The EDN Designer's Companion

The EDN Designer's Companion

Edited by Ian Hickman and Bill Travis

BUTTERWORTH
HEINEMANN

Butterworth-Heinemann Ltd
Linacre House, Jordan Hill, Oxford OX2 8DP

A member of the Reed Elsevier group

OXFORD LONDON BOSTON
MUNICH NEW DELHI SINGAPORE SYDNEY
TOKYO TORONTO WELLINGTON

First published 1994

British Library Cataloguing in Publication Data
EDN Designer's Companion
I. Hickman, Ian II. Travis, Bill
621.3815

ISBN 0 7506 1721 7

Library of Congress Cataloguing in Publication Data
The EDN designer's companion/edited by Ian Hickman and Bill Travis.
p. cm.
Includes bibliographical references and index.
ISBN 0 7506 1721 7
1. Electronic circuits. 2. Electronic circuit design.
3. Electronic systems – Design. I. Articles reprinted from EDN, 1989–1993.
II. Hickman, Ian. III. Travis, Bill. IV. EDN.
TK7867.E175
621.3815–dc20 93–50687
CIP

Composition by Genesis Typesetting, Laser Quay, Rochester, Kent
Printed and bound in Great Britain

Contents

Foreword

For almost 40 years, *EDN* magazine's mission has been to provide the critical, up-to-date information that design teams in the electronics industry need to get their jobs done. In the course of delivering that information, *EDN* has become the most popular electronics trade magazine. People from all over the world write to say how much they appreciate the information *EDN* delivers.

More importantly, however, engineers, designers, and technical experts from around the world have shared their hard-won design knowledge with *EDN*'s readers by contributing thousands of 'Design Ideas' and numerous longer articles. It is the valuable knowledge contained in these articles, plus the added perspective provided by staff-written articles, that has given *EDN* its worldwide popularity.

Ian Hickman (who is certainly one of *EDN*'s biggest fans), with the assistance of Bill Travis, has done an outstanding job of collecting some of the magazine's best gems in his books. This book, the *EDN Designer's Companion*, reprints some of the wealth of information that has appeared in EDN over the past five years. You will most certainly find answers to some of your toughest design problems between the covers of this volume.

Steven H. Leibson
Editor in Chief
EDN Magazine

Introduction

Following the success of *Electronic Circuits, Systems and Standards*, the *EDN Designer's Companion* presents a further selection of articles from *EDN*. This volume has the same goal as its predecessor: to provide a useful collection of circuits, design techniques, and standards in a handy, easy-to-look-up format.

The most recent articles to be included in the earlier volume appeared in the February 15 1990 issue of *EDN*. Since then a great many more interesting and useful articles have appeared, filling two wallet files to overflowing and causing a third file to bulge alarmingly! Consequently, most of the articles in this book date from 1990 onwards, although not quite all. It was very difficult to pare down the selection of articles to be included in the first volume to fit the available space and many which would have merited inclusion regretfully, had to be omitted. I have found space for just one or two of them in this volume, although again, paring down the amount of material worthy of inclusion has proved a very difficult task.

The general layout differs somewhat from that of the earlier volume. The articles are arranged by category, for the convenience of those with limited time to browse. A subject index also appears at the end, to direct the interested reader to any particular topic he or she wishes to address.

Like the previous book, many of the articles are shorter items from the immensely popular 'Design Ideas' section of the magazine, but a good sprinkling of longer items are included as well. The editors would have liked to include more (there is no shortage of worthy candidates) but given the space constraints, the total number of different items that could be included would have been substantially reduced. Inevitably, not every reader will be interested in every item included, but it is hoped that all of the items will interest some readers and many of them will be of interest to most.

It is our hope that designers will find this book a valuable and rewarding companion.

Part One

Digital Techniques

Introduction to Part 1

Digital technology continues to advance at a dizzying pace. PCs, not to mention workstations, have more computing power than the mainframes of only a few years ago. A large part of this rapid evolution concerns itself with high-resolution imaging. A worrisome problem in this context is the staggering number of bytes a high-resolution image requires.

In this part, a staff-written three-part series investigates ways in which you can stuff ten pounds of imaging information into a five-pound (or smaller) bag. The series covers both the hardware and software products available for image compression. Next, a contributed article sheds more light on the algorithms used to compress high-resolution images.

Imaging is one hot topic in the digital community; DSP (digital signal processing) is another. Handy, versatile DSP chips take onerous burdens away from CPUs, and perform complex computational tasks speedily and efficiently. Two contributed articles in this chapter show you how to use two different DSP chips to generate random numbers and noise for use in test applications.

1 Image compression (part 1)

Introduction to image compression techniques

This first part of a three-part series covers the basics of color image information content, as it applies to CD-ROM, ISDN etc., whilst the second part deals with the three main image-compression techniques – JPEG, MPEG and H.261.

Richard A Quinnell
Technical Editor

Computers' emerging ability to manipulate and display images and video increases their information-handling potential by orders of magnitude. As the computer's utility as an information source grows, it opens a host of applications opportunities in communications, education, and image archiving. But hardware limitations can stand in the way of implementing these applications. The key that unlocks the potential of computer-based video is image compression.

When you look at the volumes of data necessary to display images and video, the need for some form of data compression quickly becomes evident. A single VGA- resolution 640 × 480-pixel image in full natural color (eight bits each for red, green, and blue color components) needs nearly 1 Mbyte of memory. An SVGA-resolution image requires 2.3 Mbytes – more than will fit on a floppy disk.

With moving images, data rate as well as data volume becomes a concern. Transmitting studio-quality television (720 × 485 pixels at 30 images/s) would require a data-transfer rate of 31 Mbytes/s (251 Mbps) – far greater than data rates available on typical PCs. CD-ROMs, for example, only have a transfer rate of 1.2 Mbps. Primary-rate ISDN (Integrated Services Digital Network) telephone lines are only sightly better, handling 1.544 Mbps in the US and 2.048 Mbps in Europe. The fastest standard local-area network (LAN), fibre distributed data interface (FDDI), only achieves 100 Mbps.

To bridge the gap between image-handling needs and PC capabilities, you have to compress image data. Ideally, the compression would not affect the image's perceived quality when finally displayed. A variety of techniques, referred to as 'lossless' compression algorithms, can reduce the required data rates without compromising the data or affecting image quality. Two of the most common are run-length coding and entropy coding.

Run-length coding replaces strings of identical data values with a code designating that value and the number of times it repeats. For example, the sequence 66666555233 could be represented by 56 35 12 23, with the first digit of each pair representing the number of occurrences and the second representing the data value. This example exhibits a 1.37:1 compression, but, as is evident with the code representing the single value '2', its success is highly data dependent. A random number sequence could easily result in a coded-data sequence longer than the original.

Run-length coding is best suited for graphics and painted animation images, where large sections of the image have the same value. Natural images, which have complex shapes and shading, are better handled with entropy coding. Entropy coding replaces a frequently occurring value with a short code word and uses longer code words for less common values. The average number of bits needed, therefore, will tend toward the shorter code-word length.

To design an efficient code, however, you need a statistical understanding of the data's patterns. That understanding can be acquired or assumed. Some entropy-coding schemes, such as the Lempil-Ziv algorithm, create the code on the fly, as they process blocks of data. Others, such as Huffman coding, make assumptions about the expected data patterns.

Making assumptions about image data patterns is not unreasonable. Images tend to be spatially correlated; pixels tend to resemble neighboring pixels. If, instead of coding raw data, you first subtract from the value of each pixel the value of its predecessor, you can modify the data to contain mostly near-zero values, regardless of the image's absolute values. This modification makes the data conform more closely to the assumptions inherent in most entropy-coding algorithms and increases the algorithm's compression effectiveness.

Even as they become more efficient, none of the lossless coding techniques can compress image information to the extent necessary to bridge the data-rate gap. These techniques typically achieve average compression ratios of around 3:1. To finish bridging the gap, you'll have to turn to 'lossy' techniques, that is, algorithms that throw away (wisely, you hope) picture information. Lossy compression methods range from simple pixel-dropping to complex motion prediction.

The most basic lossy compression techniques simply reduce the number of bits used to quantize the image. For example, you can achieve a 4:1 data reduction for an image by subsampling your picture at half the original horizontal and vertical resolution, then replicating pixels during display to restore the image's original size. Too aggressive an application of this approach, however, and you wind up with pixelation: a picture that looks like it was made out of Lego blocks. You can reduce the impact of pixelation by interpolating between pixels rather than replicating them, but the result can be a somewhat blurry image; the method eliminates sharp edges.

To avoid pixelation, you could display a compressed full-screen image as smaller than full-screen. Small-image formats for video telecommunications are part of the CCITT's H.261 standard, commonly called P × 64 because it uses data rates that are multiples of 64 kbps. The image sizes include the CIF (common interchange format, 352 × 288 pixels) and the QCIF (quarter CIF, 176 × 144 pixels). A third small-image format is the SIF, at 352 × 240 pixels.

Because subsampling to reduce picture size is a viable method of reducing image-data requirements, you need to be careful when evaluating manufacturers' claims involving compression ratios. If you don't know the size and color depth (number of bits each for red, green, and blue) of the original image and the size of the final image, the compression ratio could be an inflated and meaningless number. A better measure of compression is the number of bits needed per pixel after compression. This bit-per-pixel value lets you compare compression schemes directly and allows you to calculate the data-storage size and transmission rate needed for any given-sized image.

Another way to reduce data is to use fewer bits to code each color component. This is the equivalent of using larger quantization steps when digitizing the image. Using eight bits to code each primary color may be overkill. With monochrome images, for example, as few as six data bits can provide adequate gray shading.

If you use too large a quantization step, the image quality degrades considerably. As shown in Figure 1.1, the image generated using 5-bit quantization (b) is almost indistinguishable from the original 8-bit image (a). On the computer case, you may be able to see the subtle shading is represented by distinct bands, an artifact called 'contouring'. Figure 1.1 (c) shows the image using 4-bit quantization; it exhibits obvious contouring.

Interframe coding for video

All of the preceding methods are applicable to both still and moving images. You simply treat each frame of a moving image as an individual still image. Such handling of moving images is called 'intraframe' compression. However, moving images often exhibit considerable similarity between successive frames. 'Interframe' compression takes advantage of this similarity.

There are two types of interframe coding: predictive and interpolative. With predictive coding, you first compress a reference frame using intraframe techniques, then encode the differences between the reference frame and successive frames. To prevent errors from propagating, however, you must periodically use a new reference frame. Interpolative coding drops some frames entirely, restoring them during decompression by interpolating between reference frames.

None of the above techniques takes into account the information contained within an image. Yet, not all picture information is valuable to the end user. In particular, the eye is relatively insensitive to fine details of light and dark (intensity) and even less sensitive to details of color. You can use this fact to reduce the amount

(a) (b) (c)

Figure 1.1 These photos of my computer system demonstrate contouring effects caused by insufficient quantization. The original 8-bit photo (a) is closely matched by the 5-bit version (b), with only slight contouring visible on the computer case. The 4-bit version (c) exhibits pronounced contouring of image data by selectively throwing away information (for more on taking advantage of 'spatial-frequency' concerns, see 'color, contrast and perception').

To take advantage of the eye's nonuniform sensitivities, you must first perform two mathematical transformations on images originally expressed as a weighted combination of red, green, and blue. The first is a coordinate transformation from the red-green-blue (RGB) color representation to one that separates intensity (luminance) from two-component color (chrominance). Several such transformations are in common use in color television systems (see 'Choose your rainbow'). For digital video, however, the most common color space is YCrCb, described in CCIR Recommendation 601–2 for digital-video studios[1]. This color space is used for all of the standardized image-compression algorithms.

The second transformation converts the normal position-based image description to a spatial-frequency description (see 'Contrast, color and perception'). This transformation is analogous to the Fourier transform used to convert time-based signals to frequencies. Instead of operating on a series of points, however, these transforms operate on a two-dimensional array of points, corresponding to a block of pixels. The transform most often used in image compression is the discrete cosine transform (DCT).

The transformations themselves are completely reversible and do not compress data. Once performed on the image, however, they make it more amenable to the selective application of lossy techniques. For example, you can take advantage of the eye's relative insensitivity to color detail by subsampling the color information without affecting the more important intensity information. By encoding only one pair of chrominance values for every two luminance values, you achieve a 1.5:1 reduction in data without noticeably affecting the image.

Because the eye's sensitivity to intensity differences varies with spatial frequency, you can afford to vary the quantization step size used with each frequency component. A larger step size (fewer bits) can adequately represent the less noticeable frequencies. If the steps are too large, however, the reconstructed image may suffer from 'blocking'. Blocking occurs in regions of subtle variation when quantization completely eliminates high-frequency information, leaving a constant value for a large block of pixels.

Figure 1.2 demonstrates blocking using an image compressed with the DCT-based JPEG compression algorithm. Compressed losslessly (a), the original image needs 7 bits per pixel. At 0.5 bits/pixel (b), the image still looks reasonable. Blocking quickly becomes evident, however, at 0.3 bits/pixel (c), and at 0.25 bits/pixel (d) the background and desktop begin to look like cartoons.

When strong high-frequency information, such as lettering or noise spikes, is present in an image, coarse quantization can have the alternative effect of amplifying those frequencies. Amplification occurs when a signal's value is just above the quantization-step threshold. The value gets rounded up to the next step's level, adding as much as half the step-size to the frequency component. With coarse quantization, the round-up error can be a significant fraction of the original value. Figure 1.3 shows a close-up of the lettering in a compressed image. The ripples around the lettering come from high-frequency amplification.

The presence of high-frequency information has another unfortunate effect in DCT-based compression schemes: it adds to the compressed image's file size. My first experiments with DCT-based image compression showed that performing a lowpass smoothing filter on the image before compression yielded as much as 10% reduction in the compressed data's file size (with a slight loss of sharpness in the image). For this reason, as well as the high-frequency amplification effect, virtually all image-compression chipset evaluation boards have smoothing filters built into the image data path.

No single technique described above will serve all of your image-compression needs. Instead, most practical compression schemes use a combination of techniques to increase efficiency. The baseline algorithm of the JPEG (Joint Photographic Experts Group) still-image compression scheme is an example. The algorithm accepts a 24-bit/pixel RGB color image, transforms it into the YCrCb color space, and subsamples the color components 2:1:1. The algorithm then uses a DCT to separate the image into its frequency components and employs frequency-dependent quantization, differential and run-length coding, and entropy coding (Huffman) for the final data reduction.

Which combination of techniques is optimum for your application depends on the nature of the images you want to compress as well as the quality level required of the decompressed image. Further considerations include the availability and cost of computational resources, the speed at which images are to be compressed and decompressed, and the amount of compression you require. As a result, a number of algorithms, both standardized and proprietary, have been developed that make different tradeoffs among the many considerations. Hardware to implement these algorithms is beginning to proliferate. The list of manufacturers gives suppliers for ICs that implement various pieces of the compression puzzle.

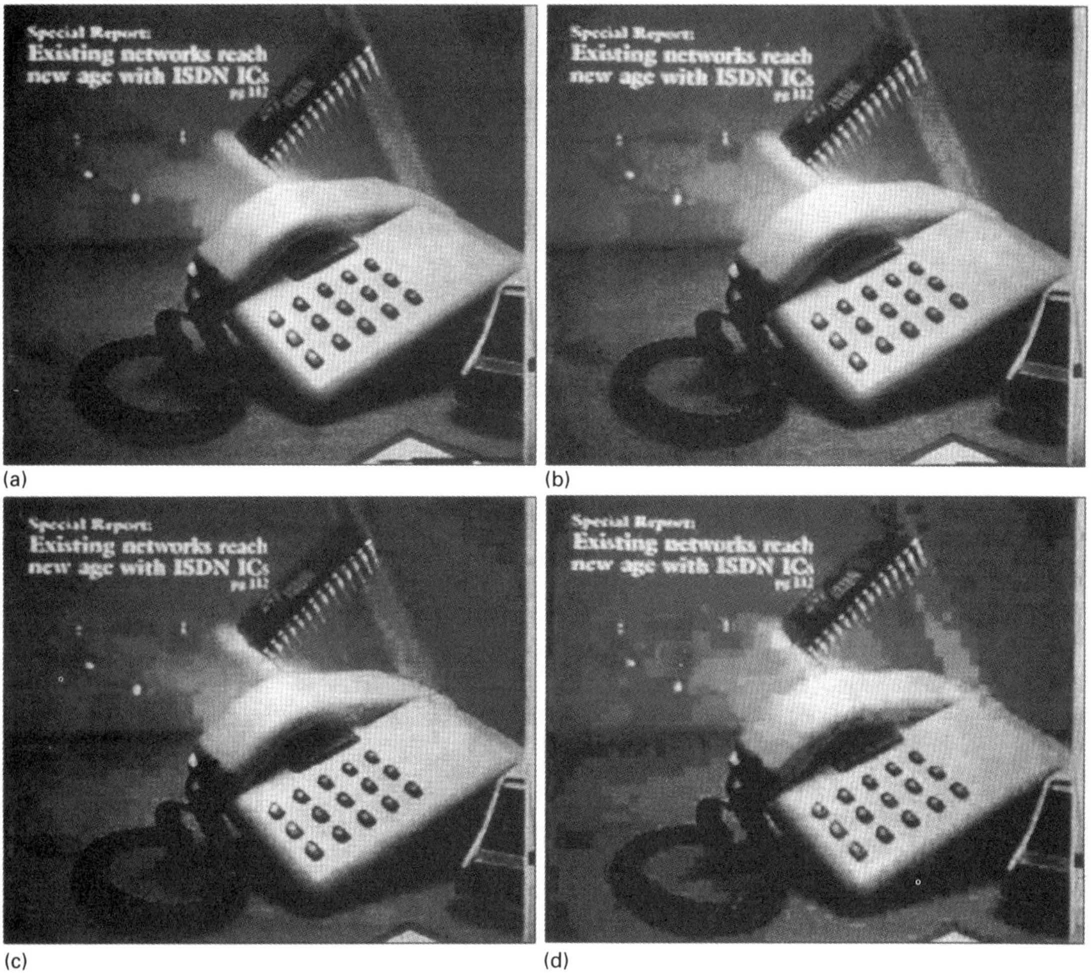

Figure 1.2 Using the JPEG algorithm to compress the original image (a) to 0.48 bits/pixel results in some degradation (b). Pushing the compression further, however, to 0.31 bits/pixel (c) and 0.25 bits/pixel (d), rapidly results in severe distortion called blocking. Blocking occurs when high-frequency information is too coarsely quantized

Figure 1.3 The ripples surrounding the lettering in this image are the result of coarsely quantizing strong high-frequency components

Contrast, color, and perception

The human eye, in conjunction with the brain's visual cortex, is a marvelous imaging system. It can operate over an intensity range spanning five orders of magnitude and detect color differences caused by a 1-nm change in wavelength. Nonetheless, our visual system has its limits, and those limits open opportunities for image compression. A camera registers information that the eye cannot see, so you can reduce digital-image data files by throwing away that information. The key to throwing away the right information is understanding the eye's ability to perceive contrast as a function of spatial frequency.

Contrast is defined using a pattern of light and dark bars where the light intensity varies sinusoi-

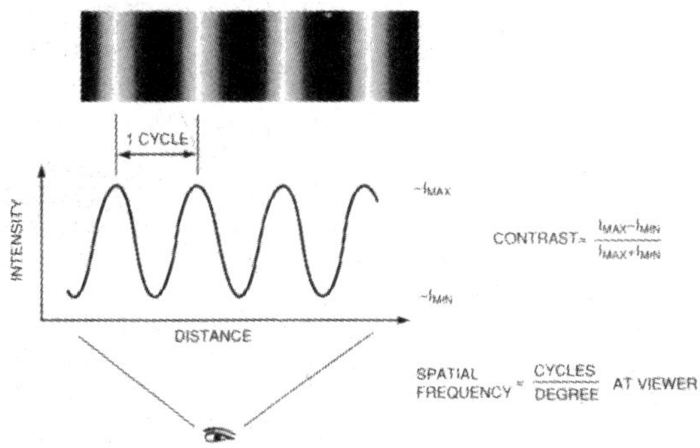

Figure 1.4 Contrast is defined in terms of the intensity variation across a sinusoidal grating. Spatial frequency is the number of grating cycles that subtend one degree of arc at the viewer's position

dally across the bars, as shown in Figure 1.4. The contrast of that pattern is defined by

$$C = (I_{max} - I_{min})/(I_{max} + I_{min}),$$

where I_{max} is the light intensity at the brightest point and I_{min} the intensity at the darkest point. Color contrast uses hue differences in place of intensity differences in the pattern.

The pattern's spatial frequency, however, does not have an absolute value. It depends on the relative positions of the pattern and the observer. Spatial frequency is defined as the number of pattern cycles that subtend one degree as seen by the observer. The spatial frequency of a fixed pattern, therefore, will decrease as the observer moves closer and increase as the observer moves away.

A variety of studies have measured the human visual sensitivity as a function of spatial frequency[2,3]. As shown in Figure 1.5, we can perceive fine detail better for intensity variations than we can for pure color variations. Furthermore, we can't perceive some spatial frequencies regardless of the contrasts involved. Patterns that fall outside our contrast/spatial-frequency range appear instead as areas of uniform color or intensity.

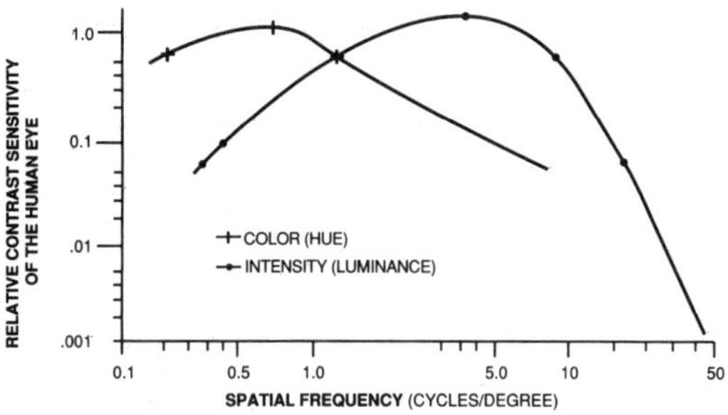

Figure 1.5 The eye's ability to distinguish subtle shading (contrast) depends strongly on the image's spatial frequency. We are less sensitive to color contrasts than to intensity contrasts

Image-compression algorithms can take advantage of these visual limitations by using fewer bits when quantizing less visible spatial-frequency components and by truncating nonvisible frequencies. Because the image's frequency content is viewer-position dependent, however, the algorithm design must make some assumptions about viewing conditions. These assumptions include the resolution of the display device as well as the viewing distance used by the observer.

For example, a 13 in (diagonal) SVGA monitor pitch can display 512 cycles (1024 points) along a 10.4-in line. At a normal viewing distance of 18 in, this corresponds to a spatial frequency of 16 cycles/degree. Spatial frequencies above that value are not displayable, so they can be removed when compressing the image.

Other limits apply for other viewing conditions. A high-quality 25-in (diagonal) color television, displaying 400 scan lines in a 15-in height, has a 15-cycle/degree capability when viewed from a distance of 6 feet. A typical color television displays only about 300 lines, or 12 cycles/degree. A printed page, on the other hand, often uses a resolution of 300 dpi for photographs. At a viewing distance of 12 in, this corresponds to 31.5 cycles/degree.

Resolution is not the only factor limiting a display's ability to reproduce spatial frequencies. The amount of contrast it can produce is equally important. If the display is set so that the minimum displayable contrast is just noticeable at a spatial frequency of 2 cycles/degree (the peak of human visual sensitivity), then the maximum contrast displayable will determine the highest spatial frequency observable. For example, a display with a contrast range of 30:1 will only be able to visibly render spatial frequencies less than 15 cycles/degree, regardless of its resolution. Attempts to display a pattern containing higher frequencies would result in the viewer perceiving a uniform gray.

Choose your rainbow

Virtually every color visible to the human eye can be created using a linear combination of three primary colors: red, green, and blue. These three colors, therefore, form the basis vectors for a three-dimensional space in which a color resides at the red, green, blue (RGB) coordinates that describe the color's composition. This RGB space is the color space used by computer displays and television monitors as well as scanners and video cameras. It is not, however, the best color space for image compression.

PAL $\begin{bmatrix} Y \\ U \\ V \end{bmatrix} = \begin{bmatrix} 0.299 & 0.587 & 0.114 \\ -0.147 & -0.289 & 0.463 \\ 0.615 & -0.515 & -0.100 \end{bmatrix} \begin{bmatrix} R \\ G \\ B \end{bmatrix}$ $U=0.493\ (B-Y)$, $V=0.877\ (R-Y)$

NTSC $\begin{bmatrix} Y \\ I \\ Q \end{bmatrix} = \begin{bmatrix} 0.299 & 0.587 & 0.114 \\ 0.596 & -0.274 & -0.322 \\ 0.211 & -0.522 & 0.311 \end{bmatrix} \begin{bmatrix} R \\ G \\ B \end{bmatrix}$ $I=V\cos 33^\circ - U\sin 33^\circ$, $Q=V\sin 33^\circ + U\cos 33^\circ$

SECAM $\begin{bmatrix} Y \\ Db \\ Dr \end{bmatrix} = \begin{bmatrix} 0.299 & 0.587 & 0.114 \\ -0.450 & -0.883 & 1.333 \\ -0.133 & 1.116 & 0.217 \end{bmatrix} \begin{bmatrix} R \\ G \\ B \end{bmatrix}$ $Dr=-1.902\ (R-Y)$, $Db=1.505\ (B-Y)$

CCIR 601 $\begin{bmatrix} Y \\ Cr-128 \\ Cb-128 \end{bmatrix} = \begin{bmatrix} \frac{77}{256} & \frac{150}{256} & \frac{29}{256} \\ \frac{131}{256} & \frac{-110}{256} & \frac{-21}{256} \\ \frac{-44}{256} & \frac{-87}{256} & \frac{131}{256} \end{bmatrix} \begin{bmatrix} R \\ G \\ B \end{bmatrix}$

Figure 1.6 A variety of color spaces are in common use, but all can be derived from the RGB information supplied by cameras and scanners. This shows the matrix transformation tables needed to convert from RGB to other color spaces

Because the eye responds differently to changes in color than it does to changes in light intensity, a color space that separates intensity (luminance) from color (chrominance) has an advantage. The luminance/chrominance color space allows compression algorithms to handle the two components differently, to better match the data to the eye's perception limits. The developers of color broadcast television understood the benefits of such a color space and developed several variations. Figure 1.6 summarizes the coordinate transformation matrices for a variety of color spaces.

The basic luminance/chrominance color space is the YUV color space. The name is not an acronym; Y is the name of the luminance vector, U and V are the chrominance vectors. The luminance vector Y derives from the RGB vectors with

$$Y = 0.299 \times R + 0.587 \times G + 0.114 \times B.$$

The RGB scaling factors reflect the eye's uneven response to the three wavelengths of light as well as normalization constants to limit the Y value range.

The chrominance vectors U and V represent the colors blue and red with the intensity information removed:

$$U = 0.493 \times (B - Y),$$
$$V = 0.877 \times (R - Y).$$

The scale factors adjust the vector magnitudes to prevent interference between the luminance and chrominance components when they are combined into a composite video signal. YUV is the color system used in the British PAL (phase alternation line) broadcast-television standard.

The color space used in the US NTSC (National Television Standards Committee) broadcast standard is called YIQ, with I and Q standing for 'in-phase' and 'quadrature', the modulation methods used to transmit the color information. I and Q are the U and V vectors following a 33° coordinate rotation. The French SECAM system (sequential color with memory) uses yet another color space, YDrDb.

Increasing interest in the use of digital signal processing for video gave rise to a color space designed specifically for digital video. That color space is the YCrCb space, defined in CCIR Recommendation 601–2[1]. It is the color space used in all the standardized compression algorithms.

The YCrCb color space starts with red, green, and blue signals quantized into 220 equally spaced values, with the value 16 representing black. It transforms those signals so that the resulting luminance and chrominance values also have 220 quantization levels. Luminance uses the value 16 to represent black; the values 0 and 255 are reserved for use as synchronization flags. Cr and Cb have black set at 128.

Manufacturers of image-compression components

Advanced Hardware Architectures
Box 9669
Moscow, ID 83843
(208) 883–8000
FAX (208) 883–8001
Entropy-coding and proprietary image-compression ICs

AT&T Microelectronics, Inc
555 Union Blvd
Allentown, PA 18103
(800) 372–2447
Programmable video-codec chip set

Brooktree Corp
9950 Barnes Canyon Rd
San Diego, CA 92121
(619) 452–7580
FAX (619) 452–1249
Color-space conversion and video-scaling ICs

C-Cube Microsystems
1778 McCarthy Blvd
Milpitas, CA 95035
(408) 944–6300
FAX (408) 944–6314
JPEG chip sets and MPEG decoders

Cirrus Logic, Inc
3100 W Warren Ave
Fremont, CA 94538
(510) 226–2200
FAX (510) 226–2240
Video-scaling and color-space conversion ICs

GEC Plessey Semiconductors
Action Desk, Lenvale House
Hollingworth Ct, Ashford Rd
Maidstone, Kent ME14 5PP, UK
(793) 518510
FAX (793) 518582
H.261 chip set

Integrated Information Technology Inc
2445 Mission College Blvd
Santa Clara, CA 95054
(408) 727–1885
FAX (408) 980–0432
Programmable video-codec chip set

Intel Corp
2200 Mission College Blvd
Santa Clara, CA 95052
(408) 765–8080
Video-compression processors

LSI Logic
1551 McCarthy Blvd
Milpitas, CA 95035
(408) 433–8000
JPEG chip set

Motorola, Inc
2100 E Elliot Rd
Tempe, AZ 85284
(602) 897–3615
FAX (602) 897–4193
Color-space decoder/converter

Philips Semiconductors
811 E Arques Ave
Box 3409
Sunnyvale, CA 94088
(408) 991–2000
Video scaling and color-space converters

SGS-Thompson Microelectronics
1000 E Bell Rd
Phoenix, AZ 85022
(602) 867–6100
DCT processors, motion estimators, and MPEG decoders

Zoran Corp
1705 Wyatt Dr
Santa Clara, CA 95054
(408) 986–1314
FAX (408) 985–1240
DCT processors, JPEG and proprietary image-compression chip sets

References

1. CCIR Recommendation 601–2, 'Encoding parameters of digital television for studios', 1990
2. Campbell, Fergus W, and Lamberto Maffei, 'Contrast and spatial frequency', *Scientific American*, November 1974, pp.106–114
3. Jacobs, Gerald H, *Comparative Color Vision*, Academic Press, New York, 1981

Acknowledgment

EDN wishes to thank Benita Kenn of Creative Labs Inc (1901 McCarthy Blvd, Milpitas, CA 94035, (408) 428–6600, FAX (408) 428–6611) for use of the Creative Labs Video Blaster card ($499), which was used to produce all of the example images in this article. Thanks also go to Media Vision (3185 Laurel View Ct, Fremont, CA 94538, (510) 770–8600, FAX (510) 623–5749), Philips Semiconductors, LSI Logic, Brooktree Corp, and Integrated Information Technology for assistance in researching this article.

2 Image compression (part 2)

Image compression standards compared

More about the Joint Photographics' Experts Group algorithm (for stills), Moving Pictures Experts' Group algorithm and CCITT H.261 – commonly known as P*64 – for video telephony.

Richard A Quinnell
Technical Editor

If you're trying to communicate compressed images between systems from different manufacturers, you'll need a standard for compression and decompression that everyone can agree on. Fortunately, three such standards now exist. These distinct international standards have been defined for image compression, each for a different application.

The three standards are commonly known as JPEG (Joint Photographic Experts Group), MPEG (Moving Pictures Expert Group), and P*64. JPEG[1] is intended for continuous-tone still images, although some users are applying it to video. MPEG[2] is intended for the distribution and interactive playback of movies and prerecorded video with an eye toward the data rates and capacity of CD-ROM drives. P*64 is the common name for CCITT H.261[3], a coder/decoder for video telephony. The name stems from the scheme's output data rate, which is an integer multiple of the 64-kbits/s basic digital telephone link. (MPEG and P*64 specify only the content and format of the compressed-image data stream; only general guidelines for the encoder are given. Implementation details of the encoder are left to the user.)

Figure 2.1 is a simplified block diagram of the encoder portion of each standard. The JPEG encoder (a) is straightforward. It converts image data to a spatial frequency representation using a two-dimensional discrete cosine transform (DCT). It quantizes the components with different degrees of granularity, reflecting the human visual system's spatial-frequency sensitivity[4], then encoded using variable-length (entropy) coding.

The MPEG encoder (b) is much more complex. Instead of simply encoding image data directly,

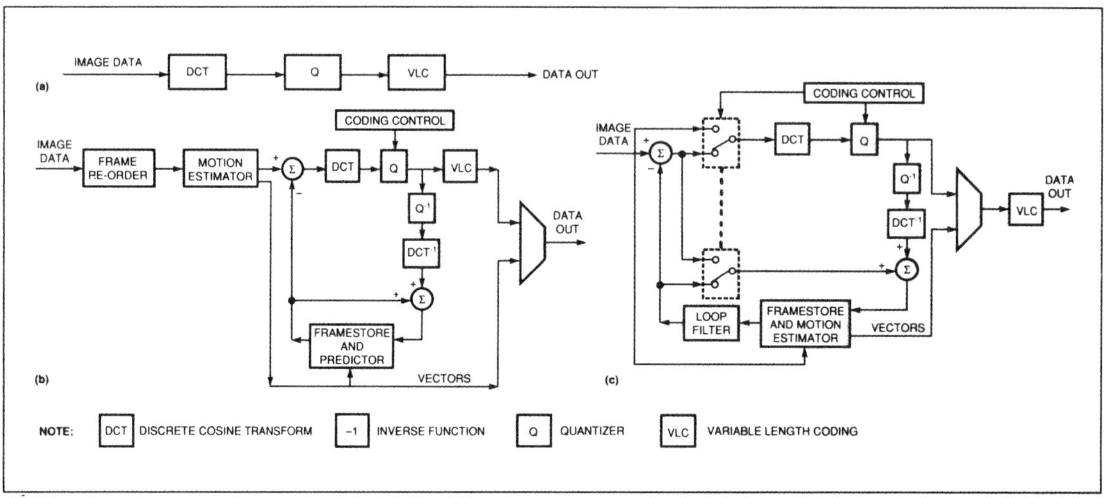

Figure 2.1 As these block diagrams show, the standard encoders have much in common. the JPEG encoder (a) uses transform coding and quantization. The MPEG encoder (b) adds motion compensation and frame prediction to the transform coding. The P*64 encoder (c) looks similar to MPEG but is block-oriented rather than image oriented

the MPEG encoder may chose to make a prediction about an image, then transform and encode the difference between the prediction and the image. The prediction accounts for movement within an image by using motion estimation. A given image's prediction may be based on future images as well as past ones, so the encoder must reorder images to put reference images before the predicted ones. The decoder puts the images back into display order.

The P*64 encoder also encodes predicted image errors, but must take them as they come. It can choose to use image data or predicted data on a block-by-block basis. The encoder's prediction is normally based on the same block of the previous image, but the encoder can include motion estimation and a filter as an option.

Despite the apparent differences, the three standards have several similarities. The most significant common element is the use of an 8 × 8-pixel DCT followed by spatial-frequency-dependent quantization. This lossy compression scheme produces two artifact types, fringing and blocking, that you may find in the reconstructed image when using any of the standards.

Fringing artifacts become more visible with increasing compression and occur at sharp edges in the image, where high spatial frequencies are present. Figure 2.2 shows an example of fringing in one frame from a video clip compressed using the JPEG standard. The fringing is particularly noticeable near the woman's throat and shoulders. In a video sequence, the fringing varies from frame to frame, giving rise to an apparent motion referred to as 'flying ants' or 'mosquito cloud' by compression experts.

Figure 2.2 This still from a JPEG-based video clip shows an artifact common to all three standard algorithms: fringing at sharp edges. The effect is most noticeable around the woman's throat and shoulders. (The image was produced using the C-Cube JPEG video evaluation board.)

Blocking artifacts occur in broad areas of colour, where low and zero spatial frequencies dominate. In such cases, too coarse a quantization causes an entire block (8 × 8 pixels) to be represented by a constant value. The effect becomes more pronounced at higher compression ratios. Some blocking is apparent in the background of Figure 2.2.

Another similarity among the three standards is the use of the YCrCb colour space with chrominance subsampling to reduce data. The subsampling causes no visible artifacts because the human visual system is less sensitive to pure colour changes. The standards, however, don't all use the same subsampling scheme. The JPEG standard calls for one chrominance pixel for every two luminance pixels. The MPEG and P*64 standards call for one chrominance sample in a block of four luminance. The subsampling patterns are given in Table 2.1 along with other parametric differences between the standards.

As Table 2.1 shows, the standards differ significantly in the image resolution they'll handle. The JPEG standard allows images as great as 65,536 × 65,536 pixels. Because they are concerned with real-time playback of moving images and with image-transmission data rates, the MPEG and P*64 standards handle much lower resolution. MPEG limits pictures to 4095 × 4095 pixels and recommends a 'constrained parameters' image of 768 × 576 pixels or smaller. P*64 specifies two image sizes, either the 352 × 288-pixel common intermediate format (CIF) or the 176 × 144-pixel quarter-CIF (QCIF).

Reducing temporal redundancy

The most significant difference, however, lies in the techniques used to reduce temporal redundancy in moving images. (Temporal redundancy is the tendency for an image to resemble the preceding image in a sequence.) JPEG has no such provision; it was designed for still images. MPEG and P*64 exploit temporal redundancy heavily, however, achieving 10× to 100× the compression achievable with transform coding alone. That improvement is not guaranteed. If the image sequence includes a great deal of motion, temporal redundancy is low and the overall compression ratio will fall. Even so, if you're using JPEG for moving images, you'll achieve much less compression for a given image quality than with the other two standards.

Both MPEG and P*64 use prediction and motion estimation to reduce temporal redundancy, but they differ greatly in approach because of their differing applications. MPEG's emphasis is on maintaining picture quality with maximum compression. P*64, because it is for video telephony, seeks to minimize encoding and decoding

Table 2.1 Image-compression standards

Standard[1]	Standards body[2]	Number	Picture block (pixels)	Maximum image size (pixels)	Motion estimation range (pixels)	Motion estimation resolution (pixels)	Transmission bit rate (bits/sec)	Color subsampling pattern[3]	Comments
JPEG (baseline sequential)	ISO/IEC	DIS 10918-1	16×8	65,536 × 65,536	Not applicable	Not applicable	Not applicable		Standard also specifies lossless, progressive, and hierarchical coding modes for special image handling needs.
MPEG (full)	ISO/IEC	11172-1	16×16	4095 × 4095	±64	0.5	100M	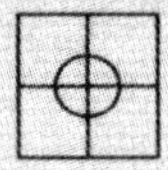	Allows VCR functions on coded data, such as fast forward, pause, and reverse. Allows 0 to 16 video and 0 to 32 audio signals multiplexed into single bit stream. Provides for editing coded sequences. Coding optimized for 350×250 image size.
MPEG (Constrained parameter)				768 × 576			1.856M		
P*64	CCITT	H.261	16×16	352 × 288 or 176 × 144	±15	1	P×64k P=1,2, . . .,30		Channel error-correction coding built in. Motion estimation and loop filter are optional.

Notes:

1. JPEG=Joint Photographic Experts Group; MPEG=Motion-Pictures Experts Group.
2. ISO=International Standards Organization; IEC=International Electrotechnical Commission; CCITT=International Consultative Committee for Telegraphy and Telephony.
3. □=Luminance sample; O=Chrominance sample.

delay while achieving a fixed data rate. The need for minimum delay stems from the interactive nature of telephony. During a conversation, a delay as small as 500 ms in receiving a response can be extremely annoying.

MPEG allows both forward and backward motion compensation, so a coded image may use information from images that followed it as well as those that proceeded it. The MPEG images are organized in groups of pictures (GOPs) that are transmitted in an order different from the display order. A GOP typically starts with a reference image, called an intraframe image or I image, compressed using just transform coding and quantization. That is followed in the data stream by a predicted image, called a P image, assembled by calculating the movement of each 16 × 16-pixel block (macroblock) in the reference image using forward prediction. Succeeding P images in the GOP use the previous P image as their reference. Between P images are one or more images (B images) assembled by calculating macroblock motion between two P images (or an I and a P image) using both forward and backward prediction. Any number of P and B images may be combined with an I image to form a group. The more B images in a GOP, the higher the overall compression.

MPEG allows editing

This group structure gives the MPEG compression standard two useful attributes not found in the P*64 standard. The first is an ability to edit the compressed data stream. Because each GOP contains all the information needed to reconstruct the image sequence, it serves as an edit point, like a movie frame, for combining prerecorded sequences. The GOP also provides some robustness in the presence of channel errors when transmitting an MPEG data stream. With predictive coding, an error in one image will be propagated as long as it serves as a reference for succeeding images. Because I images are coded without reference to other images, error propagation stops at the next I image.

P*64 uses a block update scheme to reduce temporal redundancy, although it also allows motion compensation as an option. In the block update scheme, the encoder first transmits a frame using only the transform-coding (intraframe) technique. It then compares each block on the next frame with the version transmitted in the previous frame. If there has been sufficient change, the encoder will transform code and transmit the difference between the new block and the previous one. If the change is not sufficient, the block is not transmitted. The definition of 'sufficient change' can vary from block to block.

To better understand the differing tradeoffs made in the various standards, I experimented with the encoders for each standard. The results of using the JPEG algorithm on a still image at various compression levels were given in Part 1 of this series[4]. In general, JPEG suffers from fringing and blocking artifacts when used to compress images below 0.8 bits/pixel. Using JPEG for video, therefore, requires that you limit the image size and frame rate to keep data rates to a manageable level. Further, no standard method exists for incorporating audio data into the bit stream. If you're planning to use JPEG for video, you must come up with your own method, taking care to ensure synchronization between audio and video (see 'Sound advice for compressed video').

Unfortunately, I couldn't directly study the MPEG encoder because the demonstration hardware was not ready. Instead, I traveled to the laboratories of several companies working on products that perform or incorporate video compression. At AT&T Bell Labs I saw demonstrations of the MPEG (and P*64) algorithms on the AT&T video-codec chip set. The MPEG images were compressed by various sources and looked nearly flawless. The only artifacts I noticed occurred at macroblock boundaries as small kinks in vertical lines. These were noticeable only because the image had been expanded to full-screen size through pixel replication. An interpolative scaling method would have removed this artifact.

OptImage, a maker of authoring tools for the Philips CD-I (compact-disk interactive) system, also serves as a contract shop for MPEG compression. The MPEG video the company produced was of excellent quality. The only visible artifacts were instances of blocking that occurred in large gray or black areas. The blocking was not temporally correlated, however, so the effect when viewing the video was like black snow falling in the dark background areas. Blocking was not noticeable in bright areas.

Engineers at C-Cube Microsystems showed me an MPEG decoder evaluation board with some video they had compressed at their in-house facility. Although the images were somewhat less sharp than the other demos I saw, due to lowpass prefiltering of the video, the display from the C-Cube board was free of artifacts (Figure 2.3).

Achievement of high-quality video using MPEG does not come easily, however; there is a tradeoff between encoding effort and image quality. The images I saw had not been compressed in real time. OptImage, for example, has an 80:1 time difference in compression, 80 seconds of computing for each second of video. As encoder technology improves, this differential will decrease.

Choose encoding priorities

When experimenting with the P*64 encoder, I had to choose my priorities for image-coding control. The P*64 standard requires you to control the compression process and resulting data bit rate to avoid filling a hypothetical data buffer at the decoder. It offers three main types of control: block significance, quantization stepsize, and

Figure 2.3 This still from an MPEG-based video is free of artifacts, despite the occurrence of rapid motion in the video sequence. (The image was produced using the C-Cube MPEG decoder evaluation board.)

frame dropping. The first two affect the image quality, the third affects the smoothness of perceived motion. How you manipulate the three is a choice left up to you as the encoder designer, although the specification gives an example encoder – Reference Model 8. Experienced compression designers all criticize the model as too simplistic, however, claiming it produces mediocre picture quality.

The choice in image-coding-control priorities boils down to a tradeoff between frame rate and picture quality. At PictureTel, the designers limit changes to picture quality and give up smooth motion. Whenever the image underwent significant change, such as large movements by someone in the picture, the frame rate dropped, making movements more jerky. When the image was relatively static, showing only small hand gestures or changes in facial expression, the frame rate increased.

At Integrated Information Technology, I used a video processor demonstration board to run P*64 using a constant frame rate. Figure 2.4 shows the image quality that resulted. Static images (a), even at low bit rates ($P \times 64$ kbits/s), are reasonably smooth. When I move, however (b), more macroblocks require updating, and the image becomes much more blocky as the encoder tries to reduce its output data rate. The effect becomes less prominent as the available data rate increases (**c** and **d**) and becomes invisible at rates $P > 8$.

One thing quickly becomes clear when working with the three compression standards – the quality of the image source will strongly determine the compression rates and image quality you'll wind up with. Before you run your own experiments with compression, or as you design your system, keep in mind three things about the image source.

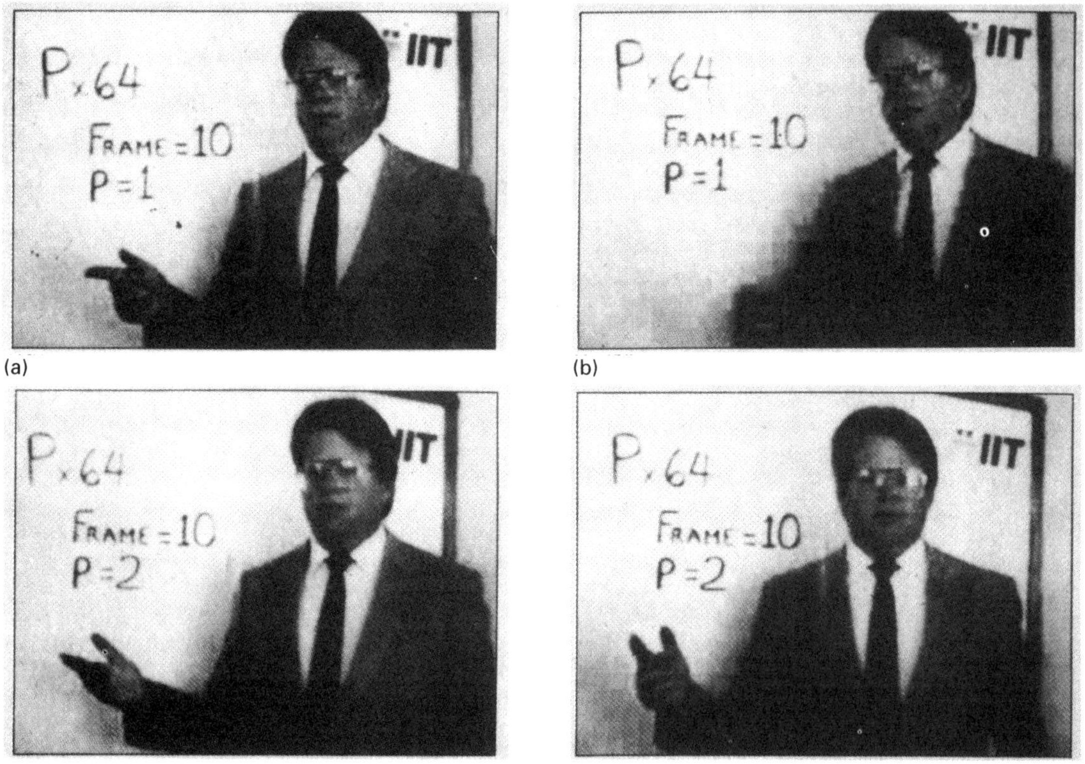

(a) (b) (c) (d)

Figure 2.4 Motion in a video sequence lowers compression efficiency and may force you to compromise image quality to keep data rates within bounds. In the P*64 standard, you can either drop frames or lower picture quality. At P = 1 (64 kbits/s) a static image (a) has acceptable quality, but large movements (b) cause a rapid degradation if the frame rate remains constant. At higher bit rates (P = 2, 128 kbits/s) the effect diminishes (c), (d). (Images created using the IIT Vision Processor evaluation board. The original images were created with a four-colour process.)

First, make sure that the image is not noisy. Noise adds high-frequency information to the image that the encoder will attempt to include in the compressed image. This causes the compression rate to decline and will increase the occurrence of fringing artifacts. A lowpass prefilter on the image source will help reduce noise as well as soften the sharp edges that also trigger fringing artifacts.

For moving images, make sure the video source has a stable time base. Motion in the image reduces compression efficiency and can cause the encoder to degrade the image if the encoder must work within a limited data bandwidth. An unstable time base causes lines in the image to appear to wriggle. That apparent movement will be captured by the encoder and dutifully transmitted, thus wasting bandwidth.

Third, watch out for global changes to the image. Such changes occur, for example, when an auto-iris camera responds to a brief reflection or when an auto-focus camera adjusts to focus on a moving target. Global changes have the same effect as movement: compression efficiency decreases and so may image quality. Remember, too, that the signal coming from a conventional video camera uses an interlaced scan. The compression standards all assume noninterlace scans. Either you'll have to drop alternate interlace fields, or combine fields in a way that doesn't introduce motion artifacts.

Sound advice on video compression

If your interest in image compression is in video applications, you'll need to consider sound. Not as an add-on, either. Sound has a significant effect on your overall system design. Three factors bear watching when integrating audio with compressed video: channel bandwidth, sound quality, and synchronization.

Sound quality and channel bandwidth affect one another. CD-quality stereo sound requires a data rate of 196 kbits/s, which can be a significant portion of your available channel bandwidth. And, the video can't recover the lost bandwidth, even temporarily. The sound data must take precedence over the video data during transmission. The eye will tolerate a momentary interruption in video, but a break in the flow of what should be continuous sound is both noticeable and annoying to the ear.

You can reduce the amount of bandwidth needed for sound by lowering the sound quality, but that may be objectionable to the end users. AT&T Bell Laboratories staff member Margaret Ann Chappell reports that customers of video-conferencing equipment expect better sound quality from the video equipment than they do from telephones.

Sound improves video

The bandwidth you spend on sound is not altogether lost to the video. Independent studies reported by AT&T, Silicon Graphics, and Dolby Laboratories show that high-quality sound will improve the perceived quality of the video. In blind studies, viewers were asked to rate two video demonstrations. The video was identical in both cases, but the sound quality varied. The demonstration with the better sound consistently received higher ratings for image quality.

Regardless of the sound quality you set, you'll have to account for synchronization of the sound and video. If you're using JPEG compression in a video application, you're on your own. JPEG doesn't have any provision for sound. One alternative is to group the sound samples in blocks, marked with the SMPTE (Society of Motion Picture and Television Engineers) time code. The code specifies the hour, minute, second, and frame number of the video to be displayed with the sound. When decoding, you can play the sound at a steady rate and adjust the video playback to match the sound.

MPEG has provision for sound synchronization in the specification. The encoder marks both audio and video packets with presentation timestamps derived from the encoder's system clock. The decoder needs to compare timestamps and ensure that matching audio and video are presented simultaneously.

The P*64 specification does not include a provision for audio because it is only part of an overall specification for video telephony. Standards for encoding audio include CCITT G.711, G.722, and G.728. Overall telephony system requirements, including audio/video synchronization, are given in CCITT standards H.221, H.230, H.242, and H.320.

References

1. ISO/IEC DIS 10918–1, 'Digital compression and coding of continuous-tone still images', January 2, 1992
2. ISO CD 11172–2, 'Coding of moving pictures and associated audio – Part 2', November, 1991
3. CCITT Recommendation H.261, 'Video codec for audio-visual services at P × 64 kbits/sec'
4. Quinnell, Richard A, 'Image compression – Part 1', *EDN*, January 21, 1993, p. 62

Acknowledgements

EDN would like to express special thanks to the companies that provided equipment and expertise for this article. These include AT&T Bell Laboratories (Holmdel, NJ), AT&T Microelectronics (Allentown, PA), C-Cube Microsystems (Milpitas, CA), Integrated Information Technology (IIT) (Santa Clara, CA), Microware Systems Corp (Des Moines, IA), OptImage Interactive Services Company (Des Moines, IA), Philips (Sunnyvale, CA), PictureTel (Danvers, MA), Production Catalysts, Inc (Manchester, NH), and Zydacron Corp (Manchester, NH). AT&T, C-Cube, and IIT make video-compression chip sets. Microware makes operating systems for real-time control of images. OptImage makes CD-I authoring tools, PictureTel and Zydacron make video-conferencing equipment, and Production Catalysts is a systems R&D consulting firm.

For more information on the image-compression standards discussed in this article, you can contact the following standards committees:

American National Standards Institute (ANSI)
11 W 42nd St
New York, NY 10036
(212) 642–4900
FAX (212) 398–0023

CCITT
Palais des Nations
CH-1211
Geneva 10, Switzerland
(22) 7305111
FAX (22) 7337256

IEC
3 rue de Varembe
Case Postale 131
CH-1211
Geneva 20, Switzerland
(22) 7340150
FAX (22) 7333843

ISO
1 rue de Varembe
Case Postale 56
CH-1211
Geneva 20, Switzerland
(22) 7490111
FAX (22) 7333430

3 Image compression (part 3)

Other standards may be more appropriate

This final installment of the series deals with some of the many other algorithms, which in certain circumstances may be more appropriate than either JPEG, MPEG or P*64.

Richard A Quinnell
Technical Editor

The use of standard algorithms for image compression helps ensure compatibility between systems from different vendors and the ability to exchange images. Yet the three international standards available, JPEG (Joint Photographic Experts Group), MPEG (Moving-Picture Expert Group), and P*64, are not always the best solution to an image-compression task. For that reason, many companies have developed proprietary image-compression algorithms.

Proprietary algorithms make the most sense when you have control over the decompression environment. For example, a stand-alone image-storage and retrieval system, such as a hospital radiology database, can readily use a custom compression algorithm, tuned to the type of images the system handles, because it doesn't need to exchange data with another system. A self-contained software application that has built-in hardware-independent decompression software is another candidate for proprietary algorithms. The correct decompression algorithm comes with the data.

Proprietary image-compression algorithms are legion, as are the reasons for their development. Some, such as IBM's Photomotion (used in the National Geographic *Mammals* multimedia encyclopedia) were commercially introduced before international standards were available. Other algorithms stem from user dis-satisfaction with the capabilities of the standard algorithms. PictureTel, for example, uses several proprietary compression schemes in its video-conference telephone systems to provide better image quality than that offered by the P*64 standard. Still other algorithms arose when users objected to the memory cost and computational requirements inherent in the standard algorithms. A W & Associates' (AWA) foursquare transform (FST) uses simpler arithmetic and single-frame storage to reduce the cost of a video-image-compression application relative to the cost of using the MPEG standard.

Standard algorithms may not be suitable for many applications. General Instruments, for example, developed its Digicipher compression scheme for compressing broadcast video to increase the channel capacity of cable and satellite television systems. The most appropriate standard alternative, MPEG, was not suitable for several reasons. First, MPEG assumes a progressive-line image scan, whereas broadcast video is interlaced. Further, users of broadcast television want to be able to change channels rapidly. The MPEG algorithm's use of predictive encoding based on an occasional reference frame would have caused unacceptable delays in the decoder's ability to switch between channels.

Because they are formulated by committee, standards don't easily adapt to technical advances. JPEG, MPEG, and P*64 are all based on transform coding using the discrete-cosine transform (DCT). Newer technologies such as fractal-based compression are invariably proprietary; they don't have an opportunity to become standardized until years after their introduction.

Compression for CD-ROM playback

A number of proprietary image-compression algorithms are available for licensing. Many of them target applications involving PC-based video playback from CD-ROMs. The Intel DVI (digital-video interactive) and the Philips CD-I (compact-disk interactive), for example, were among the first commercially available algorithms for compressing PC-based video. Philips ceased promotion of its proprietary algorithm in 1991, however, switching its support to MPEG for its CD-based consumer product. DVI is still in the running and comprises two algorithms: RTV (real-time video) and PLV (production-level video).

Both the RTV and PLV algorithms use 9-bit/pixel YUV images and compress them to 1 to 2 bits/pixel. The algorithms obtain 9-bit/pixel images from 24-bit RGB (red-green-blue) images by changing the colour space then subsampling the chrominance components. Using one 8-bit sample of each colour component within a 4 × 4-pixel block yields an average of 9 bits/pixel over the image. The algorithms can handle images as large as 768 × 480 pixels, although for CD-ROM playback and PC-based applications a smaller image size is more common.

The RTV algorithm typically works with 128 × 120-pixel images during compression, although it can rescale the decompressed image to 256 × 240 pixels using interpolation. It uses interframe coding; that is, each compressed video frame is a self-contained image. The PLV algorithm, on the other hand, uses intraframe coding. It encodes an occasional full frame as a reference, then for succeeding frames encodes only the changes from the reference frame, thus achieving a greater overall compression rate. The PLV also differs from the RTV algorithm in that PLV typically uses a 256 × 240-pixel image that it scales to 512× 480 pixels during playback.

Intel manufactures the i750 processor for engineers designing DVI compression into their hardware. It also offers PC add-in cards to capture and compress 30 frame/s RTV video sequences onto hard disk for those developing video software. If you want to use the PLV algorithm for your software, however, you'll need to go to a service bureau. Horizons Technology Inc offers a PLV compression service as does NB Engineering. In both cases, you need to provide the company with your edited video and audio on a 1-in tape marked with SMPTE time codes.

PC-based video

Apple Computer's Quicktime and Microsoft's AVI (audio video interleave) in the Video for Windows software package contain more recent examples of compression algorithms for PC-based video. Both programs provide frameworks for handling digital video and provide video codecs for the end user. Apple's Quicktime offers

Table 3.1 Quick look at video-compression algorithms

Algorithm	Maximum useful compression[1] (bits/pixel)	Comments
Microsoft RLE	7.8	Not suited to natural images. Best used for graphics or animation where greater compression is achievable
Microsoft video (Motive)	1.1	Pixellation increases rapidly with further compression.
Indeo	2.1	Color shifting becomes pronounced at higher compression rates.

Note: 1. These values were observed using a single-test video sequence. Compression capability is highly image dependent and your results will differ.

JPEG compression and a proprietary compression algorithm developed by SuperMac Technology. Video for Windows offers a run-length encoder, the Motive algorithm developed by Media Vision, and Intel's Indeo, an enhanced version of RTV. Table 3.1 provides an abbreviated comparison of the three Video for Windows algorithms.

Video playback video on the PC is not the only application that has prompted algorithm development. Advanced Hardware Architectures (AHA), for example, offers a lossless compression algorithm suitable for medical images, which cannot tolerate any data loss or the introduction of compression artifacts. The algorithm uses adaptive Huffman coding to achieve a 2.5:1 compression ratio and is available from AHA in the form of a 2-chip set (AHA3370/1, $118 (10,000)). Similarly, Zoran Corp offers its compression algorithm for still-video cameras as a 2-chip set (Zoran 031, $38 (10,000)).

The FST compression algorithm, developed by Ultech Corp (Tokyo, Japan), provides broadcast-quality video (640 × 480 pixels) on the PC at low cost. The technology is offered in the US through AWA in the form of ASICs and Commotion PC add-in cards ($7900). AWA also provides a software developer's kit ($895) and software to turn a PC into an ISDN video phone ($725). The algorithm makes no provision for audio, but AWA has defined a methodology for incorporating audio into its compressed-video data.

The FST algorithm uses 15-bit RGB colour images directly in its computations; it bypasses the RGB-to-YUV conversion step found in many other image-compression schemes. The algorithm encodes 4 × 4 pixel blocks into a 2 × 4 array, correlates neighbouring blocks, and passes a reference block together with block differences (differential coding) to a variable-length statistical encoder. Figure 3.1a shows that images compressed to 1.6 bits/pixel look like conventional broadcast video. Compression below 1 bit/pixel, however, begins to demonstrate streaking artifacts (Figure 3.1b) in broad areas of low-contrast colour.

The FST algorithm also has the option of frame-to-frame correlation. It transmits and stores a reference frame, or succeeding frames, then transmits only those blocks that change. Using correlation increases compression an additional 12 to 30%, depending on the target-image quality. At higher compression rates, however, using correlation may tend to leave ghostly after-images on screen. The after-images occur when a coloured object in the image moves against a neutral background possessing the same luminance as the coloured object. The high compression factor causes the algorithm to miss updating the colour of background blocks that get covered and re-exposed.

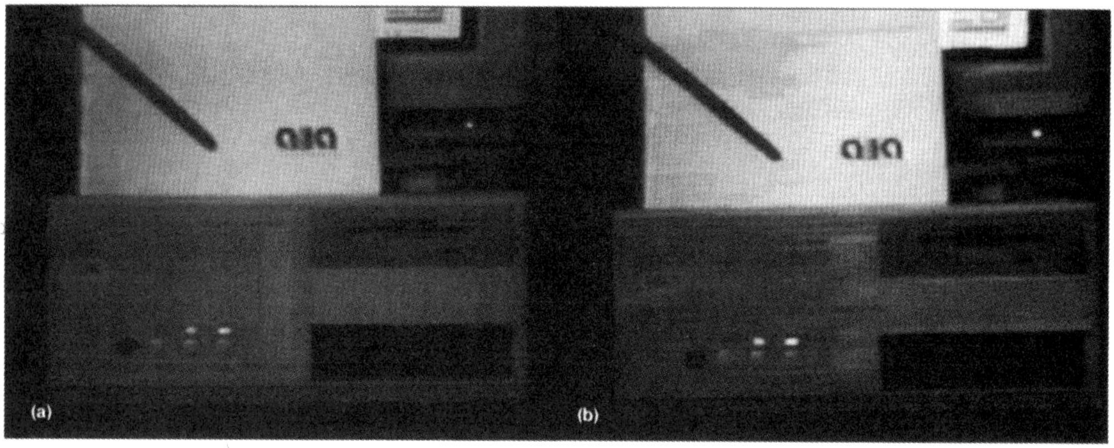

Figure 3.1 The simple four-square transform (FST) performs well in its intended broadcast-video application, compressing images to 1.6 bits/pixel (a). At about 1 bit/pixel (b), artifacts in the form of long streaks appear in areas of subtle shading

Making pictures by formula

Most image-compression algorithms share a common set of techniques: variable-length coding, differential coding, DCT transforms, and subsampling. The Fractal Transform (FT) offered by Iterated Systems, however, is a radical departure from conventional image-compression techniques. The FT is based on a nonintuitive result of topological manipulation: the existence of attractors. It lets you generate a picture from a formula.

You can create a new image (although an ugly one) by distorting – warping, shrinking, folding, flipping, sliding, rotating – an arbitrary starting image, then adding to the result a second, different distortion of the starting image, and a third, and so on a number of times. A given distortion could map all or only a section of the starting image into a section of the new image. If all the distortions you've used are members of a group called contractive operators, you get a special benefit. Repeated application of the distortion sequence – using the result of one sequence as the starting image for the next – will eventually generate an image that doesn't change with further applications. That final result is called the attractor for the sequence. Any arbitrary starting image operated on by that distortion sequence will converge to the same attractor. Thus the sequence, not the starting image, defines the final result.

The FT algorithm seeks to find a distortion sequence, or contractive operator, that has an associated attractor that closely resembles the image you wish to compress. When the algorithm has found the appropriate operator, it describes that operator in mathematical terms and uses lossless compression on the description. To decompress an image, simply decompresses the operator, then apply it repeatedly to an arbitrary starting image. The result will converge to the attractor, which resembles the original image. FT reconstruction is mathematically simpler than DCT-based reconstruction and typically executes faster.

Most attractors are fractal; that is, a portion of the attractor resembles the whole of the attractor regardless of the scale you use. The attractor's fractal nature has a beneficial attribute: it allows resolution-independent scaling. Just as the formula for a polynomial lets you reconstruct that polynomial to any scale you desire, the formula for the operator lets you reconstruct the attractor to an arbitrary scale with more detail visible at larger scales. Scaling conventional images adds no new detail and requires careful interpolation to avoid introducing artifacts.

Fractals meet JPEG

How closely the attractor resembles the original image depends on how complex an operator you let the FT algorithm find and the precision used in calculating coefficients. Both factors affect the size of the compressed file. Even with small files (simple operators, less precision), though, the decompressed image will be free of identifiable artifacts. At worst the image will resemble an Impressionistic painting – sharp edges softened and fine detail only hinted at.

Figure 3.2 shows a comparison between JPEG and FT images at the same degree of compression, 0.4 bits/pixel. As printed in the magazine, the full JPEG image (a) and FT image (b) look similar, although the lettering in the JPEG exam-

(a) (b) (c) (d)

Figure 3.2 As printed here, full-scale images compressed to 0.4 bits/pixel using JPEG (a) and Iterated Systems' Fractal Transform (b) techniques look about the same, although the JPEG image has sharper letters. A magnified view (c and d, respectively) shows the differences. Highly compressed JPEG images are blocky and have halos around sharp edges such as the phone cord. Fractal images are smooth

ple is sharper. At high magnification (c and d) the differences become more obvious. The JPEG image comprises large blocks, and there are halos around sharp edges such as the telephone cord. The FT image is smooth.

Iterated Systems licenses its patented methodology of finding suitable operators for images. In addition, it offers several software-only tools for fractal compression and decompression, a compression add-in card for ISA-bus computers, and several books on the topic of fractals.

If your system design plans call for image compression, you have plenty of proven algorithms to choose from. You may still want to develop your own, though, if you have specific system constraints that force a unique solution or if your images are restricted to a set type that can benefit from an algorithm tuned to that type.

As with circuit design, algorithm design can benefit from the use of simulation tools. At least two companies offer simulation tools suitable for developing image-compression algorithms: Comdisco Systems and Cadis Software. Although the tools are intended for general DSP design, the companies are preparing special libraries of image-compression functions. Both the Comdisco Signal Processing Worksystem and the Cadis Cossap let you design your algorithm in block-diagram form, then simulate the algorithm's operation. You can also use the tools to automatically generate C-language code for DSP processors so you can test your algorithm at a higher speed. If a silicon implementation of your algorithm is your final goal, you can use the tools to generate a VHDL (VHSIC Hardware Description Language) netlist so that you can use circuit-synthesis software to design your custom device.

Manufacturers of image-compression components

A W & Associates Inc
Long Beach, CA
(310) 420–9696
Fax (310) 420–2875

Advanced Hardware Architectures Inc
Moscow, ID
(208) 883–8000
Fax (208) 883–8001

Apple Computer Inc
Cupertino, CA
(408) 996–1010

Cadis Software Ltd
South San Francisco, CA
(415) 615–7789
Fax (415) 588–5472

Cadis GmbH
Herzogenrath, Germany
(241) 16 20 81
Fax (241) 16 14 97

Comdisco Systems Inc
Foster City, CA
(415) 574–5800
Fax (415) 358–3601

Comdisco Systems
Almondsbury, Bristol, UK
(454) 614–256
Fax (454) 614–700

General Instruments
San Diego, CA
(619) 535–2575
Fax (619) 535–2485

Horizons Technology Inc
San Diego, CA
(619) 277–7100
Fax (619) 292–9439

IBM Corp
Atlanta, GA
(800) 426–9402 × 60

Intel Corp
Chandler, AZ
(602) 554–8886
Fax (602) 554–7347

Iterated Systems Inc
Norcross, GA
(800) 437–2285
Fax (404) 840–0806

Media Vision Inc
Fremont, CA
(800) 348–7116
(510) 770–8600
Fax (510) 770–9592

Microsoft Corp
Redmond, WA
(206) 936–4188
Fax (206) 883–8101

NB Engineering
Crofton, MD
(410) 721–5725

PictureTel
Danvers, MA
(508) 762–5178
Fax (508) 762–5245

SuperMac Technology Inc
Sunnyvale, CA
(408) 245–2202
Fax (408) 524–4797

Zoran Corp
Santa Clara, CA
(408) 986–1314
Fax (408) 986–1240

Acknowledgment

EDN would like to thank A W & Associates, Intel, Iterated Systems, and Microsoft for loaning the equipment and software used in researching this article.

4 Compression algorithms reduce digitized images to manageable size

Compression algorithms for B&W images or text

This informative article deals with compression algorithms designed for black-and-white images, such as used to store technical drawings, or those used to transmit FAX messages, such as the CCITT Group 3 standard.

William C Warner
Consultant

In a computer, engineering drawings, artwork, and other nontext documents exist as digitized images. An image of a C-size engineering drawing (17 × 22 in.) commonly contains more than 14 megabits of data; an image of an E-size drawing (34 × 44 in.) contains more than 57 megabits. Storing or transmitting this much data per image is a problem. Using the algorithms presented in this article, you'll be able to significantly reduce the size of digitized images.

Scanners and video cameras digitize documents, coding them as ones and zeros. These devices examine a large number of spots, of very small area, on the face of a document. They render spots that are sufficiently dark as ones and spots insufficiently dark as zeros (Figure 4.1). In this way a document on paper or microfilm becomes an array of bits – something a computer can store, manipulate, and transmit.

The number of bits in an image depends on the size of the original document and the resolution at which you scan it. Resolution is a measure of precision. The unit for resolution is dots per inch (dpi). The resolution squared is the number of spots a scanner or camera examines and digitizes in one square inch of the original document. Common resolutions are 100, 150, 200, 240, and 300 dpi. Industry is standardizing on 200 dpi for engineering documents.

The greater the resolution, the more like the original the digitized image will appear when the computer displays or prints it. Because the quality goes up as the resolution goes up, increasing the resolution seems attractive. But the amount of data you get rises with the square of the resolution, so practical limits come up quickly.

Therefore, designers look for ways to reduce the amount of data in images without reducing the resolution and losing information. They solve this problem by encoding the data in special ways. The encoding, called compression, results in less data – often much less. The amount of data before compression divided by the amount after compression is the compression ratio. You decompress, or expand, compressed data to recover the original, raw-image bits.

The simplest compression algorithms yield compression ratios of 2:1, sometimes 5:1. The most sophisticated compression algorithms achieve 10:1, often 20:1. And nothing gets lost. Hardware or software can recover the full-blown image, bit for bit, from the compressed data.

The better algorithms work so well and are in such wide use that CCITT has issued standards for them, and IC makers offer chips that compress data in accordance with these standards[1]. A fast CPU can also compress images using software. Compressing by software is slower than by hardware, but designers sometimes offer software compression as an option for applications where reduced speed is acceptable.

```
Line
 #
 0    ...0000000000000000000000000000000000000000000000...
 1    ...0000000000000000001110000000000000000000000000...
 2    ...0000000000000000110011100000000000000000000000...
 3    ...0000000000000001110001100000000000000000002000...
 4    ...0000000000000011100001110000000000000000000000...
 5    ...0000000000000111011111111100000000000000000000...
 6    ...0000000000000111111111111100000000000000000000...
 7    ...0000000000001110000000111100000000000000000000..
 8    ...0000000000011110000000111100000000000000000000...
 9    ...0000000000111100000000011110000000000000000000...
10    ...0000000000000000000000000000000000000000000000...
```

Figure 4.1 In this raster image of the letter 'A', a scanner has rendered dark places on the page as 1s and light places as 0s

Like any specialized field, data compression has its own vocabulary. For example, common terms for uncompressed image data are *bit mapped, raster*, or *raw*. Think of the uncompressed image data as an array of bits or pixels (Figure 4.1). One row of the array corresponds to one line of a document. At 200 dpi, a document has 200 lines per vertical inch, and each line has 200 pixels per horizontal inch. Usually, the number of pixels in each line divides evenly by eight so that lines end at byte boundaries. The pixels' values are colour coded: ones are black and zeros are white (although some systems reverse the colour assignments).

A 'run' is a succession of pixels all of one colour. Runs alternate in colour and stop at the end of lines. 'Run length' is the number of pixels in a run. A busy image has a small average run length, and a clean image has a large average run length. Compression is greater and occurs faster on images with a large average run length. Very busy portions of an image may suffer negative compression; that is, compression may produce more data than it replaces (a compression ratio less than one).

Hardware will compress or expand a typical 300-dpi A-size image in three to four seconds, using the most efficient algorithms. Software takes somewhat longer. With a 20-MHz 80386 CPU, software compresses or expands the same image in six to seven seconds. (These times include writing or reading the compressed data – typically 55 k bytes – to or from a hard disk.) Larger images compress and expand in proportionally longer times, unless they are significantly more or less busy.

Average run length, or busyness, has the greatest influence on compression and expansion time for software methods (also on the degree of compression). The shorter the average run, the more runs the software must handle. Converting between raw data and run lengths accounts for a large part of processing time. The routines that perform these operations should execute efficiently; you'll probably have to write them in assembly language. The total time to process an image is, essentially, the product of the time to process a run, the average number of runs per line, and the number of lines in the image. Hence, you should strive to process each run as quickly as possible.

Simple run-length encoding is the easiest way to compress image data and is the basis for more sophisticated methods. No standards exist for this simple method, so what follows describes typical implementations.

Run-length encoding replaces runs by their lengths (Figure 4.2). Instead of saving all the pixels in a line, you calculate the lengths of all the runs and save those values. You do this calcula-

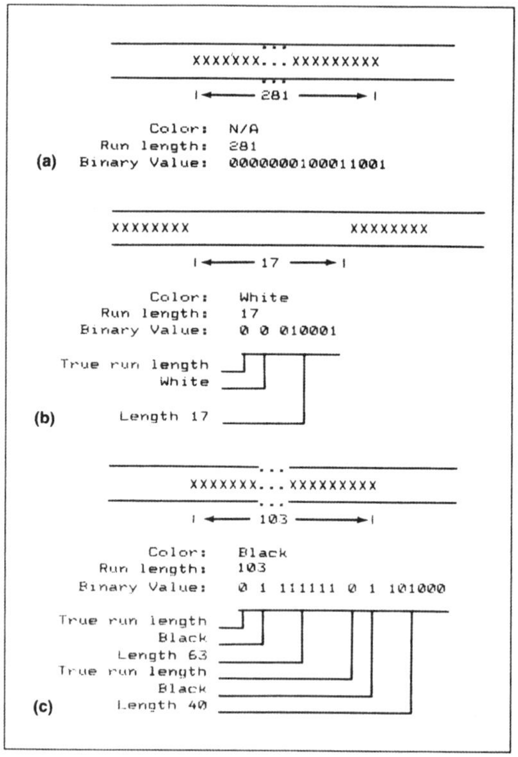

Figure 4.2 Two-byte, run-length encoding (a) reduces any pixel run to a 2-byte substitute code. On the other hand, 1-byte encoding can produce 1-byte substitute codes for short runs (b) and sometimes multiple 1-byte codes for long runs (c)

tion separately for each line. Some schemes store run lengths as 1-byte values; others use two bytes.

The only advantage of 2-byte run-length values is that two bytes can express the largest runs arising in practice. Consequently, your encoded data will be regularly spaced, and successive values will correspond to successive runs. If you assume that every line begins conventionally with a white run, perhaps of zero length, you can infer the colour of a given run from its position in the sequence of coded data.

One-byte schemes, on the other hand, must sometimes actually devote two or more values to a single long run, so successive values don't necessarily correspond to successive runs. Consequently, 1-byte schemes must use one bit of each run-length value to denote the colour of the run. The sacrifice of another bit means that even more runs may be too long for a single byte to express.

Both methods waste bits. Whenever you store a value with leading zeros, you are using more bits than you need to express that value. You can't eliminate the wasted bits by running values together because, later, you wouldn't be able to find where one value stops and the next begins. Nevertheless, the number of bits that a run-length value takes up – 8 or 16 – is usually less than the number of pixels it replaces.

Literals handle exceptions

Busy areas of an image sometimes suffer negative compression. Therefore, when doing so will avoid negative compression, some run-length schemes include bytes of literal, unencoded pixels amid the run-length codes. Typically, such schemes use the MSB of a run-length value as a flag to indicate whether the value is a true run length. When it's not a true run length, the value specifies the number of literal, unencoded bytes of pixels that follow.

The study of digitized documents reveals that some run lengths occur much more frequently than others. Researchers determined the probabilities, which are not the same for white and black pixels, and developed tables of codes to substitute for certain run-length values. The codes vary in size, and the tables assign shorter codes to more likely run lengths. More importantly, the codes eliminate wasted bits because you may splice the codes together without respecting byte or word boundaries.

If you assume that a line begins with a white run, perhaps of zero length, then you can analyze the concatenated codes and always find where one code ends and the next begins.

The substitute codes in Tables 4.1 and 4.2 are the basis of the CCITT Group 3 image-data compression algorithms.

Group 3 1-dimensional (G3–1D) encoding is a two-step process: run-length encoding followed by substituting codes for run-length values. G3–1D is one-dimensional because, like simple run-length encoding, it treats each line separately. (Two-dimensional schemes encode one line with reference to its preceding line.) G3–1D typically yields compression ratios of from 5:1 to 10:1.

G3–1D replaces runs of as many as 63 pixels with a single 'terminating' code (Table 4.1); it replaces longer runs with a combination of 'make-up' codes (Table 4.2) and, even if the make-up codes express the run exactly, a final termina-

Table 4.1 Terminating codes

White codes	Run lengths	Black codes
00110101	0	0000110111
000111	1	010
0111	2	11
1000	3	10
1011	4	011
1100	5	0011
1110	6	0010
1111	7	00011
10011	8	000101
10100	9	000100
00111	10	0000100
01000	11	0000101
001000	12	0000111
000011	13	00000100
110100	14	00000111
110101	15	000011000
101010	16	0000010111
101011	17	0000011000
0100111	18	0000001000
0001100	19	00001100111
0001000	20	00001101000
0010111	21	00001101100
0000011	22	00000110111
0000100	23	00000101000
0101000	24	00000010111
0101011	25	00000011000
0010011	26	000011001010
0100100	27	000011001011
0011000	28	000011001100
00000010	29	000011001101
00000011	30	000001101000
00011010	31	000001101001
00011011	32	000001101010
00010010	33	000001101011
00010011	34	000011010010
00010100	35	000011010011
00010101	36	000011010100
00010110	37	000011010101
00010111	38	000011010110
00101000	39	000011010111
00101001	40	000001101100
00101010	41	000001101101
00101011	42	000011011010
00101100	43	000011011011
00101101	44	000001010100
00000100	45	000001010101
00000101	46	000001010110
00001010	47	000001010111
00001011	48	000001100100
01010010	49	000001100101
01010011	50	000001010010
01010100	51	000001010011
01010101	52	000000100100
00100100	53	000000110111
00100101	54	000000111000
01011000	55	000000100111
01011001	56	000000101000
01011010	57	000001011000
01011011	58	000001011001
01001010	59	000000101011
01001011	60	000000101100
00110010	61	000001011010
00110011	62	000001100110
00110100	63	000001100111

Table 4.2 Make-up codes

White codes	Run lengths	Black codes
11011	64	0000001111
10010	128	000011001000
010111	192	000011001001
0110111	256	000001011011
00110110	320	000000110011
00110111	384	000000110100
01100100	448	000000110101
01100101	512	0000001101100
01101000	576	0000001101101
01100111	640	0000001001010
011001100	704	0000001001011
011001101	768	0000001001100
011010010	832	0000001001101
011010011	896	0000001110010
011010100	960	0000001110011
011010101	1024	0000001110100
011010110	1088	0000001110101
011010111	1152	0000001110110
011011000	1216	0000001110111
011011001	1280	0000001010010
011011010	1344	0000001010011
011011011	1408	0000001010100
010011000	1472	0000001010101
010011001	1536	0000001011010
010011010	1600	0000001011011
011000	1664	0000001100100
010011011	1728	0000001100101

White or Black Run lengths	Codes
1792	00000001000
1856	00000001100
1920	00000001101
1984	000000010010
2048	000000010011
2112	000000010100
2176	000000010101
2240	000000010110
2304	000000010111
2368	000000011100
2432	000000011101
2496	000000011110
2560	000000011111
EOL	000000000001

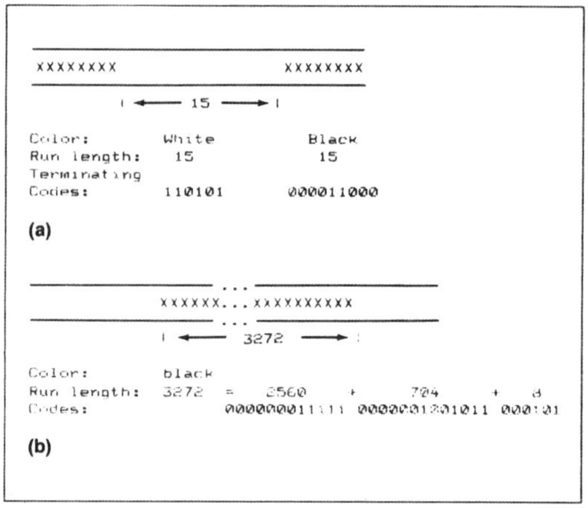

Figure 4.3 G3–1D encoding of short white and black runs uses only the terminating codes from Table 1 (a). A long run (b) uses both make-up codes from Table 4.2 and a terminating code from Table 4.1

ting code. Figure 4.3 gives examples of G3–1D encoding. Images begin and end on byte boundaries. End-of-line (EOL) codes (Table 4.2) mark the start of an image and the end of each line, and an end-of-page (EOP) code (comprising six contiguous EOLs) marks the end of an image. The first run in each line is white, by convention, even if its length is zero.

A few options for this particular standard algorithm are available. You may pad each line with 0-value bits between the end of data and the start of the final EOL code. Such padding makes the EOL codes, and hence each line, end on a byte boundary. You may also specify a minimum line length, in bytes of compressed data, to guarantee that compressed lines, when transmitted, will meet some minimum transmission time. Fill bits (zeros) flesh out lines that are too short (Figure 4.4)

Another option is the order of bits within bytes (Figure 4.5). Image-data bits reside in bytes of computer memory, starting at some address, and filling successive bytes upward into higher memory. The stream of bits (which is the image) can enter each byte at bit seven and leave at bit zero, or enter at bit zero and leave at bit seven. Designers can make the choice of bit order independently for raster and compressed data.

Obviously, to expand compressed images, you must know whether the compressed data includes padding or filling, and you must know the bit order of the compressed and raster data.

The big payback in image compression comes with the step up to two-dimensional (2-D) encoding. Two-dimensional schemes encode each line with reference to the preceding line. Two-dimensional encoding achieves high compression because, in images of real drawings, many lines differ very little from the lines just above and below them. (A glance back at Figure 4.1 will show you that each line differs only slightly from the line above it.) Compression ratios of 10:1 are common, and 20:1 is not uncommon.

Research shows that 75% of transitions from one color to another occur directly below or within one pixel to the left or right of a transition on the line above. Two-dimensional encoding exploits this regularity by employing very short codes to express a 'coding' line in terms of the line above – the 'reference' line. In places where the coding and reference lines aren't sufficiently similar, 2-D schemes use G3–1D run-length substitution codes to encode that portion of the coding line. Each raw coding line becomes the reference line for its successor.

The important features in a line of pixels are the points at which the pixels change colour. These

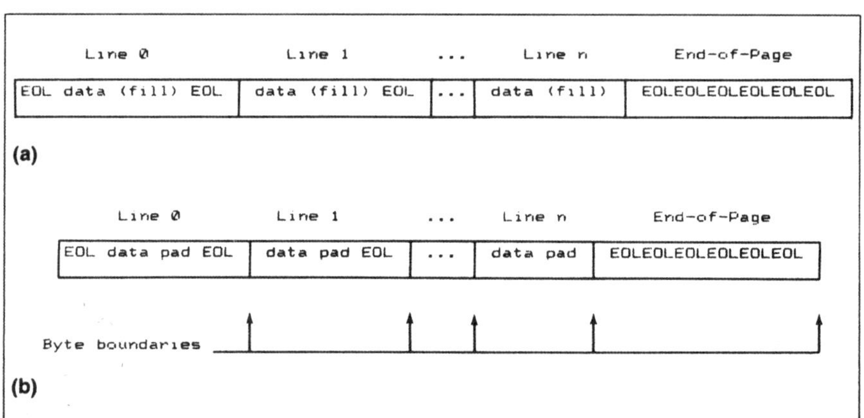

Figure 4.4 G3–1D encoding allows several options, such as encoding with optional fill but no byte alignment (a), or encoding with byte alignment but no fill (b)

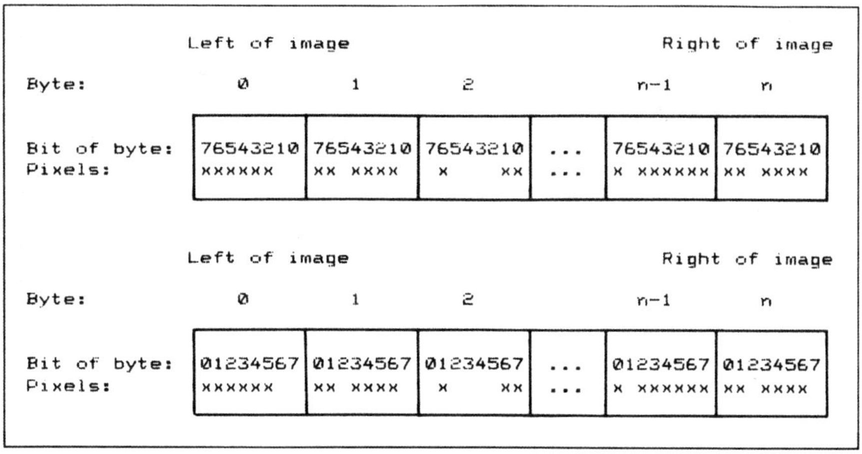

Figure 4.5 Hardware or software that will work with compressed images must know which bit-ordering scheme to use

are transition points, or changing points; every pixel that isn't the same colour as its immediate predecessor is a changing pixel. Compression algorithms look for and then classify these changing points before deciding how to encode a line.

For 2-D compression, five changing pixels figure into the encoding of each segment of a line. Their labels and meanings are

- A0: The starting pixel on the coding line. At the start of each line, A0 rests on an imaginary white pixel to the left of the first real pixel of the line. A0 moves to the right as you encode each segment of the line.
- A1: The next changing pixel on the coding line to the right of A0.
- A2: The next changing pixel on the coding line to the right of A1.
- B1: The first changing pixel on the reference line to the right of A0 but of opposite colour.
- B2: The next changing pixel on the reference line to the right of B1.

To use the 2-D algorithm, label changing pixels on the coding line and on the reference line (Figure 4.6a). Depending on how the changing pixels on the two lines relate, select one of seven types of 'vertical' codes, a 'pass' code, or a 'horizontal' code.

A portion of a line qualifies for vertical encoding when pixel A1 is directly below B1, or within three pixels right or left of directly below. In this case, store the appropriate vertical code (Figure 4.6b), advance A0 to A1, and place the other labels accordingly (Figure 4.6c).

If pixel A1 is located to the right of B2 (Figure 4.7), store a pass code, advance A0 to the pixel below B2, and place the other pixels accordingly.

If a section of a line doesn't meet the criteria for vertical or pass encoding, store the horizontal code, followed by the G3–1D codes to express the run from A0 to A1, followed by the G3–1D codes to express the run from A1 to A2. Then move A0 to A2, and place the other labels accordingly (Figure 4.8).

Occasionally, pixels will qualify for more than one type of encoding. When that happens, pass takes precedence over vertical, which takes precedence over horizontal. Note that you should always splice each new code to what you have encoded before, without respecting byte boundaries.

Two-dimensional encoding offers an additional, little-used option: the K (error-immunity) factor. Designers may choose to embed 1-D lines at regular intervals amid the 2-D data. If a transmission error occurs, the receiver can use the 1-D line to regain sync. The value of K specifies the number of 2-D lines that will occur between each 1-D line. If K is not infinity, the compressed image begins with a line in 1-D codes, and straight 1-D lines appear every K lines thereafter.

One further standard, CCITT Group 4 compression, yields the greatest compression ratio. Group 4 is 2-D encoding with all the details fixed to yield the greatest compression (Figure 4.9). It allows no embedded, 1-D lines (K = infinity). The first reference line is an imaginary all-white line above the top of the image. No EOLs mark the start of the image or the end of lines. Group 4 has no padding and no filling and does not respect byte boundaries, except that the compressed image starts and ends on byte boundaries. An EOP mark, comprising two concatenated EOLs, follows immediately on the end of the compressed data, with padding to the end of the final byte.

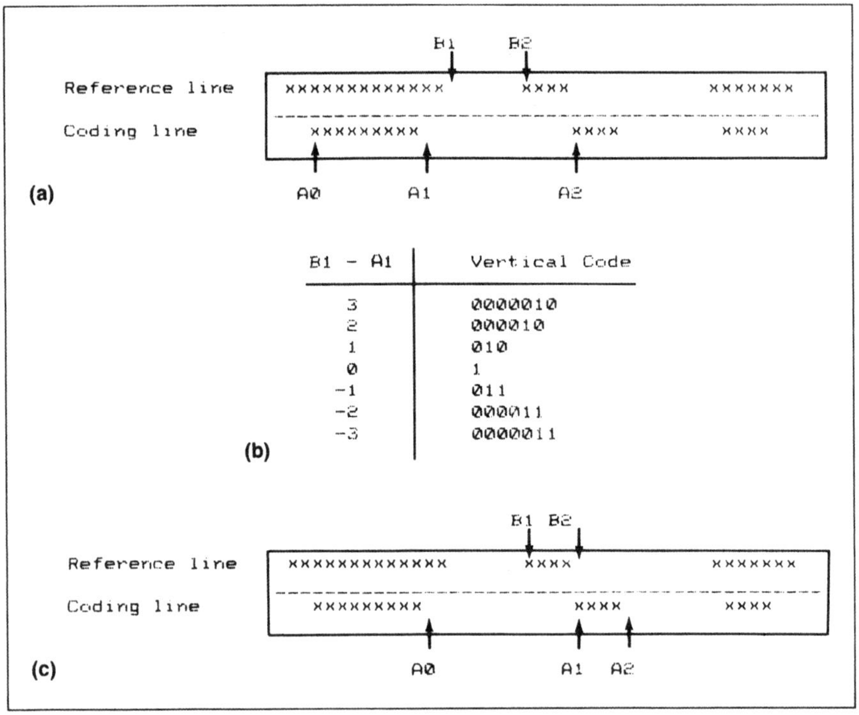

Figure 4.6 Two-dimensional-encoding algorithms find and label certain transition points in the reference and coding lines (a). Then, if the A1 transition is within 3 bits of the B1 transition, the algorithms replace the run in the coding line up to A1 with the applicable vertical code from (b). Next, they reposition the transition labels and repeat the process (c)

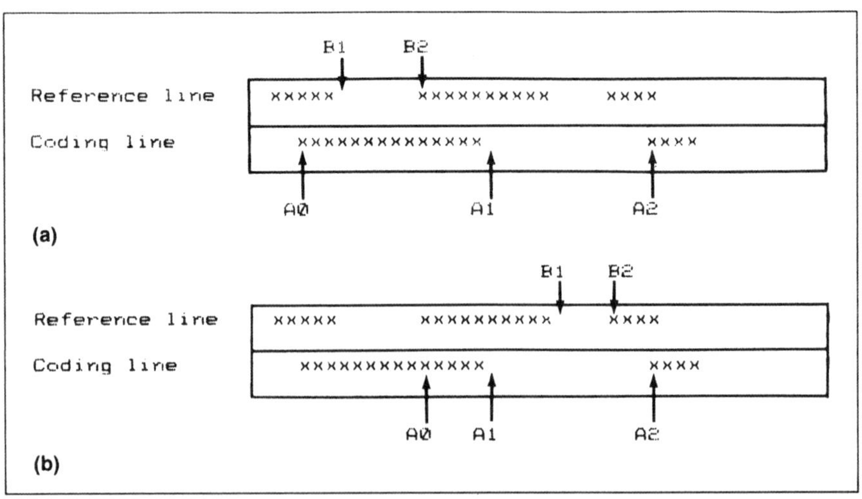

Figure 4.7 In cases where pointer A1 is to the left of B2 (a), insert a pass code (0001) instead of one of the vertical codes from Fig 4.6b. Then reposition the transition labels and move on to encode the next portion of the line (b)

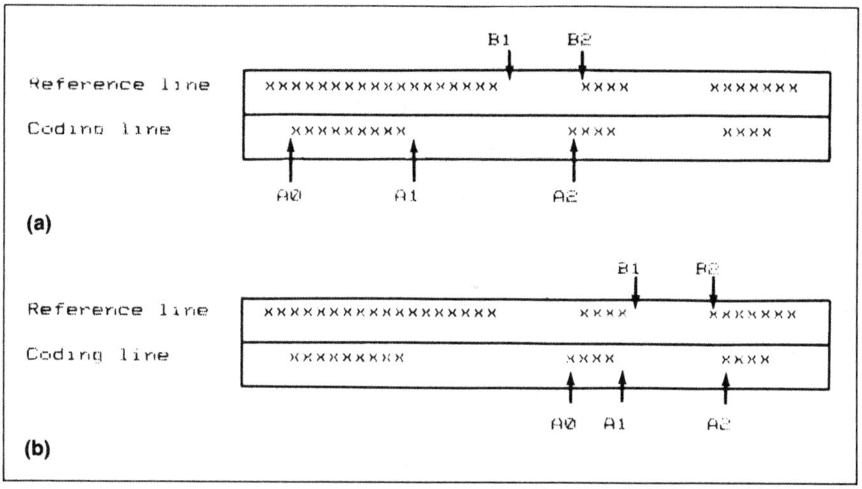

Figure 4.8 In many cases, neither vertical nor pass codes are applicable to the disposition of bits on two successive lines. When A1 is to the left of B1 but not near it (a), insert code 001. Then encode both the run from A0 to A1 and the run from A1 to A2 with G3–1D substitution codes, finally repositioning the labels (b)

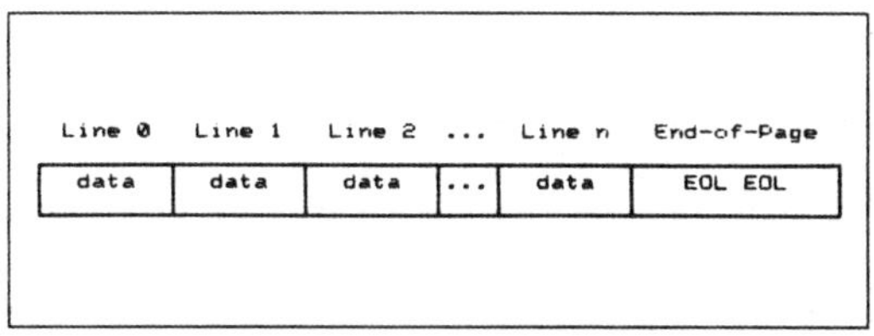

Figure 4.9 CCITT G4 encoding allows no start-of-page code, no EOLs, no padding or filling, and no byte alignment to achieve the most compact compression possible

Glossary

Bit-mapped data: Uncompressed image data.
Busy: A busy image has a small average run length.
Clean: A clean image has a large average run length.
Compression: Encoding bit-mapped data in a more compact form.
Compression ratio: The amount of data before compression divided by the amount of data after compression.
Decompression: Recovering the original raw-image bits from compressed data.
dpi: Dots per inch, a measure of resolution or precision.
EOL: End-of-line indicator.
EOP: End-of-page indicator.
G3–1D: CCITT Group 3 one-dimensional. Run-length encoding followed by substituted codes for run-length values.
Horizontal codes: A code that replaces a horizontal run in a 2-D coding scheme.
K: The spacing between 1-D lines of pixels included in a 2-D encoded image for error-recovery purposes.
Negative compression: Compression that produces more data than it replaces (a compression ratio of less than one).
Padding and filling: Extra bits inserted between the end of data and the start of the final EOL code that make the EOL codes, and hence each line, end on a byte boundary. Fill bits flesh out lines that are too short to guarantee a minimum transmission time.
Pass codes: A code used in 2-D encoding schemes.
Raster data: Uncompressed image data.
Raw data: Uncompressed image data.
Run: A succession of pixels all of one color. Runs alternate in color and stop at the end of lines.
Run length: The number of pixels in a run.
Run-length encoding: Replacing runs by their lengths.
Transition points: Every pixel that isn't the same colour as its immediate predecessor is a transition, or changing point.
Two-dimensional (2-D) encoding: 2-D encoding exploits normal raw data's regularity by employing very short codes to express a coding line in terms of the line above – the reference line.
Vertical codes: A vertical code used in 2-D encoding schemes.

Reference

1. Fuchs, Peter M, 'Compressing data conserves memory in bit-mapped displays', *EDN*, October 30, 1986, p. 173.

5 DSP chips can produce random numbers using proven algorithm

> ## *Uses for a random number algorithm*
>
> Not only is the application (speedy system-response testing) interesting, but the math behind it (which is well explained) is equally intriguing.

Paul Mennen
Tektronix Inc

If you incorporate mathematically proven methods into your programming, your DSP μP can generate reliable random-number sequences. Many of the properties of the 'linear congruential sequence' method of random-number generation have been proven, making it an obvious choice. Without too much work you can translate this method from mathematical theory to general DSP-μP pseudocode to executable code for your DSP μP.

A random-number generator must satisfy two conflicting requirements. First, the generator must be able to repeat the same random sequence many times. This ability allows you to repeat a test exactly. It also helps, when averaging the results of several measurements in the time domain, to reduce the effects of external disturbances. The repeatability requirement is the easiest to meet because any computational algorithm will produce the same sequence every time you execute it if it begins with the same initial conditions.

Second, and almost opposite, the sequence must be truly random. A truly random sequence minimizes the effect of external disturbances by averaging many frequency-domain measurements. This requirement is impossible to satisfy with a deterministic computational device such as a digital signal-processing (DSP) chip. The best you can do is to make the sequence long enough so that it seldom repeats during a measurement. Consequently, the term for such sequences is 'pseudorandom'.

Usually you would want the numbers in the sequence uniformly distributed over the available range. In this case, however, a uniform distribution at the DAC input would produce a Gaussian distribution at the output because of the averaging effects of the DAC's smoothing filter. For frequency-response applications you should seek spectral flatness so that all frequencies can excite the DUT evenly.

Of the hundreds of methods in use for generating random sequences, perhaps the best known method is the 'linear congruential sequence' (LCS) method. One advantage of the LCS is that mathematicians have been able to prove many of its properties. Such proofs reduce the amount of testing required and ensure well-characterized outputs.

The proven method explained

The defining equation for the LCS is

$$R' = (aR + c) \bmod m$$

where R is the random number, a and c are constants, and mod m signifies taking the remainder after dividing by m (modulus). To use this simple algorithm, first you must pick a starting value for R, called 'the seed'. Then multiply R by a constant, add a different constant to the product, divide the sum by m, and assign the remainder resulting from this division to be the next value of R. Repeat these steps every time you require a new random number.

The numbers produced lie between 0 and $m - 1$, and the sequence cannot be longer than m (assuming that a, c, m, and the seed are positive). The number you choose for the seed is unimportant, but you must pick appropriate values for a, c, and m – not a trivial task. Make your choices to satisfy the following six criteria (where $n = \sqrt{m}$):

1. c and m are relatively prime.
2. $a - 1$ is a multiple of every prime factor of m.

3. $a - 1$ is a multiple of 4 if m is a multiple of 4.
4. a and c are both greater than n.
5. $a \div$ n (truncated) is a power of 2.
6. a mod n is less than n $\div$ 2.

The first three criteria ensure that the algorithm's output will be a maximal-length sequence; that is, the sequence will hit all of the numbers between 0 and $m - 1$, in random order, before repeating. (This result is provable, but that is beyond the scope of this article.) Criterion 4 ensures sufficient spectral flatness and distribution uniformity. Criteria 5 and 6 will simplify the computation without compromising the sequence's quality.

Computations for a specific DSP µP

You may want to use the AT&T WE-DSP16 because it is the only DSP µP fast enough to compute random numbers and interpolation filters in real time. These filters vary our Fourier analyzer's random-noise bandwidth because our analyzer's DAC's output rate and smoothing-filter bandwidth are fixed.

Note that choosing m to be a power of 2 simplifies computation because shifts can accomplish the division. The WE-DSP16 includes a 16-bit multiplier, so $m = 2^{16}$ would be a logical choice. Assuming a maximal-length sequence and a 512 kHz output rate (required for a 200 kHz bandwidth), the sequence will repeat every 0.128 seconds (65536 $\div$ 512000). Unfortunately, this sequence is too short for practical purposes. The next logical choice then is $m = 2^{32}$, resulting in a sequence length of 2.33 hours (65536 $\times$ 0.128 seconds). This time is long enough to satisfy demanding test situations.

For this choice of m you need to do a 32-bit multiply. Like other fixed-point DSP chips, the DSP16 has only a 16-bit multiplier. Therefore the 32-bit multiply requires multiple-precision arithmetic, which complicates the code considerably. To formulate a multiple-precision algorithm, define a_{lo} and a_{hi} *to be the lower and higher 16 bits of the constant a.* Similarly, denote the two halves of the random number as R_{lo} and R_{hi}. *Now the LCS algorithm, using multiple-precision arithmetic, becomes*

$$R = c + a_{lo} * R_{lo} + ((a_{lo} * R_{hi} + a_{hi} * R_{lo} + zz) << 16) \quad (1)$$

where
* multiplies two 16-bit numbers (32-bit result),
\+ sums two 32-bit numbers,
<< is a logical left-shift operation, and
zz = a_{lo} if R_{lo} is greater than 32768, or
zz = 0 otherwise.

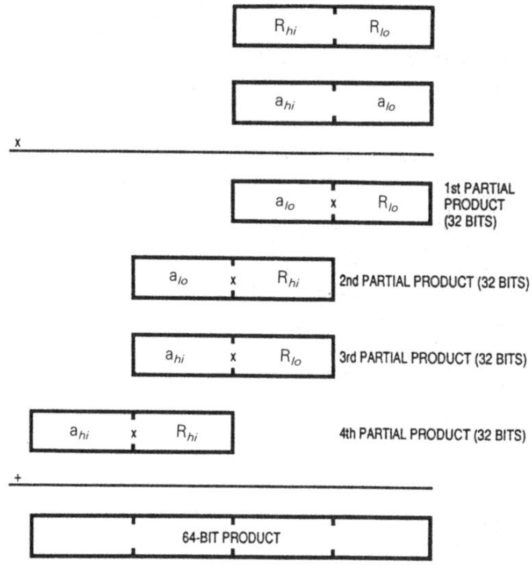

Figure 5.1 Computing a 64-bit product of two 32-bit numbers with a 16-bit, 2's-complement multiplier complicates the algorithm

To see where this algorithm comes from, first look at the multiple-precision product of the 32-bit constant a and the 32-bit random number R. Figure 5.1 shows this operation.

Because the remainder after dividing by 2^{32} (m) is simply the low-order half of the 64-bit product, you don't need to compute the high-order half. Thus you need save only the lower half of the second and third partials, and you don't need to compute the fourth partial at all. Because of criterion 5, a simple left shift computes the third partial, leaving only two required multiplications.

By now you should be able to verify Equation 1 except for the mysterious 'zz' operator. This term arises because partial products require unsigned multiplication for the 64-bit multiple-precision product to be correct. However, most multipliers (including the DSP16s) are 2's-complement, signed multipliers. Using a signed multiply for the second, third, and fourth partial products produces errors only in the unused, high-order half of the product, so only the first partial remains a concern. Although the multiplier will always interpret a_{lo} as a positive number (because of criterion 6), R_{lo} can be any combination of 16 bits. Whenever the most significant bit of R_{lo} is 1, you need to correct the result of the signed multiply (with the zz operator) to fool the multiplier into producing an unsigned result.

For this example, the actual values chosen for constants a and c are:

$$a = 107465_{HEX} \text{ and } c = 234567_{HEX}$$

Listing 5.1

```
TEMP=7465*Rhi          ; second partial product (TEMP is 32 bits)
TEMP=TEMP+Rlo<<4       ; add third partial product ahi=10 (hex)
if Rlo>7FFF            ; if Rlo>32767, correct the first partial
   TEMP=TEMP+7465      ;    product for using a signed multiply
TEMP=TEMP<<10          ; shift left 16 (discard 16 high-order bits)
TEMP=TEMP+7465*Rlo     ; add first partial product
R=TEMP+234567          ; 32-bit add produces the new random number
```

Listing 5.2

```
.ram
ALO:    int                         /* storage for a(lo) - Multiplier       */
CHI:    int                         /* storage for c(hi) - Add constant     */
CLO:    int                         /* storage for c(lo)                    */
.endram
alo:    int 0x7465                  /* a(lo)                                */
chi:    int 0x0023                  /* c(hi)                                */
clo:    int 0x4567                  /* c(lo)                                */
        i = 0                       /* used as table increment              */
        a0 = i                      /* initialize a0 (seed)                 */
        r0 = ALO                    /* points to alo                        */
        rb = ALO                    /* beginning of circular memory buffer  */
        re = CLO                    /* end of circular memory buffer        */
        pt = alo                    /* point to table of constants          */
        do 3 {
              y=a0  x=*pt++         /* save constants alo, chi, and clo to  */
              *r0++ = x             /* memory locations ALO, CHI, and CLO   */
             }
        pt = alo                    /* point to alo                         */
        y=a0  x=*pt++i              /* x will always contain constant alo   */

/* NOW WE ARE SET UP, AND READY TO ROLL OUT THE RANDOM NUMBERS QUICKLY      */

do n {  /* Execute the loop in the instruction cache (needed for speed) */
        y = a0                      /* y = Rhi                              */
        a1l = *r0++                 /* a1l = alo (zz for Rlo >= 32768)      */
        a0 = a0 << 16               /* a0h = Rlo                            */
        if pl a1=a1<<16             /* clear zz if Rlo < 32768              */
        p=x*y  y=a0  x=*pt++i       /* p = alo * Rhi;  y = Rlo              */
        a1 = a1 + p                 /* 2nd partial product + zz             */
        a1 = a1 << 16               /* save only 16 LS bits in a1h          */
        a0 = a0 << 4                /* a0 = ahi * Rlo (3rd partial product) */
        p=x*y  y=a0  x=*pt++i       /* p = alo*Rlo;  yh = ahi*Rlo = 16*Rlo  */
        a0=a1+y  y =*r0++           /* add 3rd partial product, yh = cH     */
        a0=a0+p  yl=*r0++           /* add 1st partial product, yl = cL     */
        a0=a0+y                     /* add c                                */
        pdx0 = a1                   /* output high part of R                */
     }
```

You can verify that these constants satisfy the six criteria. (Note, however, that many other equally fine choices exist.) Using these values for a and c, you can write Equation 1 in psuedocode as follows. All constants in the psuedocode are in hexidecimal. As before, R_{lo} and R_{hi} *are the lower and upper 16 bits, respectively, of the random number* R. (See Listing 5.1.)

Listing 5.2 contains the AT&T WE-DSP16 code segment, which computes this LCS sequence. The computations' order maximizes the number of parallel operations. Note that you have computed a 32-bit random number, but you only need 16 bits to send to the DAC. You should choose the upper 16 bits because they are more random than the lower half.

Generating each random number takes only 15 cycles. Therefore, if you were to use the fastest version of the DSP16A (25 ns clock), and the DSP chip was doing nothing else, you could generate a new random number every 375 ns. In our case however, the DSP chip has additional work. The LCS sequence must go through a series of digital interpolating filters to lower the bandwidth when required. The filters' output then goes to the output DAC at a fixed rate of 500 k samples/s, providing random-noise bandwidths to 200 kHz.

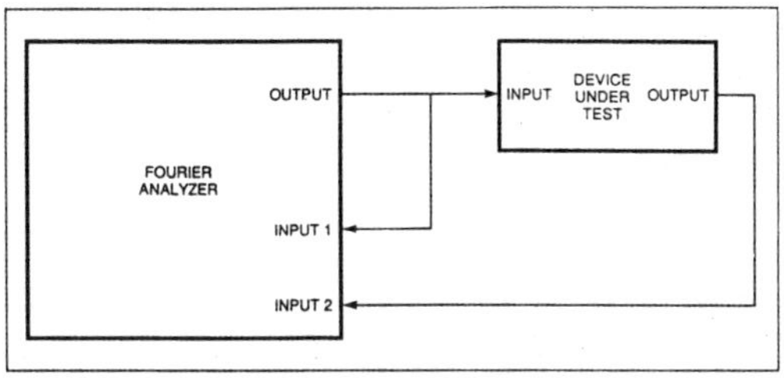

Figure 5.2a A Fourier analyzer derives a frequency-response plot from the Fourier transforms of the input and output waveforms of the device under test

Random noise speeds testing

Before the days of Fourier analyzers, engineers measured frequency response by exciting the system under test with a series of sine waves and measuring the system's output level at each excitation frequency. Random-noise excitation allows you to make the same measurement much

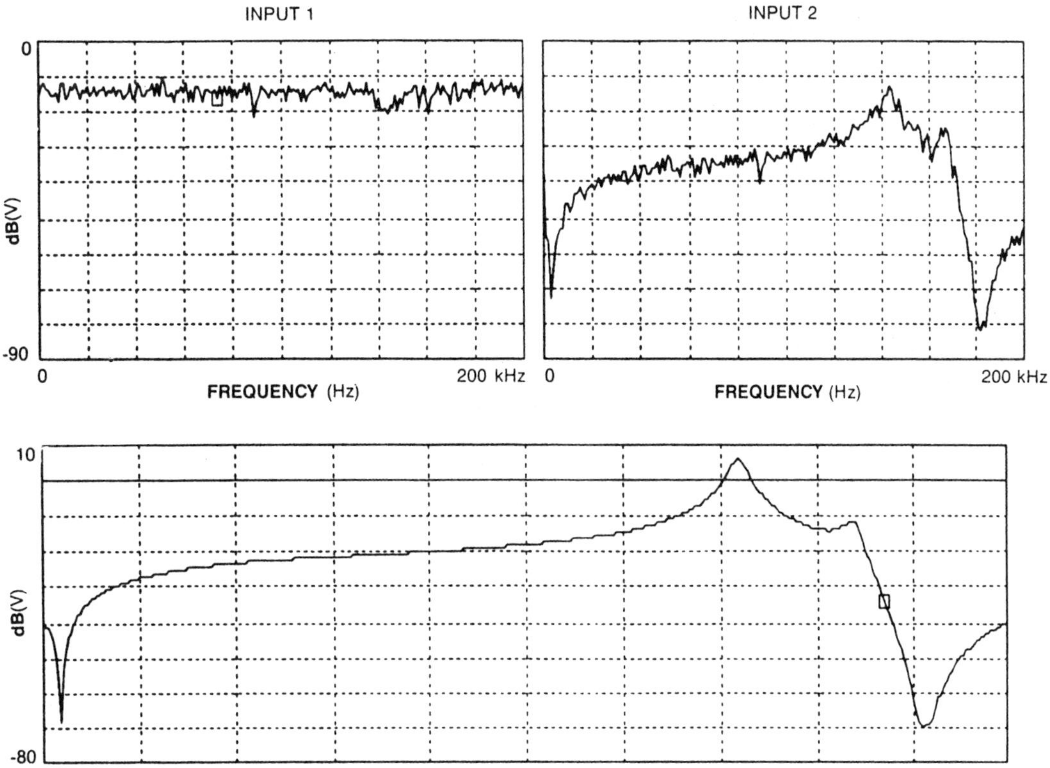

Figure 5.2b A Fourier analyzer derived the lower frequency-response plot from the upper two input and output waveforms

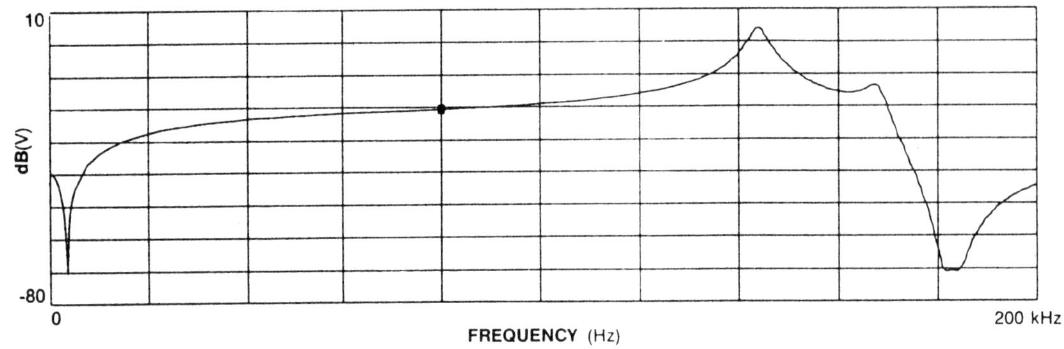

Figure 5.2c A swept-sine analyzer's result is nearly identical to the Fourier analyzer's but takes much longer to derive

faster. The idea is to excite the system with a signal having all frequencies of interest simultaneously instead of just one frequency at a time. You then compare the Fourier transforms of the input and output to generate a frequency-response plot.

Figure 5.2a shows a test setup for measuring the frequency response of an electrical network. Figure 5.2b shows the results from a test between 0 and 200 kHz using a Fourier analyzer. The analyzer computed the two spectra on the upper left and right from the Fourier transforms of Inputs 1 and 2, respectively. The lower trace is the resulting frequency-response curve. This measurement required 1.5 seconds. Repeating the measurement using the swept-sine technique yielded the same result (Figure 5.2c), but took 3 minutes.

6 Fixed-point DSP chip can generate real-time random noise

DSP random noise generator using the LCS algorithm

A description of a DSP implementation of the random number algorithm described in the preceding article, used to generate either uniform- or Gaussian-distributed noise, and its application to testing the performance of MF (multi-frequency) decoders (dial-tone detectors in telephony) in the presence of noise.

Bill Salibrici
TeleSciences Co Systems

As device costs drop and development support grows, DSP chips are finding use in diverse applications, such as in real-time digital waveform synthesis. Increased DSP use raises two questions: can you exploit the fast multiply-accumulate capability of single-chip DSP devices to generate real-time random noise samples, and, if so, how would the performance compare to a commercially available noise generator? To answer those questions, we wrote and tested a DSP application. This application required a signal plus random noise, as is often needed in testing digital filters used to implement tone detection. We developed the digital noise generator (DNG) algorithm to generate real-time, pseudorandom, uniformally or Gaussian-distributed signals.

We used the μPD77P25, a 16-bit fixed-point DSP chip from NEC Electronics, to test the DNG algorithm. The chip is a second-generation, medium-speed, single-chip device. We chose the chip out of convenience; the hardware platform from an existing application was already available. However, the concepts presented in this article are applicable to other fixed-point, as well as floating-point, DSP chips. Figure 6.1 shows a block diagram of the hardware used to test the algorithm. The μPD77P25 has a serial output port that can interface to a simple, low-cost codec, which we used as a DAC.

The codec also contains internal antialiasing filters, which further simplifies the hardware. Other fixed-point DSP chips that include a serial output port are the Analog Devices ADSP-2101, AT&T DSP16, Motorola DSP56000, and Texas Instruments TMS320C25. The sample rate used here is 8 kHz, which means that the highest frequency component contained in any synthesized output waveform must be band limited to 4 kHz or less to prevent aliasing errors.

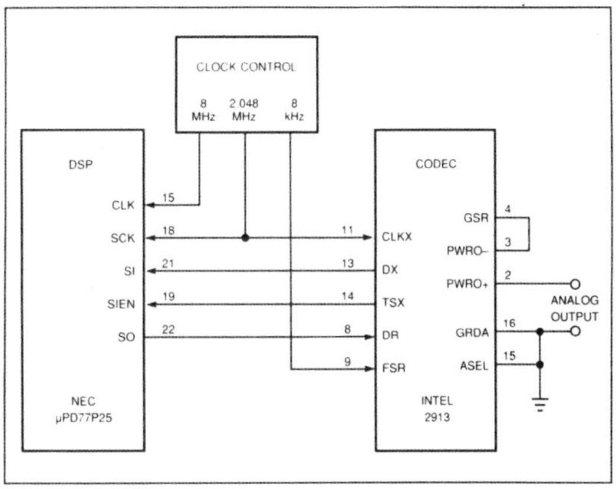

Figure 6.1 To test the noise produced by the DNG algorithm, we used an existing platform that included the DSP and codec

The linear congruential generator[1,2] was used to generate a uniformly distributed random-number sequence given as follows:

$$x(n + 1) = [a \times x(n) + c] \bmod m, \quad (1)$$

where a, c, and m are constants; mod is the modulus operator; $n \geq 0$; and $x(0)$ is the seed or starting value.

For properly chosen constants and a given seed value, this algorithm produces a uniformly distributed, pseudorandom sequence of positive integers between 0 and m. The term pseudorandom simply means that you use a deterministic method to generate a sequence of numbers that appear to be independent draws from a uniform distribution. Equation 1 basically states that you can obtain the next value in the sequence from the previous value through several operations, including the multiply-add combination for which DSP devices are well suited. The modulus operator gives the remainder of one number divided by another. If m is chosen to be the word size of the DSP chip, then the modulo m operation becomes quite simple. The operation (X) mod m is accomplished by taking m LSBs (least significant bits) of X. For example, ($a \times$ X) mod m is simply the product ($a \times$ X) truncated to m LSBs.

One method to generate a Gaussian (or normal) distributed random-number sequence is to apply the Central Limit Theorem to batches of uniformly distributed random numbers[3,4]. You can mathematically express this idea as follows:

$$g(n) = \frac{1}{B} \sum_{i=0}^{B-1} x(i), \qquad (2)$$

where the x(i) are uniformly distributed, independent random numbers; B is the batch size; and g(n) is a member of a set of random numbers approximately Gaussian distributed.

This algorithm produces a pseudorandom sequence whose distribution is close to Gaussian, provided that the batch size B is greater than 10. If you choose a power of 2 as your batch size, then you can use a shift-right operation to simplify the required division within the DSP chip.

The data representation internal to the 77P25 is the 16 bit, 2's-complement, fractional format. This means that the most significant bit (MSB) is the sign bit followed by an implied binary point. The remaining 15 bits contain the fraction magnitude. You can represent the range of numbers –1.0 to (1 – 2^{-15}) in this format, or you can express them with hexadecimal numbers as 8000H to 7FFFH.

To apply the linear congruential generator in this case, we used $m = 2^{15}$, which is the size of the fraction magnitude. Then Equation 1 will generate positive integers between 0 and 32768.

However, the interpretation of these integers is x(n)/m, so that the end result will be a sequence of numbers that lie between 0 and 1. References 1, 2, and 5 discuss some criteria for choosing the four constants of Equation 1. The modulus m should be as large as possible so that the sequence it generates has a long period. The multiplier a should lie between 0.01 and 0.99 m, such that a mod 8 = 5. For example, a value of a = 9821 satisfies this criteria. The increment c is immaterial when a is a good multiplier, except that it must have no factor in common with m. For example, a value of c = 6925 satisfies this criteria.

You may arbitrarily choose the seed value x(0). Notice that a given seed value will exactly reproduce the same stream of random numbers. This feature may be useful in those applications where identical random numbers would yield a more precise comparison of the performance of different systems. For most of this article's data, 11186 is the seed value, although several other values were also tested. The linear congruential generator now becomes:

$$x(n + 1) = [9821 \times x(n) + 6925] \bmod 2^{15}. \qquad (3)$$

The LSBs of the numbers Equation 3 generates are not very random, but cycle with regular patterns. This nonrandom behavior is characteristic of all generators of Equation 1's form, where m is some integer power of 2. However, the MSB cycles with a full period. Therefore, the sequence Equation 3 generates has a full period of length m (i.e. the sequence repeats after m values).

Figure 6.2 shows the algorithm used to implement Equation 3 for uniformly distributed pseudo random numbers. Notice that Equation 3 is partitioned so that [$a \times$ x(n)] mod m is calculated first, and then the modulo m addition to c is done. Figure 6.3 shows the algorithm used to implement Equation 2 for Gaussian-distributed pseudorandom numbers. The batch size for Gaussian noise was 16, and the uniformly distributed random numbers come from Equation 3. Notice here that each element of the summation is divided by the batch size before adding it to the running sum. You do this to prevent overflow

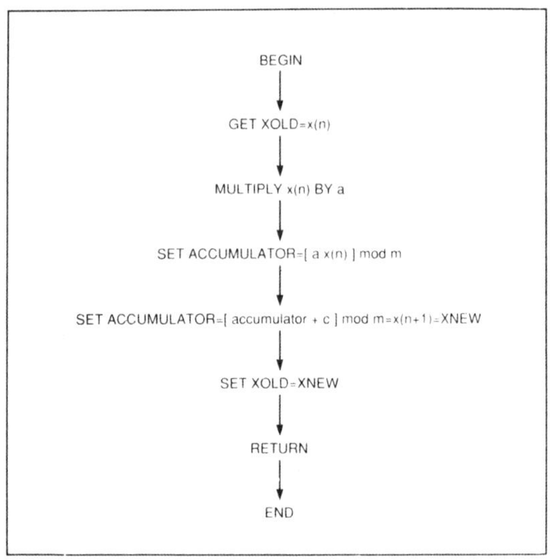

Figure 6.2 You can generate uniformly distributed, pseudorandom numbers with the program illustrated here

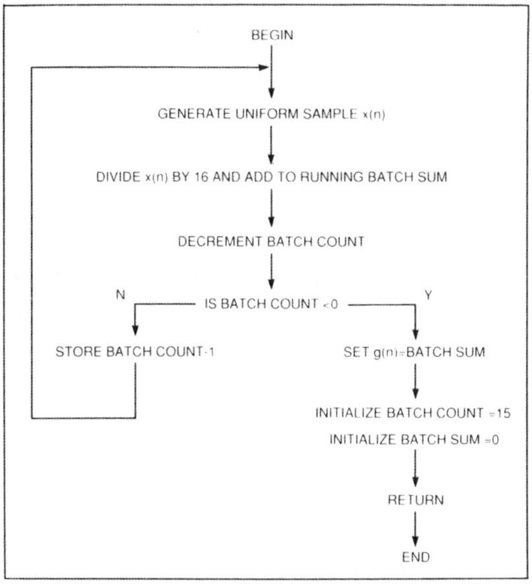

Figure 6.3 You can generate Gaussian-distributed, pseudorandom numbers by taking the output of the uniform algorithm described in Figure 6.2 and following this flowchart

errors. You can easily apply these algorithms to other single-chip DSP devices with perhaps minor changes due to architectural differences.

Listing 1 shows the source code used to implement the Uniform algorithm on the 77P25. Listing 2 shows the source code used to implement the Gaussian algorithm. The Listing Table shows the first five numbers, in integer format, of the Uniform and Gaussian output sequences. (Both Listings and the Listings Table are located on the *EDN* BBS.)

Evaluating the results of several simulation programs addressed the question of how random the DNG is. These simulations showed that the raw DNG noise distributions have the expected shape. The results of two statistical tests for proper fit indicate that the hypothesis of normality for the Gaussian case is accurate. The autocorrelation function of the DNG compared to the drand48 function (a function Sun Microsystems Inc's C Library provides) are slightly different, but not radically so.

We performed further tests using the hardware design shown in Figure 6.1. We scaled the internal DNG random numbers for a zero mean by subtracting ½ and expanded them to full scale by multiplying by 2 so that the DSP chip's output range was between –1 and +1. In order to use the codec DAC, we performed a linear-to-μlaw software conversion before outputting the noise sample. A lowpass filter inside the codec band limits the analog noise output signal to approximately 3 kHz. The codec output has a full-scale excursion of ±3 V, which introduces a voltage gain factor of 3 between the DSP chip's output and the final analog output.

These lab tests showed that the measured values for the mean and variance are close to theoretical values. Also, the Gaussian frequency spectrum of the DNG compares well to that of a commercially available noise generator for the 3 kHz band-limited case.

We needed the Gaussian noise source in order to test the detection of multifrequency (MF) tones in the presence of Gaussian noise. The Bell Communication Research LSSGR[6] defines the required MF decoder performance. We modified our internally developed DSP-based waveform synthesizer to incorporate the DNG algorithm. Thus, we had a composite real-time signal consisting of MF tones plus Gaussian noise to use as a source. We used an already existing user interface to vary the levels of the tones and noise.

Table 6.1 shows the results of the MF decoder performance for the DNG versus the GenRad 1390B, which is a commercially available random noise generator. In all cases, we used nominal MF frequencies, and the level of each tone was set to –22 dBm at 600 Ω. The MF tone-switching rate was 70 ms ON and 70 ms OFF.

Both noise sources performed equally well for a noise level of –30 dBm, which actually exceeds the LSSGR requirements. We then raised the noise level to cause the MF decoder to fail (i.e. miss digits). The two noise sources differ by 8.5% for the noise level required to produce the same MF decoder-error rate. After accounting for variations in the hardware test configurations, approximately a 5% difference remains between the DNG and 1390B noise levels. The internal structural differences of the noise sources themselves appear to be the cause of this difference. For this test case, the MF decoder appears to tolerate a slightly higher level of Gaussian noise from the DNG than the 1390B, for a given detection error rate.

The 16-bit, fixed-point DNG real-time performance compared favorably to that of a commer-

Table 6.1 DNG output sequence values

	Uniform x(n)		Gaussian g(n)	
Iteration, n	Hex Value	Decimal Value	Hex Value	Decimal Value
1	66B7	26295	347F	13439
2	1588	5512	340F	13327
3	1D75	7541	379F	14239
4	2C8E	11406	372F	14127
5	5EA3	24227	4ABF	19135

cially available noise generator for moderate DSP sample rates. The fixed-point DNG performance is probably adequate for those applications where the accuracy of the random distribution is not of paramount importance. The DNG algorithms can offer a flexible and cost-effective solution for random noise generation, especially in those applications where you are already using DSP techniques.

You could increase the DNG's noise bandwidth if you increased the sample rate. However, this increase impacts hardware cost, especially for the DAC and the output lowpass filter needed. The commercial unit used for comparison here can generate noise up to 5 MHz in bandwidth.

A major deficiency of the fixed-point DNG is the relatively small word length. Reference 1, for example, recommends that the word length be at least 30 bits. Another deficiency has to do with the dynamic range (overflow and scaling considerations). You could address both of these deficiencies by using a 32-bit, floating-point DSP chip. However, floating-point DSP devices are typically slower and more expensive than their fixed-point counterparts.

Increasing the batch size would improve the distribution of the Gaussian sequence at the expense of increased computational load. The tradeoff is between the accuracy of the output distribution and the DSP chip's speed. This tradeoff ultimately limits the bandwidth of the resultant noise signal.

References

1. Knuth, D E, *The Art of Computer Programming, Volume 2 / Seminumerical Algorithms*, Addison-Wesley, 1981 Second edition, pp. 9–24 and 170–173.

2. Law, A M, and W D Kelton, *Simulation Modeling and Analysis*. McGraw-Hill, 1991 Second edition, pp. 424 to 427.

3. Davenport, W B, and W L Root, *An Introduction to the Theory of Random Signals and Noise*. IEEE Press, 1987, chapter 5.

4. Brown, A D E, 'Generate Gaussian Noise with a DSP Chip', *EDN*, June 23, 1988, pp. 183–184.

5. Park, S K and K W Miller, 'Random Number Generators: Good Ones are Hard to Find', *Communications of the ACM*, Volume 31, Number 10, October 1988, pp. 1192–1201.

6. Bell Communications Research, TR-TSY-000064 LSSGR, Revision 2, June 1985, Section 6.4.

Acknowledgement

The author would like to express appreciation to Dr Ken Boudwin of Exxon Research and Engineering, Florham Park, NJ, for his help in understanding the statistical tests for randomness, and to Sherrie Gumienny, English Department Head for Life Center Academy, Burlington, NJ, for her review and helpful comments of the article.

Part Two

Analog Technology

Introduction to Part 2

As digital technology advances, real-world analog electronics must keep pace. After all, what's the good of having a sophisticated process-control computer if you can't sense and properly condition the linear signals to feed the computer? This part presents a variety of techniques to help you get the most out of your linear-circuit designs. Much of the material has to do with eliminating problems – those inherent in single-supply designs, boosting common-mode rejection, the need for filtering, and squashing instability problems through compensation, for example. And, lest you succumb to an analog overdose, an article on Spice-based Monte Carlo analysis shows you how to measure the effects of component tolerances on your linear designs.

You'll note that this part contains articles by two of the all-time greats of the electronics trade press: Jerry Graeme and Jim Williams. These analog gurus have produced a staggering volume of useful articles, on a a mind-boggling variety of topics. For your amusement: note the pock-marked, burnt-phosphor oscilloscope traces in Jim Williams's article. I was struck by Jim's reliance on his 'Old Faithful' scope when I first started editing his material for *EDN* in 1982; I'm sure we'll still be seeing the same phosphor-burn patterns at the turn of the century and beyond.

7 Op-amp selection minimizes impact of single-supply design

How to cope with analog when only a +5 V supply is available

At a time when much interest in the digital world is shifting to systems operating at a supply voltage of 3.3 V or lower, much analog processing is only now moving to single supply 5 V working; A/Ds and DACs with 12, 14 or more bits are now available for this supply voltage. For many older op-amps, 5 V working puts severe limits on their capabilities, but more and more new types specially developed for this application are now appearing, some of them particularly aimed at very low curent operation for battery powered systems.

Walt Jung and James Wong
Analog Devices Inc

As system designers begin using single-supply power not just in digital circuits but also in analog circuits, performance often suffers. Yet, the demand for higher performance continues unabated. When designing a single-supply application, you should employ amplifiers expressly designed for the job. Among the performance specs you'll enhance are the supply-power and operating ranges, dynamic range, and input and output ranges. You'll also achieve more linear operation overall.

When most op-amps operate from ± 15 V supplies, they accept an ample input common-mode (CM) range and provide a wide output range – both typically being ± 10 V or more. Single-supply systems, on the other hand, often handle much smaller voltage ranges: 12 V down to 5, 3, or even 1.5 V. At these voltages, most available dual-supply op-amps simply cease to function; a few will operate but with performance degradation. In general, your choice of single-supply amplifiers is much more limited than dual-supply amplifiers.

General-purpose amplifiers such as 741s and 1558s, as well as most FET-input amplifiers, can operate from supplies as low as ± 5 V, or 10 V total. Precision-amplifier families such as the OP-07 and OP-27 do not work below 5 V. Some selected family types may be useful for some regions of this voltage spectrum; for example the OP-97 proves useful down to ± 2.25 V (4.5 V). Typically however, only amplifiers specifically designed for single-supply or low-voltage operation can function without major performance compromises below 10 V total.

Amplifiers designed for ± 15 V operation typically require 2 to 5 V of headroom, with respect to both supply rails, at both inputs and the output. Even the more flexible dual-supply amplifiers approach only within 1 V or so of the rails.

In single-supply applications, this need for headroom limits linear operation over wide signal swings. These limitations are most acute at lower supply voltages. For this reason, single-supply amplifiers' input stages usually remain linear even when the common-mode voltage applied is right at the negative rail (ground) or even slightly below (~ – 200 mV). Op-amp input stages that operate this way are pnp bipolars, CMOS or PMOS types, and N-channel JFET types.

Obviously, devices with these various input stages will have differing input-bias currents, noise voltage and currents, and offset voltage and drift. Spec sheets delineate these characteristics, but a subtle (and potentially important) behavior for a given device is how it reacts to common-mode stresses.

Overdriving many amplifiers below the negative rail (pnp inputs in particular) will usually cause a nondestructive output-phase reversal. The overdriving may not actually damage the device if the input-fault current is limited, but the circuitry around the amplifier may react violently. If the circuitry is a polarity-sensitive servo, for example, it may lock up or oscillate.

Few single-supply amplifiers have internal protection against overdrive. Fortunately, preventing overdrive is rather easy. Simply clamp the input voltage to no more than 300 mV below the negative rail with a low-threshold Schottky diode[1].

You must protect CMOS- and PMOS-input-stage amplifiers more carefully because exceeding their supply rails can trigger a parasitic SCR within the device – with destruction possible. As with any CMOS device, you should keep possible transients in mind and follow data-sheet recommendations.

At the positive end of input-CM range, some devices can operate to within 1 V of the rail; others may need 1.5 V of headroom. Generally, overdriving an op-amp's common-mode inputs above the positive rail causes nonlinearity (as opposed to phase reversal). Again, do not drive CMOS- or PMOS-input devices beyond the positive rail.

With the exception of rail-to-rail output-stage devices, most op-amps can swing only to within 1 to 2 V of the positive rail. Consequently, when these devices operate from supplies of 5 V or less, output swing is greatly reduced. Even if the input's noise floor were to remain constant (which it seldom does), the S/N ratio and dynamic range suffer for devices operating from low voltages. 'Output-stage swing makes or breaks single-supply operation', covers some salient points on output-stage designs for single-supply low-power amplifiers.

Ultralow quiescent current drain

Another fundamental reason for using single-supply designs is to conserve power. For this reason, most single-supply op-amps draw low standby current. Very low-current designs involve fundamental tradeoffs. In general, low-current designs tend to sacrifice bandwidth, slew rate, and input noise voltage.

Standby-current drain per amplifier channel can often make or break a device when a system's power consumption is critical. The industry has no standard definition of 'low power'. We suggest that devices drawing quiescent currents of ≤ 1 mA/channel are low power, and those drawing <100 µA/channel are 'micro-power'. Most of the devices discussed in Chapter 8 meet one of these definitions.

Given a requirement for low-power designs, engineers are forced to face not just slower speeds but also bandwidth restrictions arising from higher circuit impedances.

Noise versus supply current is an issue where tradeoffs are likely. Table 7.1 lists three representative op-amps and shows that lower noise is attainable at the price of increased current.

Table 7.1 Noise-vs-supply-current tradeoffs

Device	Noise (nV/√Hz)	Supply current/channel
OP-295	51	160 µA
AD820/AD822	12.5	750 µA
OP-213	6	1.45 mA

Table 7.2 Current-vs-bandwidth tradeoffs

Device	Current drain (µA/channel)	Bandwidth	Slew rate (V/µsec)
OP-295	160	75 kHz	0.03
AD822	750	2 MHz	3.5
324/358	500	1 MHz	0.6

Bandwidth, too, is another area where tradeoffs are likely, as Table 7.2 shows. The dual amplifiers in Table 7.2 bracket the industry-standard 324 and 358, which feature 500-µA/channel current drain, 1-MHz bandwidth, and a 0.6 V/µsec slew rate. Greater speed also comes at the price of increased current.

Assigning ground is a task that can become quite important in single-supply, ac-coupled op-amp circuits. You can make this 'false' or 'pseudo' ground anything you desire. The best choice depends on the individual application, but maximum amplifier flexibility helps make signal referencing less problematic, particularly if the design has substantial dynamic current into an elevated ground.

Possible choices for false ground range from simple ac-bypassed resistive dividers to fully buffered op-amp follower stages for the lowest wideband dynamic impedance. A well-bypassed, noise-free divider is suitable for high-impedance loads when the dynamic current is low. When dynamic currents are high, use a simple op-amp follower such as the one in Figure 7.1. This circuit

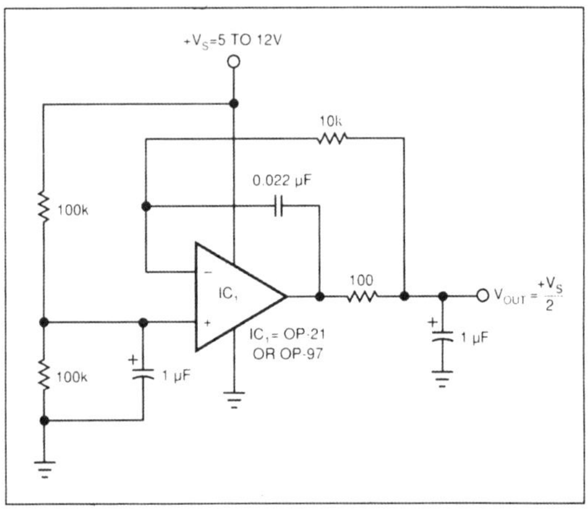

Figure 7.1 You can make your 'false' or 'pseudo' ground any value desired. When dynamic currents are high, use this simple op-amp follower

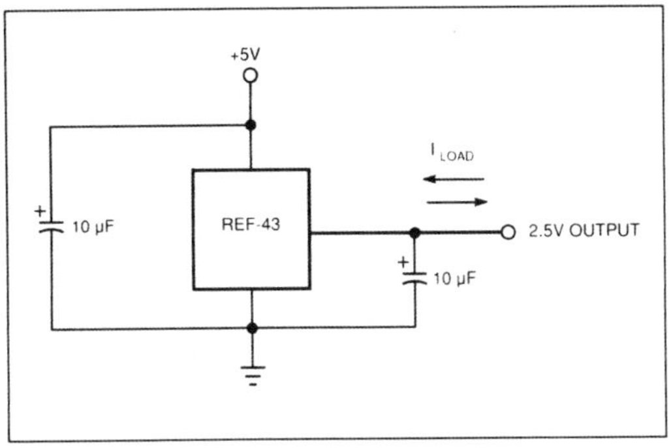

Figure 7.2 Use a reference IC that can source and sink current if you need a precise 'pseudo-ground' voltage. This circuit can handle load currents both into and from the 2.5 V source. Low-ESR bypass capacitors help keep the circuit's ac impedance low

works with a wide range of supplies to produce a low-noise output of $V_{SUPPLY}/2$. Your choice of IC_1 determines the circuit's standby current.

If you need a precise 'pseudo-ground' voltage, use a reference IC that can source and sink current, such as that in Figure 7.2. Operating from a 5 V supply, this circuit can handle load currents both into and from the 2.5 V source. Low-ESR (equivalent series resistance) bypass capacitors help keep the circuit's ac impedance low.

Power-supply noise, if not dealt with correctly, can shred even the best paper design. Low-power circuits tend to have poorer supply rejection, and the noise from commonly used 5 V digital supplies is nearly worst case. Simple power-line filters may not be adequate for high-performance analog stages. Although a common analog and digital supply may be attractive for size and cost reasons (or simply decreed by your boss), you should avoid a common supply if at all possible.

First, power taken from the middle of a logic layout contains huge amounts of high-frequency noise – 100 mV or more. Worse yet, logic supplies are typically switching types, which have large output spikes. Thus, using a separate, linear-mode, low-noise supply for the sensitive analog circuits is better whenever possible.

If you must use a 5 V logic supply, isolation and circuit partitioning as well as optimized decoupling and filtering can help greatly[1]. First, make sure that you tap the supply for analog power right at the supply terminals – not from the logic stages. This routing avoids the voltage drops of the logic-supply runs and allows maximum supply regulation.

Bypass capacitors alone are usually not adequate for filtering switch-mode glitches, so you need to take additional steps. Figure 7.3 shows a more effective technique[2]. This balanced LC filter comprises two high-frequency inductors – each made from a three-turn, toroidal ferrite bead – and a large, composite output capacitor. The capacitors should be low-ESR switching types for best performance. The filter as shown suppresses glitch noise by 40 dB or more.

Even given a device that has a wide output swing, the application's configuration can limit the swing. For example, the 2 op-amp instrumentation amplifier in Figure 7.4 normally uses dual supplies. But in this case, the instrumentation amplifier uses a single supply and has a gain of approximately 100.

At first glance, this instrumentation amp will faithfully amplify a 10 mV differential input to 1 V at V_{OUT} with a low common-mode voltage. However, a close examination proves otherwise. Amplifier IC_1's output must be a *negative* 94 mV to satisfy the loop's requirements and produce 1 V at the output. Obviously, IC_1 cannot produce a negative output from a single supply. Indeed, if the circuit attempts such operation, IC_1 saturates, and V_{OUT} becomes nonlinear.

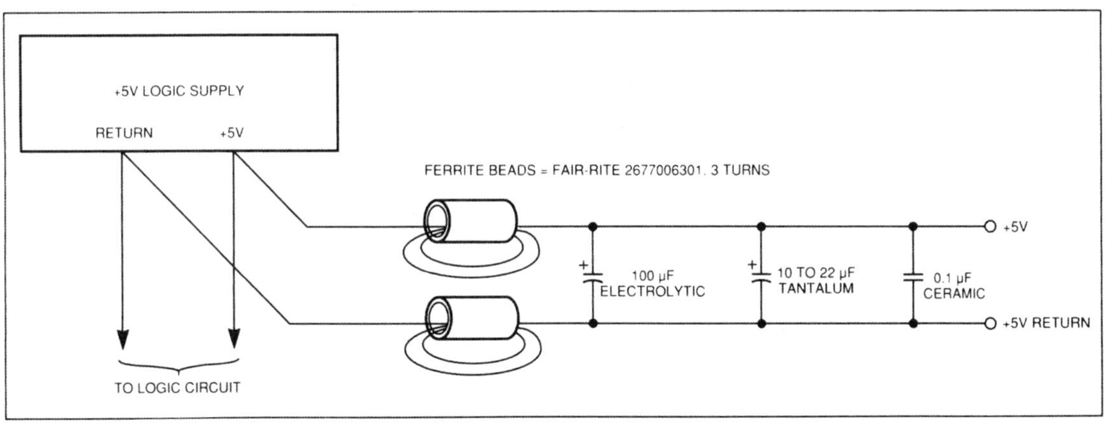

Figure 7.3 This balanced IC filter suppresses glitch noise by 40 dB or more

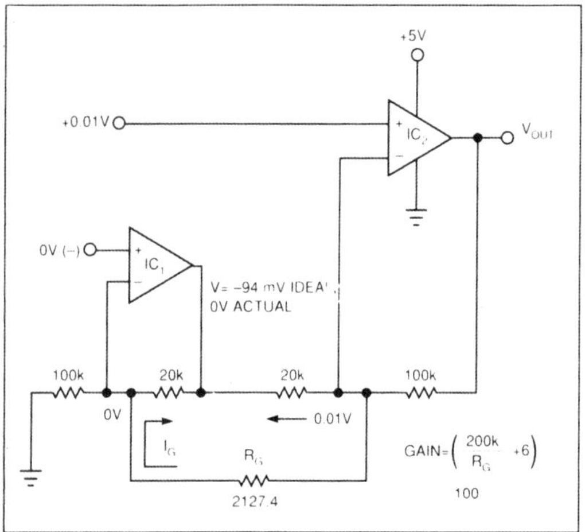

Figure 7.4 This 2 op-amp instrumentation amplifier normally uses dual supplies. Using a single supply constrains the design's common-mode range

As it turns out, this input-CM limitation is a function of the instrumentation amp's gain as well as the output swing of the op-amps within it. Therefore, low negative-V_{SAT} or rail-to-rail output-stage op-amps help improve common-mode handling. In Figure 7.4 the worst-case minimum-CM voltage is 0.4 V (assuming a 5 V output swing and each op-amp swinging to within 100 µV of each rail). However, op-amps that cannot swing to 0 V will have a worse input CM-voltage minimum. The point here is that even though an op-amp could be designed for single-supply operation, your configuration can still constrain the op-amp.

Despite its basic limitations, this topology can still be useful if you apply it carefully. For example, Figure 7.5 shows an accurate strain-gage bridge amplifier circuit in which the bridge and amplifier are powered from 5 V. The amplifier produces linear outputs to as low as 600 µV from ground. In this case, the bridge itself supplies the 2 V common-mode bias, which keeps the op-amps in the middle of their common-mode range. Driving the bridge with a low-noise 4 V supply further minimizes noise. (Part 2 of this series details a low-noise 4 V supply.)

Figure 7.6 shows another instrumentation-amplifier approach. Compared with the Figure 7.4 circuit, the Figure 7.6 circuit's primary advantage is that it doesn't exhibit extreme common-mode limitations. Also, the topology in Figure 7.6 switches divider taps for gain changes, so moderate switch resistances (10 to 100 Ω) do not cause major errors. You can easily set the circuit's gain with jumpers or dpdt CMOS switches. However, adding the same gain-setting switches in series with R_G in Figure 7.4 would produce serious errors.

In Figure 7.6, amplifier IC_{1A} is a follower for signals at terminal − V_{IN}, and IC_{1B} is a follower for signals at + V_{IN} and an inverter for signals at its negative input. The linear subtraction of CM signals and amplification of differential signals provides the circuit with a precise gain that you can vary between 10 and 100.

The performance keys to this circuit are the resistor network and the amplifier. For best

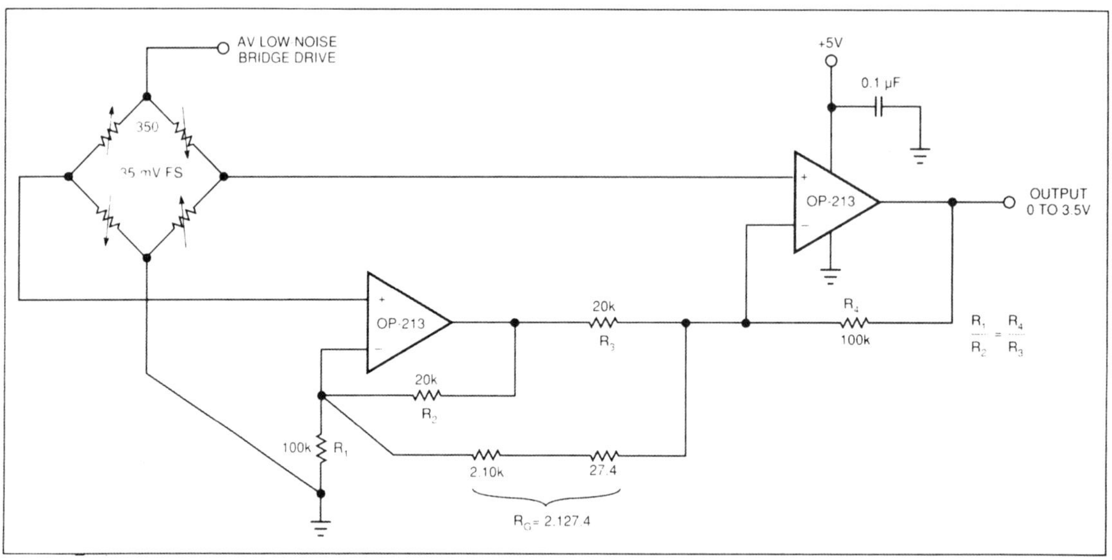

Figure 7.5 This accurate strain-gage bridge amplifier operates from 5 V and produces linear outputs to as low as 600 µV from ground

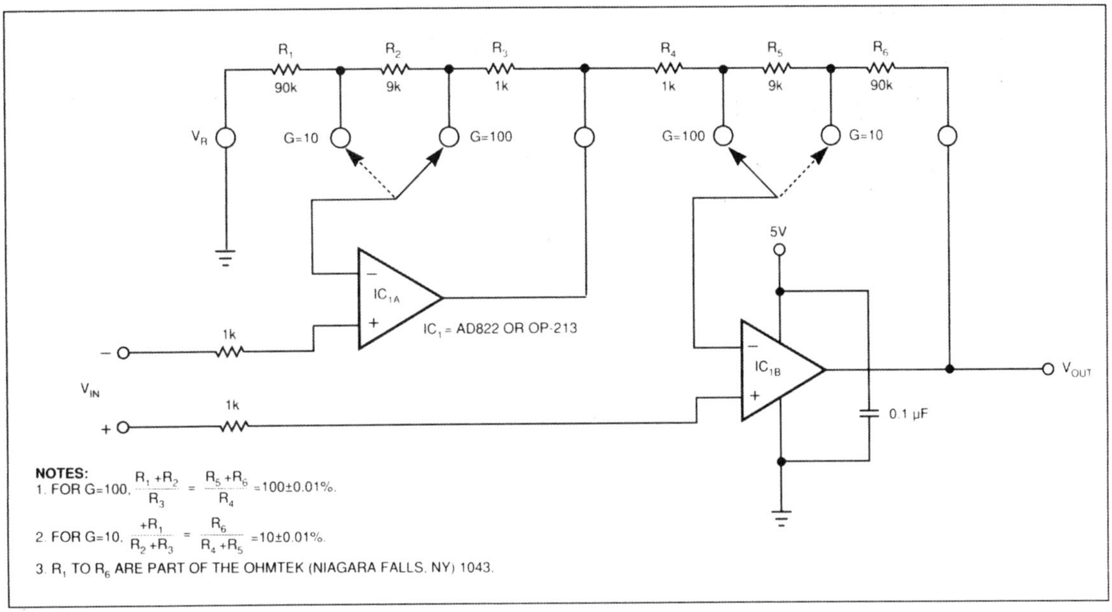

Figure 7.6 This instrumentation amplifier shrugs off switch resistances, enabling you to switch gains

performance with premium amplifiers, the resistor network should have a ratio-match specification of 0.1% minimum with 0.01% as a goal. For ideal amplifiers and 0.01%-match gain resistors, the common-mode error at the output will be on the order of – 100 and – 120 dB for gains of 10 and 100, respectively.

You must observe the common-mode limitations of the configuration if you use a single supply. The circuit produces a V_{OUT} that's referred to the potential applied to the resistor network's V_R pin. This ground assignment implies that if you want the circuit to exhibit high linearity for even small V_{IN} difference voltages, the op-amps should be able to swing close to ground.

Output-stage swing makes or breaks single-supply operation

Not all single-supply amplifiers' outputs can swing to the negative rail (ground). Some may swing to within a diode drop (0.6 V); others may swing within a few tens of millivolts. Only a very few swing to less than 1 mV from the negative rail.

To understand which op-amp functions best for output-swing-critical applications, you must have some detailed knowledge of output stages. Figure 7.7 shows a sampling of output-stage topologies used in single-supply op-amps.

The bipolar complementary emitter-follower stage in Figure 7.7a is only active to within a diode drop of each rail, at best. This stage may or may not use a Darlington for transistor Q_1 but usually has a single pnp for Q_2. This output stage cannot swing to ground without outside help, such as the added resistor $R_{PULLDOWN}$. When used, this resistor allows output linearity to ground for source-type loads only. It also raises overall dissipation for sustained high-level outputs. Amplifiers with this type of output stage can be useful for single-supply applications, but you must carefully apply them to optimize all tradeoffs.

The NMOS bipolar emitter-follower common-source output stage in Figure 7.7b has, by nature, an asymmetrical voltage swing. However, the stage features low saturation voltages to ground because of the NMOS pulldown transistor N_1. Saturation voltage to the positive rail is about 1 V (or more). The advantage of this stage is that the NMOS pulldown transistor can drive current-sink loads to within a few millivolts of ground, which enhances linearity in many single-supply applications.

The CMOS stage of Figure 7.7c is, by definition, fully complementary and offers a resistive connection to the supply rail for a high or low output, that is, rail to rail. With appropriate low-R_{ON} transistors for P_1 and N_1, saturation drops to either rail can be a millivolt or less at low currents. Because this stage is inherently Class A, the amplifier design must carefully control the static currents in P_1 and N_1 for low quiescent current. For output currents of a few milliamps,

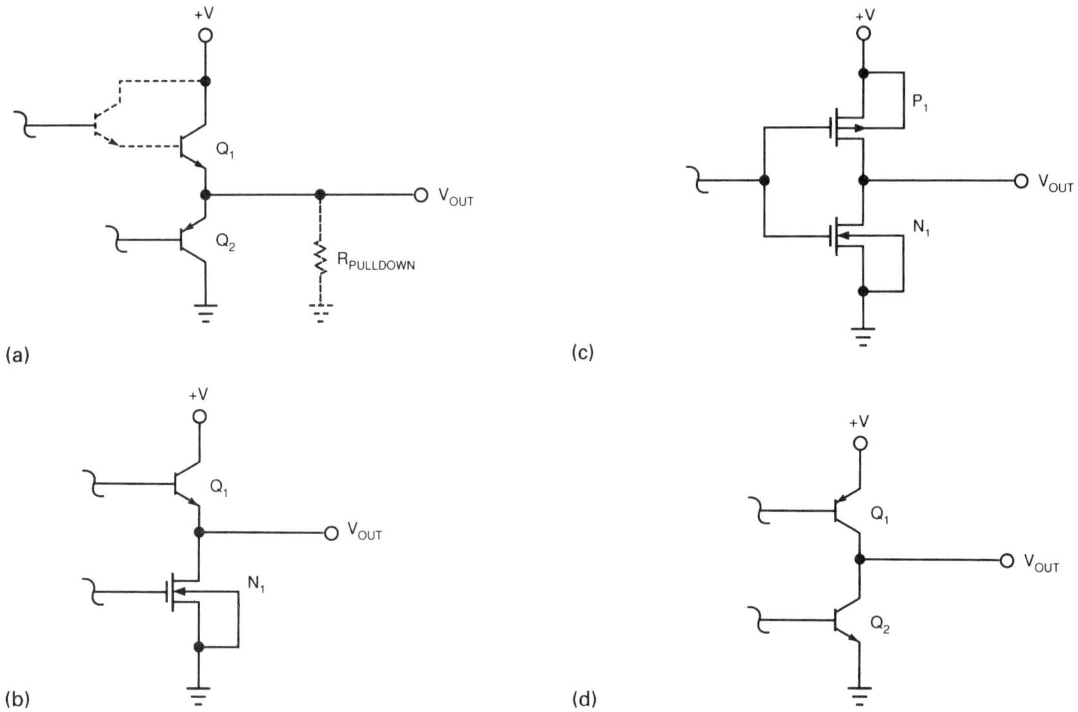

Figure 7.7 Among this sampling of output-stage topologies used in single-supply op amps, the bipolar complementary emitter-follower stage (a) is active only to within a diode drop of each rail at best. The bipolar emitter-follower, NMOS common-source output stage (b) has, by nature, an asymmetrical voltage swing. The CMOS stage (c) is, by definition, fully complementary. The complementary-bipolar common-emitter stage (d) is another rail-to-rail output stage

this type of output stage is effective and quite versatile because of its rail-to-rail nature.

The complementary-bipolar common-emitter stage of Figure 7.7d is another rail-to-rail output stage. Saturation drops to the rails range from a few millivolts to hundreds of millivolts over current ranges up to 20 mA. Like the CMOS rail-to-rail output stage, this bipolar counterpart is both effective and versatile, but avoiding punitive tradeoffs in power is critical to the design of the bipolar stage.

References

1. Kester, Walt, *Amplifier Design Guide*, Analog Devices, Norwood, MA, 1992.

2. Wong, James 'Simple filter quiets power line', *EDN*, June 4, 1992, p. 143.

8 Analog circuits bypass single-supply design constraints

As well as op-amps, other analog ICs now feature +5 V working

In addition to op-amps, analog signal processing requires a whole range of other devices that will work at a supply voltage of 5 V or below. This article looks at just some of them.

Walt Jung and James Wong
Analog Devices Inc

Designing analog circuits using a single-supply voltage places unfamiliar constraints on designers who typically work with ± 15 V supplies. This article presents a collection of applications – including references, regulators, 4- to 20-mA loops, DACs, low-noise amplifiers, temperature sensors, and current monitors – that illustrate some general guidelines for single-supply, low-power design. All circuits in this article work at 5 V and up, and several operate down to 3 V.

Low-power references

You will encounter many problems trying to make stable dc voltage references work from 5 V and lower power supplies. These problems include

- minimizing quiescent power consumption
- maximizing overall power efficiency
- operating down to 3 V
- achieving low dropout
- obtaining minimum noise output.

Because supplies ≤ 5 V can't support zener devices, low-voltage references must be bandgap types. Operating from 3 V, for example, dictates a reference diode of appreciably lower voltage. One solution is a 1.2 V reference diode and appropriate low-power support circuitry (Figure 8.1a).

On the surface, this circuit appears trivial. But if you want to minimize power drain, your design options narrow. D_1 must be a low-operating-current diode. In this case, D_1 is a 1.235 V device with a minimum current of 50 μA. With a value of 27.4 kΩ for R_1, the circuit can work with a supply voltage as low as 2.7 V and still produce the minimum diode current. Obviously, you must minimize loading on the unbuffered diode, but you can account for loads of a few microamps as long as they are static in nature.

Dynamic loads present a larger problem, and a power-conscious designer will not want to burn excess current in a reference diode for occasional load changes. IC_1 alleviates this problem by buffering the diode to allow higher source and sink currents, but at the expense of 160-μA quiescent current. You can reduce power consumption by using a lower-current op-amp, but make sure the amp's dynamic performance is adequate.

Without gain-scaling resistors R_2 and R_3, the output of the Figure 8.1a circuit is 1.235 V. If you use these resistors, you can set the output anywhere between the supply rails because of the rail-to-rail output swing of IC_1. Output-voltage drift is primarily a function of the diode grade and can be as low as 10 ppm/°C.

Low-standby-current reference

The low-standby-current reference in Figure 8.1b exhibits maximum power efficiency. This self-biasing circuit maximizes current economy by using the quiescent current of a low-power op-amp to both bias and simultaneously buffer the diode[1]. With this arrangement, the circuit needs no separate diode-bias current, and the otherwise wasted amplifier supply current helps stabilize the diode and quiescent current.

To implement this circuit, the minimum diode and amplifier currents should be compatible. In this case the op-amp current adds to the current through R_3 to make the sum ≥ 50 μA and allow circuit start-up. R_1 and R_2 scale up the output voltage from 1.235 V, as noted. Standby current drain is typically 55 μA.

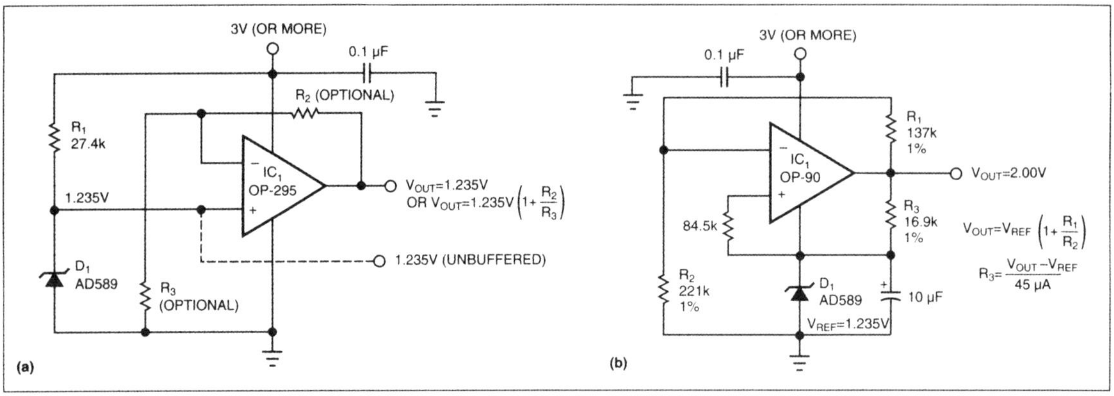

Figure 8.1 Reference circuits may appear trivial, but single-supply operation constrains device choices. In (a), D_1 must operate with low currents. The self-biasing circuit in (b) maximizes current economy by using the quiescent current of a low-power op-amp to both bias and simultaneously buffer the diode

Of course, V_{OUT} must be compatible with the output swing of the amplifier and the minimum supply voltage. In this example, IC_1 can supply several milliamps of current, and there must be at least 1 V between the supply and V_{OUT}. The circuit can generally work with many amplifiers, but your choices narrow for low-voltage and low-current operation. Note that the 1.235 V reference subtracts from the supply voltage available to the amplifier, so only a device capable of working at 1.8 V or less can be used with a 3 V source.

Low-dropout references

Low-dropout references are of interest when DACs or ADCs require a reference potential close to the supply; for instance, 4.5 V with a 5 V supply. The low-power circuit in Figure 8.2 illustrates how you can use a standard 2.5 V, 3-terminal reference (IC_2) and a scaling amplifier with a rail-to-rail output. Resistors R_1 and R_2 scale the 2.5 V reference up to 4.5 V; trim is available with optional resistor R_{TRIM}. Capacitor C_1 provides a low ac output impedance. This circuit has a dropout voltage of about 160 mV for a 1-mA load and a standby current drain of 610 µA. IC_2's resistive saturation drop of approximately 150 Ω determines the dropout voltage.

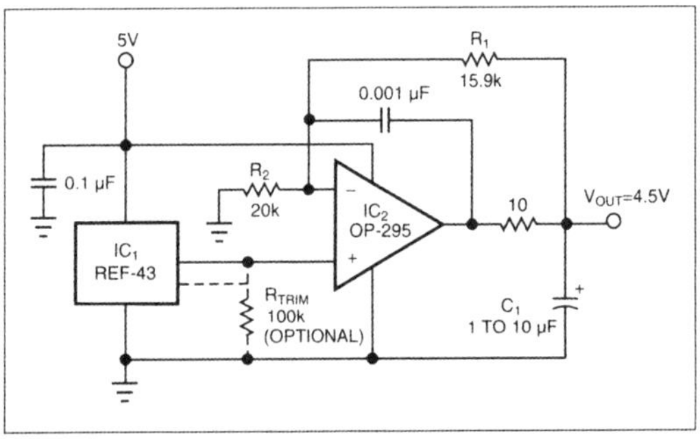

Figure 8.2 Providing a 4.5 V reference from a 5 V supply requires a low-dropout reference circuit. This circuit has a dropout voltage of about 160 mV for a 1-mA load and a standby current drain of 610 µA

By adding a boost transistor to low-dropout reference circuitry, output currents up to 1 A become possible. The circuit in Figure 8.3 retains the low standby current and dropout voltage of the previous designs. The low-dropout regulator features a 800-µA standby current and suits a variety of voltages and output currents up to 100 mA. The circuit achieves a 100-mA current using controlled-gain pass transistor Q_1, an MJE170 type. Using series resistor R_3 to constrain Q_1's base drive provides output-current limiting. This scheme limits the base current to about 2 mA. Then, the max h_{FE} of Q_1 will allow currents no more than 500 mA, which limits short-circuit dissipation to safe levels.

You can adapt this circuit for different voltages by changing the value of R_1, as the table in Figure 8.3 shows. Dropout voltage with a 100-mA load is about 200 mV, thus maintaining a 5 V output for inputs above 5.2 V. You can attain lower output noise by including the optional reference filter comprising R_2 and C_1. Output voltages down to 3 V or less are possible as long as V_{IN} is 3.2 V or more.

Adding a second amplifier to the loop allows foldback current limiting, and high-current PMOS control transistor IRF9531 allows outputs up to 1 A (Figure 8.4). In this circuit, IC_{1A} is the loop's voltage-control amplifier. It controls the output voltage as established by R_1, R_2, and the 2.5 V reference.

IC_{1B} is the current-control amplifier. For output currents less than 1 A, the output is low, which reverse-biases D_1. As the output current exceeds

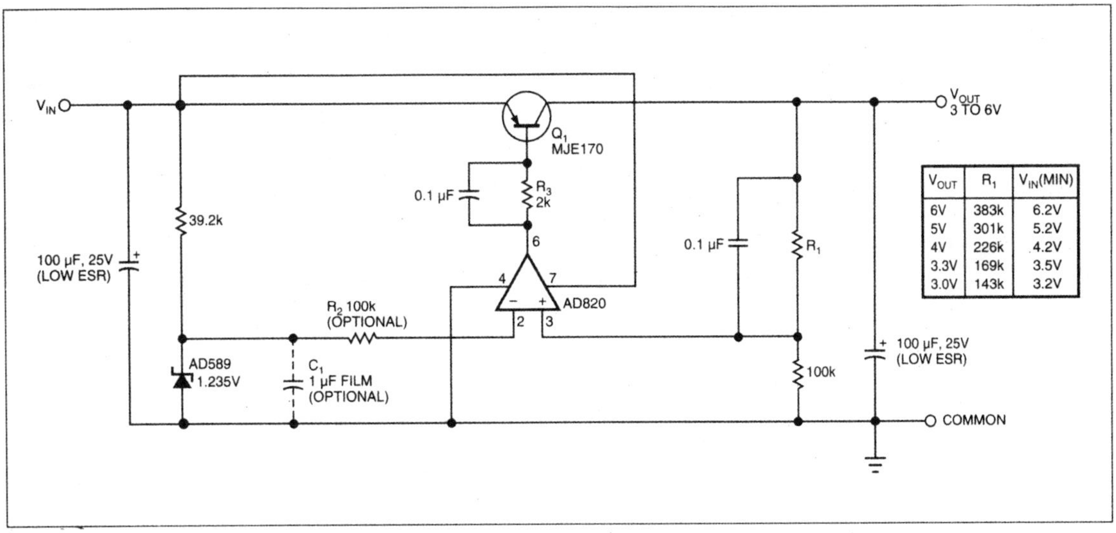

Figure 8.3 Adding a boost transistor to a low-dropout reference results in a low-dropout regulator that provides a 100-mA output current, 800-µA standby current, and 200-mV dropout voltage

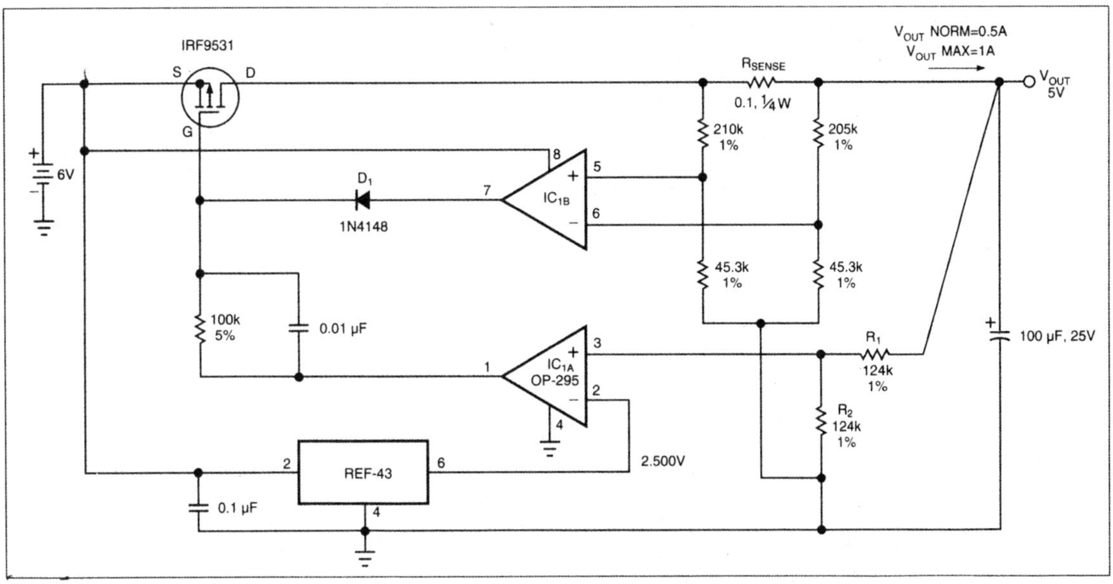

Figure 8.4 Adding a second amplifier to the regulator loop allows foldback current limiting. A high-current PMOS control transistor supplies output currents up to 1 A

1 A, the drop across R_{SENSE} rises, which forces IC_{1B}'s output high and forward-biases the diode. This action closes the current-control loop, which overrides the drive from IC_{1A} because of the much higher impedance in that leg of the circuit.

If an output current greater than 1 A persists, the current loop forces a reduction of output to the load, thus causing a corresponding drop in voltage. As the output voltage drops, the current-limit threshold also drops fractionally. The result is a decreasing output current as the voltage decreases; the limit is 0.2 A at 1 V of output. This foldback effect reduces power dissipation in the control device, which lets you use simple heat sinking.

When operating from a 6 V raw supply, the rail-to-rail output drive from IC_{1B} can produce the full gate-source voltage and fully enhance (turn on) the PMOS transistor. The dropout voltage is 0.2 V at 500 mA and 0.6 V at 1 A.

4- to 20-mA loop circuits

Amplifiers whose outputs swing close to the negative rail enhance and simplify the design of 4- to 20-mA loop transmitters. Amplifiers that can't swing close to this rail have saturation-voltage limits that reduce the accuracy of the amp's zero-scale signal range. The output – and many times the input – of an amplifier often operates at or near the negative rail, yet the amplifier must remain linear. A case in point is the circuit in Figure 8.5, a loop-powered strain-gauge sensor. The amplified 50-mV full-scale (FS) bridge output is calibrated for a 4- to 20-mA transmitter output.

IC_1 linearly amplifies the bridge signal by a gain of about 40. IC_1's output range includes the negative rail, so the IC can amplify a 0- to 50-mV bridge signal to 0 to 2.008 V referred to the loop's common bus (pin 5 of IC_1). Because all negative-supply device pins refer to this point, the bulk of the loop's quiescent current flows into R_6 and the external loop and termination, R_{LOAD}.

With no output from the bridge, IC_1's output will be at the negative rail. No current flows through R_1 or R_2 into the summing point of IC_2 because IC_1 servos the summing point to the negative-rail potential. For this zero-scale signal condition, R_3 (4-mA NULL) calibrates the loop for a 4-mA output current or for 0.4 V across R_{LOAD}. Because no current flows through R_1 in this zero-scale condition, R_1 has no effect on nulling, which ensures that the NULL and SPAN trims don't interact.

R_3 and R_4 effectively appear across the 5 V reference output, so the current the reference injects into the loop is constant. The loop's output summing resistors, R_5 and R_6, scale the current. The expression for this current is

$$I_{NULL} \approx \left(\frac{5V}{R_3 + R_4}\right)\left(1 + \frac{R_5}{R_6}\right),$$

where 5 V represents IC_3's reference voltage.

When the bridge output is 50 mV FS, IC_1 amplifies the output to the 2.008 V FS level and supplies signal current to IC_2's summing point. Like the reference current through R_3 and R_4, the loop scales up the signal current in R_1 and R_2. The current appears through R_6 as I_{SPAN}:

$$I_{SPAN} \approx \left(\frac{V_{BRIDGE} \times A}{R_1 + R_2}\right)\left(1 + \frac{R_5}{R_6}\right),$$

where V_{BRIDGE} is the output of the bridge, and A is IC_1's gain. For a 50-mV FS bridge output, R_1 (20-mA SPAN) trims the circuit for a 20-mA output current (4-mA I_{NULL} plus 16-mA I_{SPAN}) or 2 V across R_{LOAD}.

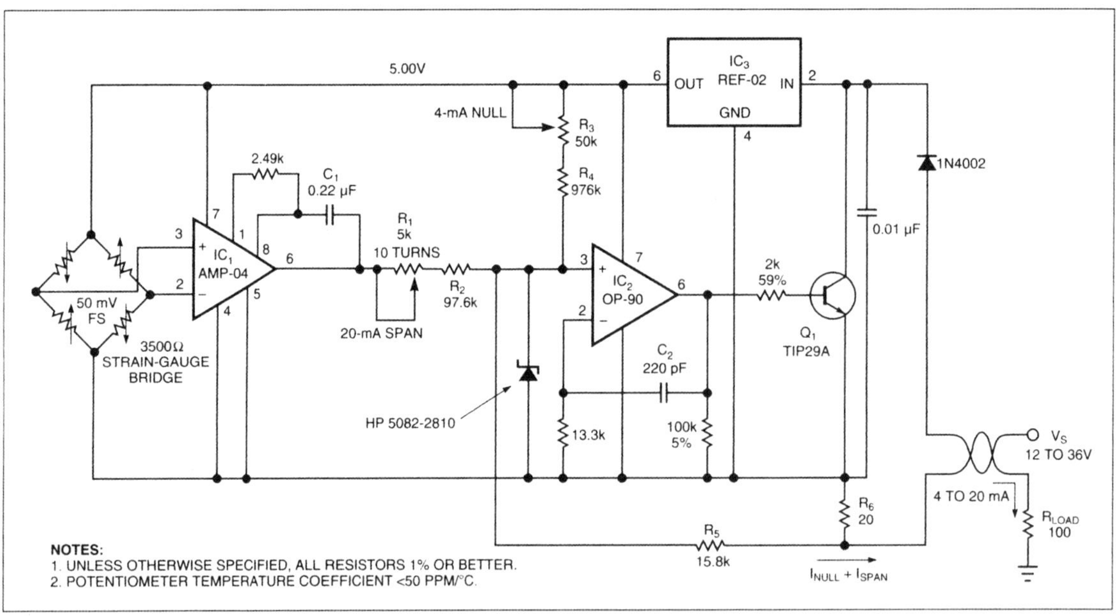

Figure 8.5 Single-supply amplifiers enhance the design of 4- to 20-mA current loops – such as this circuit, which features noninteractive trims – because the amplifiers can swing close to the negative rail and accurately amplify zero-scale signals

In this circuit, the three active devices and the 3500 Ω bridge draw 3.75 mA worst case, which is safely below the system's 4-mA zero scale. C_1 provides a 7-Hz lowpass filter to limit noise, and C_2 stabilizes IC_2's output loop. Q_1, a TO-220 device, conducts the 0- to 16-mA I_{SPAN} portion of the loop output.

Most DACs require dual supplies. Among those that don't, most work from 12 to 15 V supplies, as opposed to 5 V only. This situation greatly reduces your component choices for single 5 V-supply systems. CMOS R/2R-ladder DACs are natural choices for low-voltage operation because many of these units work from 5 V supplies. However, when you use DACs in their standard multiplying mode with a positive reference and an inverting-output amplifier, the DACs also require a negative supply so the amplifier can output negative voltages. Or, conversely, if you use a negative reference, the output will be positive. In either case, you will need a negative supply.

To make such a DAC operate in a totally unipolar fashion, you can turn it around and operate it in what is known as its inverted, or voltage-output, mode. In Figure 8.6 the DAC's standard V_{REF} input to the ladder becomes the voltage-output node (pin 1), and the normal I_{OUT} node becomes the reference input (pin 3). This voltage-output-mode CMOS DAC circuit works from 5 V using two 8-pin ICs to achieve serial-input 12-bit operation. The DAC produces a 0 to 4.095 V output (1 mV/LSB).

The circuit uses a 12-bit multiplying DAC, so the output voltage equals $(D \div 4096) \times V_{REF}$, where D is the 12-bit digital word's value, which ranges from 0 to 4095. With a 1.235 V V_{REF}, the unbuffered output at pin 1 of IC_1 will be $(4095 \div 4096) \times 1.235$ V at full scale, or approximately 1.2347 V (~ 300 μV/LSB). The output amplifier brings the overall scale at V_{OUT} to the $(D \div 4096) \times V_{REF} \times A$ level, where A is the amplifier gain. With the appropriate choice of A, the output will be 1 mV/LSB. The amplifier gain required for this weighting is about 3.3. The gain is variable (see R_2 and R_3), so you can trim out tolerances in both the reference voltage and gain resistors. For bipolar amplifiers, making the equivalent resistance of the gain divider equal to the DAC's output resistance of 11 kΩ proves helpful.

The circuit is clean and straightforward, but some application points are worth discussion. Note that the enhancement voltage of the DAC's internal NMOS switches equals the difference between the supply and reference voltages. With a 5 V-only supply, the enhancement voltage of the switches is 3.765 V with a 1.235 V reference. Thus, a V_{REF} of 1.3 V is about the highest practical for this DAC operating in the voltage-output mode. If you use a higher reference voltage, the MOSFET's gate drive can become starved, which would lead to a higher on-resistance and output-voltage nonlinearity.

The amplifier strongly affects dc accuracy, speed, and power, and some tradeoffs may be necessary for the best overall performance. For a

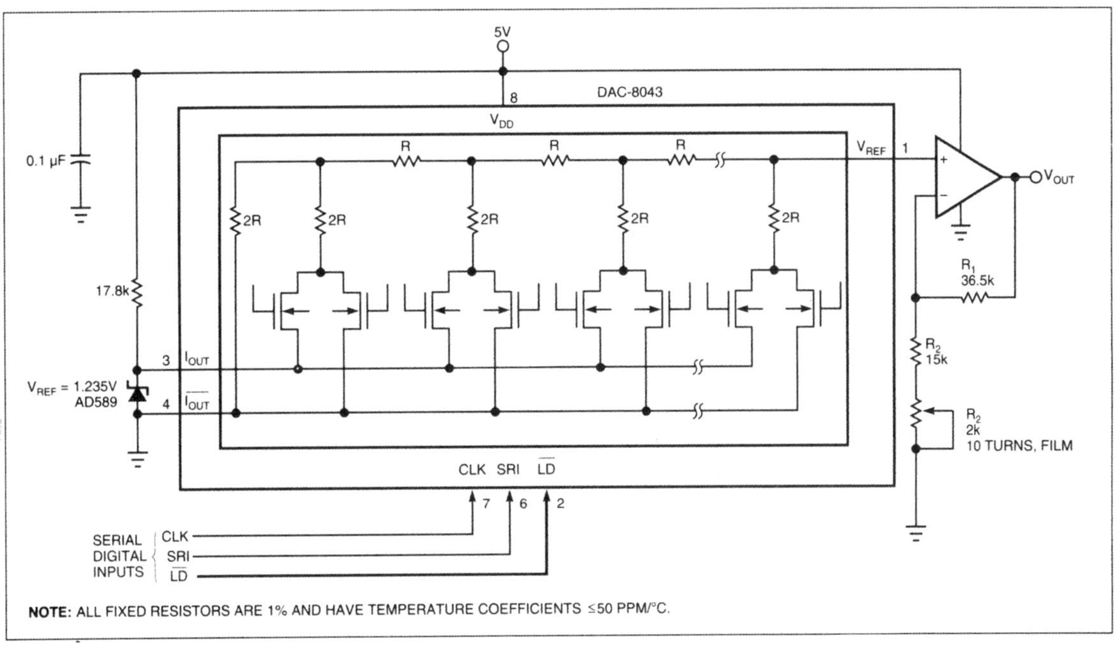

Figure 8.6 To run a standard CMOS R/2R-ladder DAC from a single supply, you have to operate the DAC in the inverted, or voltage-output, mode

linear V_{OUT} range of 0 to 4.095 V, the amplifier must be a rail-to-rail output-stage type. The worst-case output offset will be about 3.3 times the amplifier's offset voltage. Relatively slow response and the best overall dc accuracy and lowest power consumption is possible with a CMOS output op-amp. For example, the OP-295 has a settling time less than 150 μs and draws about 660 μA. A faster bipolar output op-amp, such as the AD820, settles in 2.5 μs but draws around 1.2 mA.

For various reasons, achieving low-noise operation from single supplies is much more difficult than with dual supplies. Both the circuit configuration and the specific devices can make designing for low noise difficult. Single-supply devices limit performance because of their low-current internal biasing, which compromises voltage noise in bipolar stages. For example, relatively high input noise of 40 to 50 nV/$\sqrt{}$Hz is not uncommon for low-power devices. Also, single-supply circuit configurations can themselves limit optimal noise performance because of the necessary biasing and their nonideal characteristics. Dual-supply designs don't have these limitations.

Figure 8.7, a single-supply ac amplifier arranged for low-noise operation from either low- or high-impedance sources, depending on the amplifier you choose – addresses these difficulties. With two sections of a dual op-amp available, the circuit is useful as a low-level preamp for multimedia and other audio applications. Because of 5 V single-supply operation, bypass capacitors and input and output coupling capacitors are necessary.

To achieve lowest circuit noise, IC_1's gain resistor, R_2, has a value of 221 Ω, which corresponds to a Johnson noise that is one third the noise of the op-amp (2n V/$\sqrt{}$Hz). This resistor value ensures that the noise of the amplifier will dominate for low V_{IN} source resistances.

For lowest noise in the surrounding circuit, biasing must also be noiseless[2] or free from noise added directly or indirectly by the biasing. This caveat means that resistors with dc across them should have low excess noise or be ac bypassed. Thus, R_1, R_3, R_4, and R_5 are metal-film resistors, and R_5 is bypassed. R_4 and R_5 provide a 2.2 or 2.5 V bias source, which biases the output of IC_1 approximately midway in its range. In addition, both C_1 and C_2 are low-ESR types.

To provide a low source impedance for low noise, C_1 must be larger than you would choose if you were considering only the time constant. Otherwise, the current noise of a bipolar amplifier would produce an equivalent voltage noise across C_1 and the source impedance. Alternately, using a

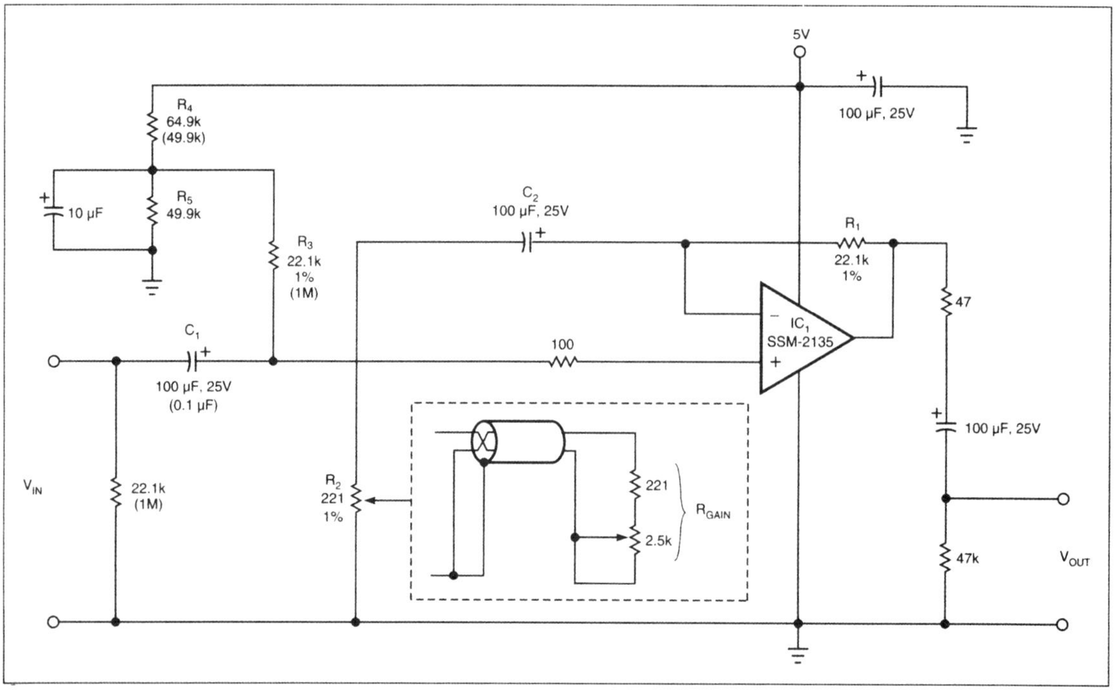

Figure 8.7 Devices and circuit topologies make achieving low noise difficult in single-supply designs. One critical design technique is to limit resistor noise by using low-noise resistors or adding bypass capacitors across them, as this circuit does for R_5

JFET-input op-amp with low current noise and bias current allows the input impedances to be much higher and C_1 to be correspondingly smaller. The gain of the stage in Figure 8.7 is 100 with the fixed value of R_2. You can replace R_2 with the optional cable and R_{GAIN} resistors to add remote gain control to the circuit.

The bandwidth of this stage varies with the amplifier you choose for IC_1; in this case, from 20 to 30 kHz at maximum gain. The IC_1 bipolar amplifier is suitable for low-impedance sources. For high-impedance sources, a FET-input device such as the AD820 will exhibit lower noise because of its lower input-current noise. (Component values for using the AD820 as IC_1 are in parentheses in Figure 8.7.) Aside from the source-impedance choice between the two amplifiers, there are power-consumption differences. A bipolar-input device may draw a higher quiescent current.

Many methods of sensing temperature are available, but high-precision sensing using single power supplies remains a challenge. Figure 8.8 shows a precision platinum resistance temperature detector (RTD) that runs from a 5 V supply. To minimize self-heating, a regulated 200-μA current, I_{BRIDGE}, drives the RTD bridge. Overall current drain is 1.3 mA. The circuit yields a 0 to 400°C output from a 5 V supply and a 0 to 350°C output from a 4.5 V supply.

R_{SENSE} senses the 200-μA current, and IC_{1A} compares the voltage drop across R_{SENSE} to the 0.2 V reference voltage across R_5. The common-mode voltage at the bridge output is 0.21 V, which is a level difficult for conventional IC or discrete 5 V instrumentation amplifiers to handle. In this circuit, IC_2, a single-supply instrumentation amplifier, amplifies the output. The gain is 100 kΩ ÷ R_{GAIN} (100 kΩ is the value of IC_2's internal resistor, and R_{GAIN} is the combination of R_8, R_9, and R_{10}). In this case, the gain is nominally 245. This gain times the bridge's output of 38 μV/°C yields a V_{OUT} scale factor of 10 mV/°C. C_1, in conjunction with IC_2's internal 100-kΩ resistor, provides a lowpass filter that forces the gain to roll off above 3.4 Hz.

Several factors in the surrounding circuit are important for best performance. All critical resistors – marked with one asterisk in Figure 8.8 – should be 0.5%, ± 25-ppm/°C types, or about 100 times better than the RTD's TC of 3850 ppm/°C. R_1 and R_2, in particular, should be from the same manufacturer and batch, and R_4 should be from the same manufacturer to help minimize differential TCs. Gain resistor R_8 should have a TC that is low compared with the 50-ppm/°C TC of IC_2. All potentiometers should be multiple-turn film types.

The bridge-amplifier combination has a small amount of nonlinearity, typically ≤ 0.5%. In

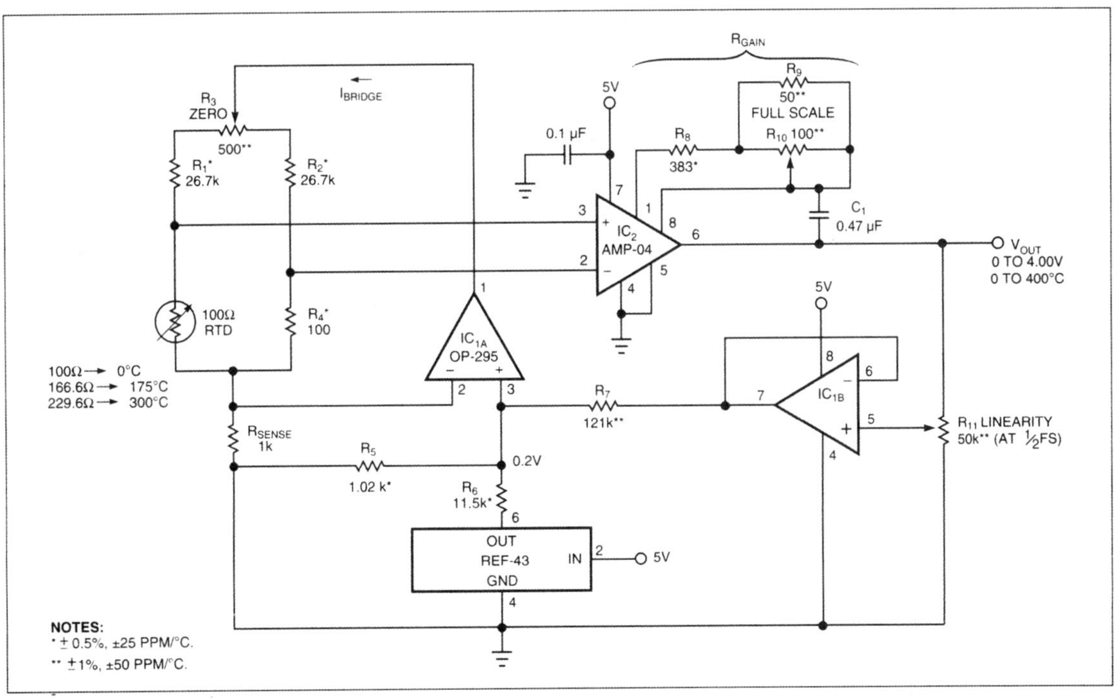

Figure 8.8 The overall current drain of this temperature sensor is 1.3 mA. The circuit yields a 0 to 400°C output working from a 5 V supply and a 0 to 350°C output from a 4.5 V supply

comparison, the RTD has a much larger nonlinearity – as much as 6°C over a 400°C span, or 1.5%. Fortunately, controlling the positive feedback, which increases the bridge drive as the output increases, corrects this nonlinearity. R_{11} picks off a fraction of V_{OUT}. IC_{1B} buffers this fractional voltage, and R_7 sums the buffered voltage with the reference voltage. With correct adjustment, this scheme cancels the RTD's nonlinearity.

Calibration takes three steps. First, carefully set the circuit for 0°C, because lack of a negative supply prevents this endpoint from going through 0 V. One technique for this step is to substitute an exact 100 Ω resistor for the RTD (or place the actual RTD in an ice bath) and adjust R_3 until V_{OUT} begins to swing positive. Then trim R_3 in the reverse direction until V_{OUT} just stops changing, which should be at 0 V.

For the second step – the full-scale trim – preset R_{11} to a midpoint and substitute a 274.04 Ω resistor for the RTD. Then trim R_9 for 4.000 V at V_{OUT} (400°C). Step three is the linearity trim: Substitute a 175.84 Ω resistor for the RTD and trim R_{11} for 2.000 V at V_{OUT}. Repeat steps two and three for best accuracy because the trimming operations interact. When you have fully trimmed the circuit, the output errors will be within ± 0.5°C over the 0 to 400°C range.

Supply-current monitors

You can arrange single-supply amplifiers to make accurate current monitors for either a negative or positive supply lead. Such circuits deliver output voltages that are directly proportional to the load current and voltages that are positive with respect to the supply ground. You can adapt the monitor circuits for wide operating-current ranges.

Figure 8.9 shows a simple way to monitor supply current with a single-supply amplifier. IC_1 senses the voltage across R_{SENSE} in the negative supply lead. The current through R_{SENSE} is I_{LOAD}, which equals the total supply current. Resistors R_1 and R_2 scale and invert the voltage across R_{SENSE}, thus producing a positive output voltage, V_{OUT}.

Because IC_1 can sense inputs close to its negative supply rail and has an output that swings rail to rail, the circuit responds linearly to small currents down to a V_{OUT} level of a few millivolts. R_{SENSE} and the R_1/R_2 ratio set the current-to-voltage output scale factor, which is 10 mV/mA (1 V = 100 mA).

You can adjust R_{SENSE} proportionally for other scale factors, such as 0.1 Ω for 1 mV/mA or 0.01 Ω for 0.1 mV/mA. The supply voltage essentially dictates full scale. For example, the 6 V supply in Figure 8.9 lets the circuit monitor currents up to 600 mA. The ratio of R_1 and R_2 should match the desired gain to within 0.1% for low gain error, and you should use stable types for both these resistors and R_{SENSE}.

For highest performance, consider some other factors. For low power losses and sense errors, the drop across R_{SENSE} should be 100 mV or less. In Figure 8.9, IC_1's 160-μA supply current is part of

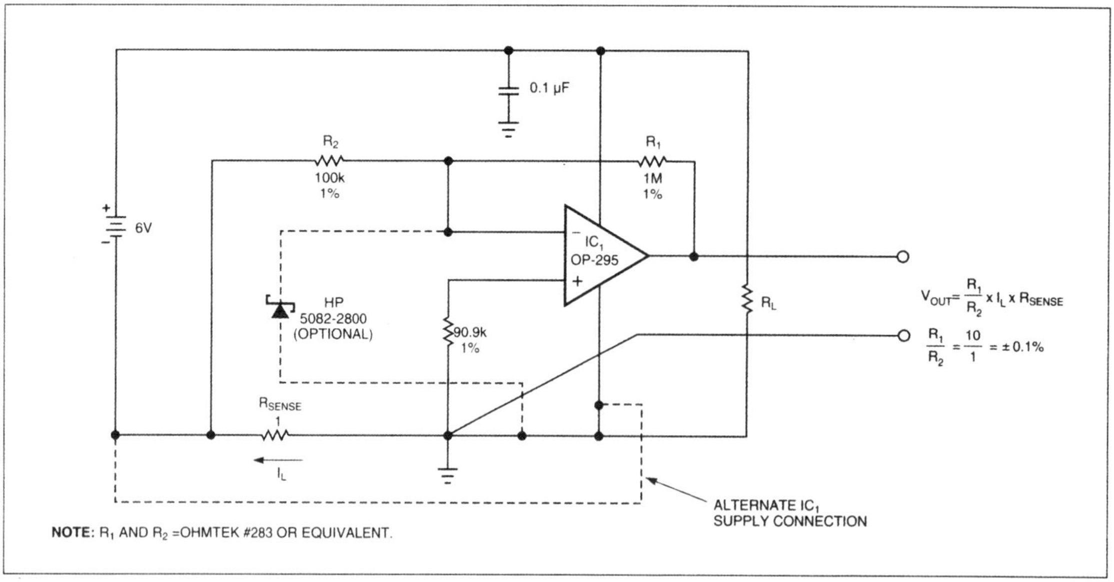

Figure 8.9 In this simple supply-current monitor, IC_1 senses the voltage across R_{SENSE} in the negative supply lead. The current through R_{SENSE} is I_{LOAD}, which equals the total supply current. Resistors R_1 and R_2 scale and invert the voltage across R_{SENSE} to produce a positive V_{OUT}

I_{LOAD}, but the supply current can bypass R_{SENSE} if necessary. If input transients can take the summing point 0.3 V or more below ground, use a Schottky diode clamp for IC_1, which Figure 8.9 shows in dashed lines.

References

1. Jung, W, and G Erdi, 'Self-biasing reference consumes low power'. *EDN*, April 27, 1989, pg 219.
2. Motchenbacher, C, and F Fitchen, *Low-Noise Electronic Design*, John Wiley & Sons, New York, NY, 1973.

9 External compensation boosts op-amp performance

How to optimize performance

Externally compensated op-amps provide the user with the opportunity to tailor the frequency response as required. But for some applications, merely ensuring stability without excessive peaking in the frequency response is not enough. Sometimes the time-domain response is more important; e.g. for use ahead of an ADC in a sampled-data system, an op-amp's settling time is crucial. This article explains the relevant details.

Charles Kitchin and Scott Wurcer
Analog Devices Inc

Although most high-speed op-amps feature built-in compensation, some new devices also provide optional external compensation. This additional compensation lets you use an op-amp to perform functions that aren't possible with standard op-amps – such as driving high capacitive loads while maintaining a fast settling time. By adding the proper external capacitor to an amplifier, you can also improve your op-amp's slew rate and bandwidth in a variety of circuit configurations. To apply external compensation to the AD744 bipolar FET (BiFET) op-amp, for example, you connect the proper capacitor between pin 5 and pins 1 or 8, depending on the application. The schematic of the AD744 (Figure 9.1) details where these pins connect internally. As shown, the external capacitor is in parallel with the 5-pF internal compensation capacitor.

Both the slew rate and the gain-bandwidth product of a high-speed amplifier are inversely proportional to the value of the external-compensation capacitor, C_{ext}. Therefore, to maximize an amplifier's speed, you should try to minimize the value of the compensation capacitance. Conversely, use large values of C_{ext} to slow an amplifier to a point at which the slew rate is perfectly symmetrical and well controlled. The table in Figure 9.2 summarizes the effects of external compensation on the AD744's slew rate and bandwidth under three different load conditions.

Amplifiers that provide optional external compensation also give you the flexibility to implement a variety of designs without sacrificing performance. If you configure the AD744 as a unity-gain voltage follower and add external compensation (Figure 9.2), for example, the op-amp can drive capacitive loads in excess of 2000 pF and maintain a 1.5-μs settling time to 0.01%. For stable operation, this configuration requires a minimum external-compensation capacitance of 5 pF.

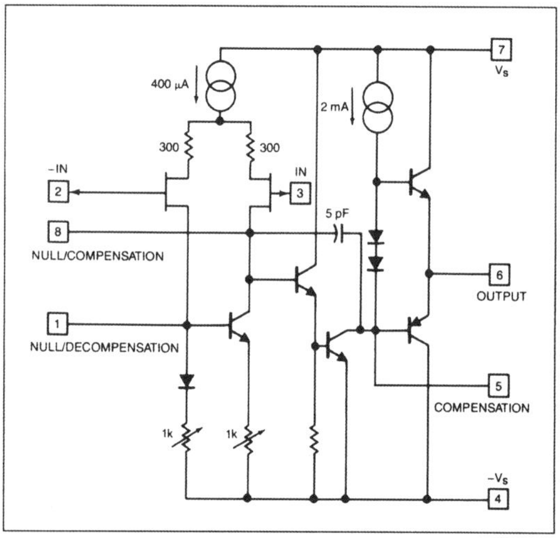

Figure 9.1 In addition to containing 5 pF of internal compensation, the AD744 BiFET op amp can accommodate external compensation

Figure 9.3 shows the AD744 as a voltage follower with gain (a) and as an inverting amplifier (b). In these applications, external compensation isn't necessary for stable operation. However, external compensation might be appropriate for driving capacitive loads in the range of fifty to several thousand picofarads.

The ability to provide external compensation is particularly useful when you are driving high capacitive loads such as long commercial shielded cable. Cable capacitance varies widely, but it generally ranges from 9 pF/ft for low capacitive audio cable to as much as 50 pF/ft for standard

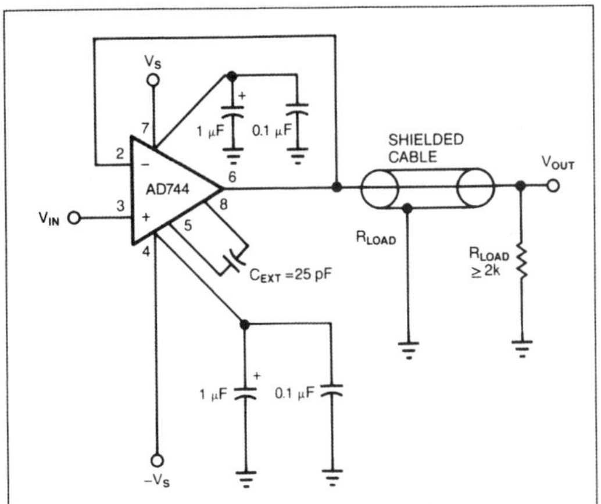

GAIN	MAXIMUM C_{LOAD} (pF)	C_{EXT} (pF)	SLEW RATE (V/μSEC)	-3-dB BANDWIDTH (MHz)
1	50	5	37	6.5
1	150	10	25	4.3
1	2000	25	12.5	2.0

Figure 9.2 Configured as a unity-gain voltage follower with external compensation, this circuit can drive long shielded cables. The table gives the recommended compensation-capacitance values and the typical performance of this circuit under three different load conditions

computer wiring. Driving a 100-ft cable with a total capacitance of 900 to 5000 pF isn't possible with most fixed-compensation op-amps; the excessively high capacitive load reduces the op-amp's phase margin, often causing oscillation.

The table in Figure 9.3 lists the recommended compensation-capacitance values for driving cable loads of various capacitances. In each case, the recommended C_{ext} is a minimum value. You can use a larger value, but doing so degrades the slew rate and the bandwidth.

Extend gain bandwidth with decompensation

When you use the AD744 in applications that have a closed-loop gain greater than 10, you can enhance the gain-bandwidth product by connecting a small capacitor between pins 1 and 5 (Figure 9.4). At low frequencies, this capacitor cancels the effects of the chip's internal-compensation capacitor, C_{int}, effectively decompensating the amplifier. The table in Figure 9.4 provides a gain-vs-performance summary of the AD744 in this decompensation configuration.

When implementing decompensation, however, remember that manufacturing conditions cause the value of internal capacitance to vary among amplifiers. You should therefore optimize

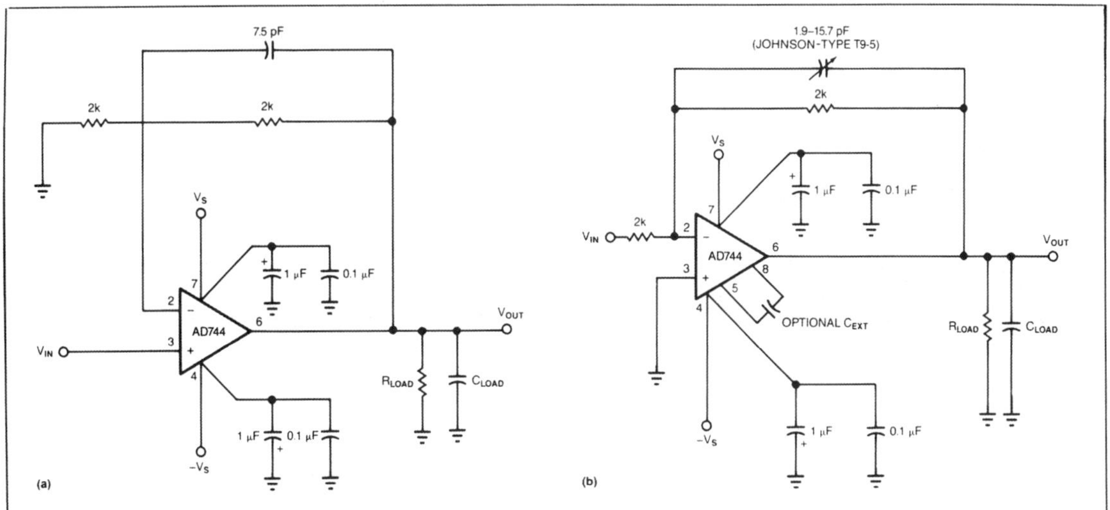

R_1 (Ω)	R_2 (Ω)	GAIN FOLLOWER	GAIN INVERTER	MAXIMUM CABLE CAPACITANCE (pF)	C_{EXT} (pF)	C_{LEAD} (pF)	SLEW RATE (V/μSEC)	−3 dB BANDWIDTH (MHz)
4.99k	4.99k	2	1	50	0	7	75	2.5**
4.99k	4.99k	2	1	150	5	7	37	2.3**
4.99k	4.99k	2	1	1000	20		14	1.2
4.99k	4.99k	2	1	>2000	25		12.5*	1.0
4.99	4.99k	11	10	270	0		75	1.2
4.99	4.99k	11	10	390	2		50	0.85
4.99	4.99k	11	10	1000	5		37*	0.60

NOTES:
1. SLEW RATE IS SPECIFIED FOR A 50-pF LOAD. IN ALL OTHER CASES, SLEW RATE IS SPECIFIED FOR THE RATED MAXIMUM LOAD. FOR HIGH CAPACITIVE LOADS, THE AD744'S 25-mA OUTPUT-CURRENT LIMIT SETS THE OP AMP'S SLEW RATE (IN V/μSEC) EQUAL TO 0.025A DIVIDED BY C_{LOAD} (IN μF).
2. The C_{LEAD} IS ADJUSTED FOR MINIMUM SETTLING TIME.

Figure 9.3 When you use the AD744 as a unity-gain inverter (b) or use it to drive high capacitive loads, you should employ the amp's optional compensation feature

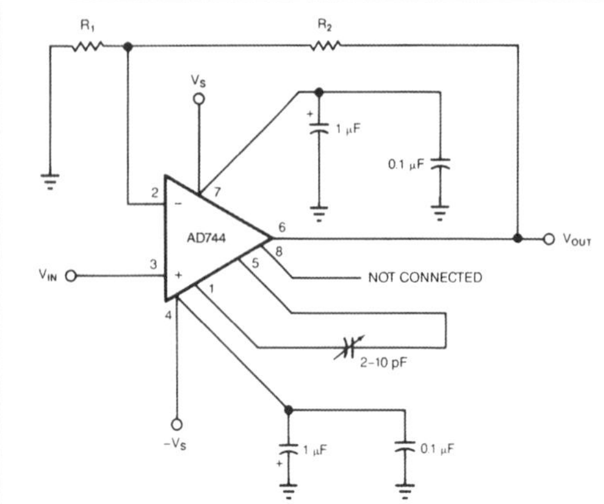

R_1 (Ω)	R_2 (Ω)	GAIN FOLLOW	GAIN INVERTER	-3-dB BANDWIDTH	GAIN/ BANDWIDTH PRODUCT (MHz)
1k	10k	11	10	2.5 MHz	25
100	10k	101	100	760 kHz	76
100	100k	1001	1000	225 kHz	225

Figure 9.4 For applications involving a closed-loop gain greater than 10, you can connect a decompensating capacitor between pins 1 and 5 on the AD744 to extend the circuit's gain-bandwidth product. The table summarizes this circuit's performance

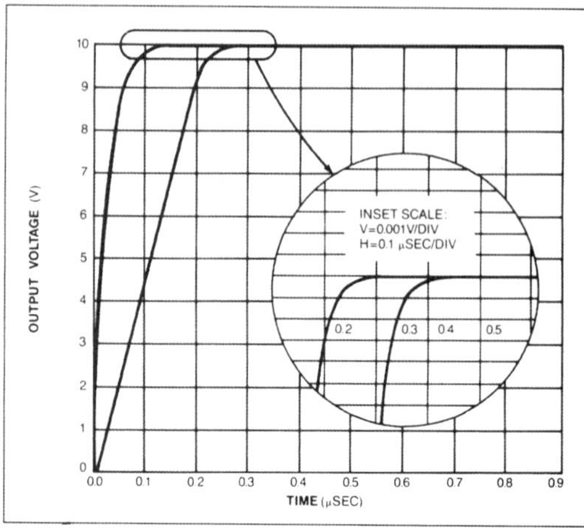

Figure 9.5 The responses of the two 13-MHz unity-gain inverting amplifiers approach their final value. The responses of the limited-slew -rate and infinite-slew-rate devices begin to look the same, but the limited-slew-rate device's response is delayed by approximately 125 ns

your amplifier's response for the desired gain by using a 2- to 10-pF trimmer capacitor, rather than a fixed-value capacitor. If you modify the commonly used 3-op-amp instrumentation-amplifier connection to include decompensation, you can substantially increase the circuit's bandwidth.

In addition to having high bandwidths and fast slew rates, some amplifiers, such as the AD744, feature single-pole responses to allow the fastest true settling times. Amplifiers that use a pole-zero mismatch in their compensation network (to increase the phase margin at the crossover point) usually have long settling tails. These tails aren't present, however, in single-pole amplifiers.

Figure 9.5 depicts the computer-generated responses of two idealized single-pole-compensated, FET-input op-amps. The inverting amplifiers have a 13-MHz unity-gain crossover frequency. One response is from an amplifier with a 50-V/µs slew rate and the other is from an amplifier with an infinitely high slew rate. As the outputs approach their final values, the two responses begin to look the same. However, the response of the limited-slew-rate amplifier is delayed by approximately 125 ns. The magnified inset in Figure 9.5 provides details of the settling time.

Build a fast-settling inverter

The fast-settling inverter circuit in Figure 9.6a combines the virtues of a precision bipolar amplifier with the rapid slew rate and the wide bandwidth of a high-speed BiFET op-amp. In addition to exhibiting a 900-ns settling time to 0.01%, this composite amplifier features the high dc precision associated with a quality bipolar amplifier.

This circuit takes advantage of a novel compensation-trim scheme that uses the AD744's external-compensation pin to null the long tails typically associated with composite amplifiers operating as integrators. Amplifier IC_1 senses IC_2's input offset voltage (via IC_2's summing junction) and nulls the input offset voltage by applying an offset voltage to IC_2's noninverting input.

Normally, this approach yields a circuit that has both a fast slew rate and a high degree of de precision but suffers from long, slow settling tails (settling times of 5 to 10 µs are typical). These long settling times occur because the circuit's net response has a pole and a zero that aren't coincident. However, in Figure 9.6a circuit, a small trimmer capacitor applies feedback via the AD744's external compensation pin. When correctly adjusted, this capacitor moves the pole and the zero back together, thereby removing the troublesome tail from the circuit's response.

The oscilloscope photo in Figure 9.6b illustrates this circuit's rapid settling characteristics. To adjust the compensation, drive the circuit's input with a 1-kHz square wave and adjust the trimmer capacitor for a minimum settling time.

Figure 9.7 presents another configuration that uses external compensation to increase a circuit's performance. This design is a variation of the

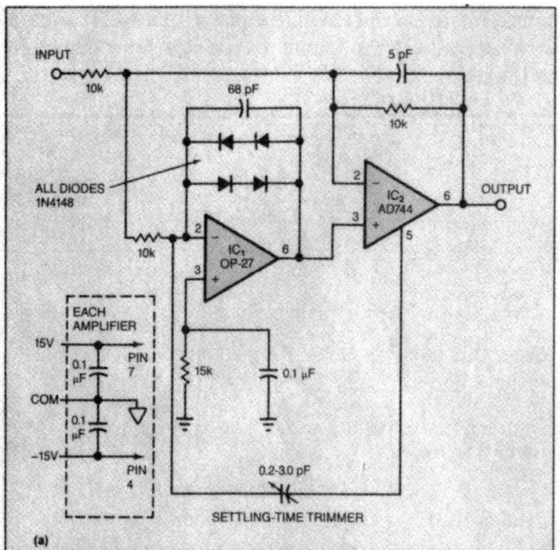

(a)

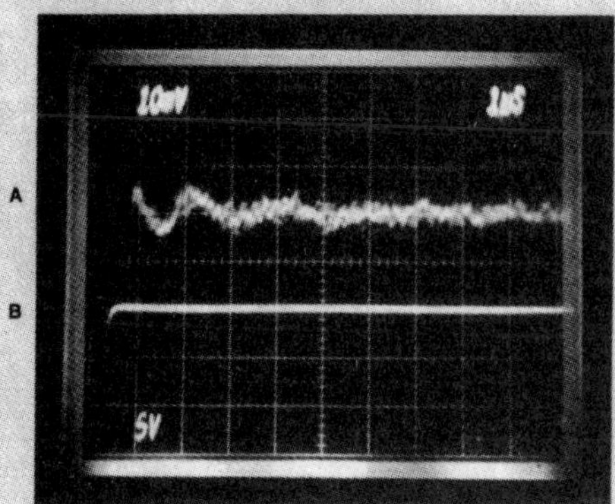

	VERTICAL	HORIZONTAL
A	10 mV/DIV	1 μSEC/DIV
B	5 V/DIV	1 μSEC/DIV

(b)

Figure 9.6 By combining the virtues of a precision bipolar amplifier with the rapid slew rate and wide bandwidth of a high-speed BiFET op-amp, you can configure a fast-settling inverter circuit (a). This circuit's rapid settling characteristics are illustrated in the oscilloscope photo (b)

1-op-amp instrumentation-amplifier principle. The capacitor in Figure 9.7 applies feedback to the FET op-amp's offset pins, rather than to its inverting-input terminal. Because feedback is applied directly to the input FETs' drains, both the inverting and noninverting inputs form a true high-impedance differential-input section.

This circuit offers significant improvements over previous designs because it lets you either enhance or degrade the bandwidth by connecting a single external capacitor between pin 5 and pin 8 or 1 of the AD744. A circuit's gain equals the

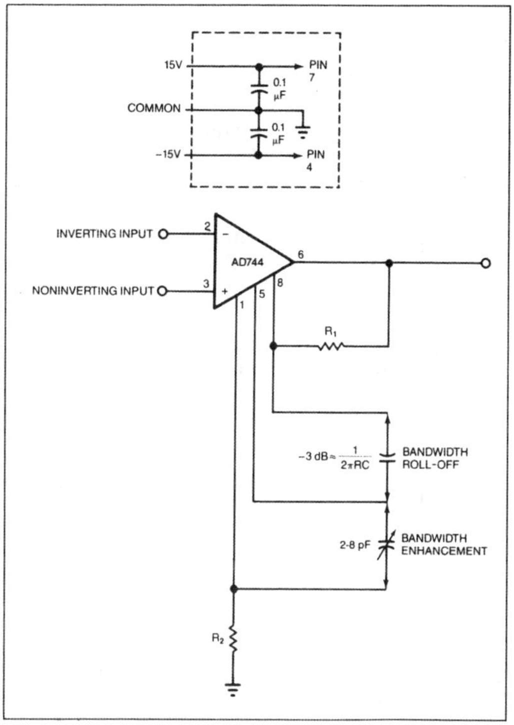

Figure 9.7 By modifying a 1-op-amp instrumentation amplifier, you can enhance a circuit's bandwidth and roll-off

ratio of the value of the two external resistors to the transconductance of the input FETs. Within an accuracy of ± 10%, the gain is:

$$\text{gain} \approx R/2430\,\Omega,$$

where R is measured in ohms. At a gain of 100 (R = 243 kΩ), the circuit's bandwidth will be 750 kHz if you connect a 2- to 10-pF bandwidth-enhancement capacitor between pins 5 and 1. Under these conditions, the circuit's total harmonic distortion is less than 0.01% with a 3 V output and a 1-kHz sine wave. The circuit exhibits little distortion because feedback is applied directly to the FETs' drains, rather than to the amplifier's offset-null pins. Also, the AD744 has an inherently low level of distortion.

Because the circuit's closed-loop gain is set between the drains of the input stage's FETs and the amplifier's output, the input stage's gain is completely independent of the ratio of external resistor values. The input stage's gain is determined solely by the input stage's transconductance. However, because the input stage's gain varies with temperature, the circuit's overall gain also varies slightly with temperature.

Choose capacitors carefully

Because some amplifiers, such as the AD744, have inherently high bandwidths, you should exercise care when selecting external components. The simple model in 'The mathematics of external compensation', explains how to deal with a high-speed op-amp's external time constants. The passive devices you use in a design can greatly affect your circuit's performance. The chief villain to guard against when selecting a capacitor is dielectric absorption; it can greatly increase an amplifier's settling time.

The capacitor in Figure 9.8 illustrates how dielectric absorption can affect a circuit. As shown, there are effectively two capacitors connected in parallel. The principal capacitor, C_1, is easily charged or discharged. However, the voltage across the smaller capacitor, C_2, can't change quickly because of the effective series resistance. The increased charge/discharge time required by a capacitor with high dielectric absorption significantly contributes to an op amp's overall settling time. For example, if the optional external capacitor in Figure 9.3b is a high-K ceramic trimmer, the amplifier's overall settling time will be much greater than the AD744's typical 500-ns settling time. You can minimize this problem by selecting external components that exhibit low dielectric absorption, such as polypropylene, polystyrene, and mica capacitors.

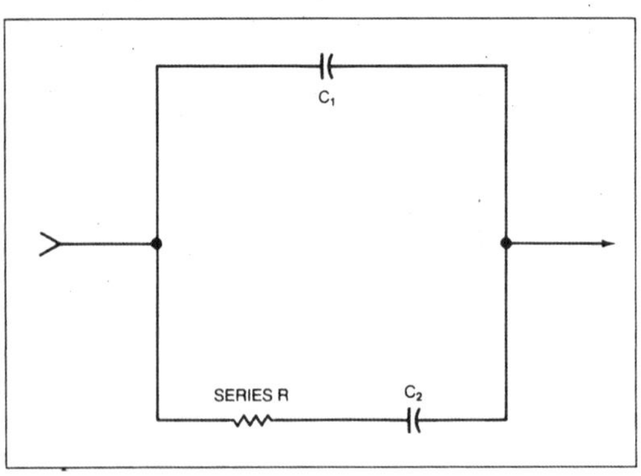

Figure 9.8 Effective series resistance can increase the time required for a capacitor to charge and discharge and therefore significantly increase an op-amp's overall settling time

The mathematics of external compensation

An amplifier with a single-pole or ideal-integrator open-loop frequency response will achieve the minimum possible settling time for any given unity-gain bandwidth. However, when this so-called ideal amplifier is used in a practical circuit, the actual settling time increases above the minimum value because the external circuity introduces additional time constants. These time constants are often caused by additional capacitance on the amplifier's summing junction. You can minimize this increase in settling time by selecting the proper value for the feedback capacitor, C_f.

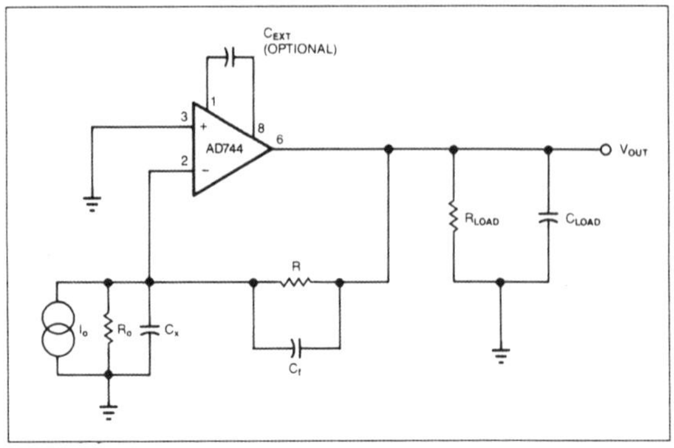

Figure 9.9 In this simplified model of an ideal integrator, an op-amp is configured as a current-to-voltage converter

The integrator circuit in Figure 9.9 shows an op-amp connected as an I-V (current-to-voltage) converter. For an op-amp configured as an ideal integrator with a unity-gain crossover frequency, use the following equation to determine the circuit's small-signal behavior.

$$\frac{V_{OUT}}{V_{IN}} = \frac{-R}{\frac{R(C_f + C_x)S^2}{2\pi f_0} + \frac{G_n}{2\pi f_0} + (RC_fS) + 1},$$

where S = the Laplace operator, F_0 = the op-amp's unity-gain crossover frequency, and G_n = the circuit's noise gain $(1 + (R/R_o))$. Solving the above equation for C_f yields:

$$C_f = \frac{2 - G_n}{R2\pi f_0} + \frac{2\sqrt{RC_x 2\pi f_0 + (1 + G_n)}}{R2\pi f_0}.$$

The first equation would completely describe the output of the system if the op-amp didn't have a finite slew rate and other nonlinear characteristics. Even if you consider these effects, the fine-scale settling to less than 0.1% is determined by the op-amp's small-signal behavior.

In these equations, the capacitance, C_x, is the total capacitance that appears at the op-amp's inverting terminal. This capacitance is the sum of the output capacitance of the current source and the input capacitance of the op-amp, which includes any stray capacitance at the op-amp's input. When modeling an I-to-V converter, you can use the Norton equivalent circuit in Figure 9.9.

Replacing R_o and I_o with their Thevenin V_{in} and R_{in} equivalents creates the general-purpose inverting amplifier in Figure 9.10. Here, C_x represents the AD744's input capacitance (5.5 pF) plus any stray capacitance due to the wiring and the IC package.

In either case, the capacitance causes the system to change from a 1- to a 2-pole response. The additional pole increases the settling time by introducing peaking or ringing in the op-amp's output. If you can estimate the value of C_x with reasonable accuracy, you can use the second equation to determine the correct value for C_f and use a small capacitor that optimizes the amplifier's response. Usually, you can adjust the response to that of a critically damped 2-pole system with a quality factor of 0.5 and a – 3-dB bandwidth equal to:

$$f(-3\text{dB}) \approx 0.64 \sqrt{(f_0)\frac{1}{2\pi R_2\ (C_F + C_x)}}.$$

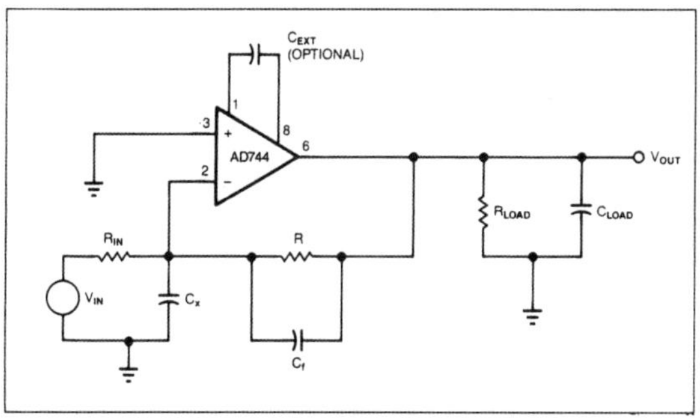

Figure 9.10 In this general-purpose inverting amplifier, the current source of Figure 9.9 is replaced with an input-voltage source

Note that when resistor R is greater than 1.6 kΩ, the amplifier's response is improved by the addition of a small capacitor, C_f, that has the proper value. For critical applications in which the maximum possible bandwidth is desired (or the value of capacitor C_x isn't known), you should make C_f a variable. C_f should be a trimmer capacitor whose range centers around the ideal calculated value. Use the previous equation to calculate the system's resulting response.

In some cases, the frequency response is greater than basic theory ($f = f_o/G_n$) predicts. Above a certain impedance level, the time constants introduced by external components limit the bandwidth to less than the theoretical value.

The AD744 lets you directly modify its open-loop frequency response via an external capacitance connected between pins 5 and 8 of the op-amp. The major benefit of this feature is an improved phase margin – and therefore greater stability – when the amplifier is driving a high capacitive load or when it's operating at a noise gain of less than two. Use the following three equations to calculate the AD744's resultant slew rate, crossover frequency, and compensation capacitance. Note that in each equation, C_{ext} is measured in farads.

$$\underset{(\text{V}/\mu\text{s})}{\text{SLEW RATE}} = \frac{3.0 \times 10^{14}\ \text{V}/\mu\text{s}}{C_{EXT} + (5 \times 10^{-12})}.$$

$$\begin{array}{c}\text{CROSSOVER}\\ \text{FREQUENCY}\\ (\text{Hz})\end{array} = \frac{1}{2\pi(2450)\ C_{EXT} = (5 \times 10^{-12})}.$$

$$\underset{(\text{FARADS})}{C_{EXT}} = \frac{1}{f_c(2\pi)2450} - (5 \times 10^{-12}).$$

10 Simple techniques provide compensation for capacitive loads

How to ensure stability with capacitive loads

Driving a capacitive load often poses a problem for an op-amp. The extra phase shift caused by the time constant of the load with the op-amp's output impedance can eat away at the phase margin, and even cause oscillation. Just a very few op-amps can cope with large capacitive loads; e.g. the TLE2027 preserves a 7-MHz small-signal unity-gain bandwidth even with a 10-nF load and will in fact tolerate any size capacitance provided the slew rate does not cause it to hit current limit – it will even drive a 2-kHz 6 Vpp sinewave into 680 nF. But most op-amps are less forgiving, needing the sort of precautions outlined below.

Sergio Franco
San Francisco State University

Capacitive loads have a notorious tendency to destabilize negative-feedback circuits because of the pole formed by the load capacitance, C_L, and the output impedance, R_o, of the error amplifier. Because this pole is located within the feedback loop, its phase lag erodes the phase margin of the system, thus leading to potential instability. You can also look at this potential instability from an alternative viewpoint: because the open-loop gain rolls-off with frequency, the load capacitance is affected by the op-amp's closed-loop output impedance, which increases with frequency and behaves inductively. Unless you provide adequate damping, resonance may occur at the circuit's output and excessive peaking and ringing may result.

Figure 10.1 shows a popular cure for this instability problem. This circuit uses a small series resistance, R_C, to decouple the error-amplifier output from the load capacitance, and a small feedback-capacitance, C_C, to provide a high-frequency bypass from the output back to the input. The phase lead introduced by C_C compensates for the phase lag due to C_L, thus restoring stability. As a rule of thumb, R_C is usually made equal to R_o, and the value of C_C is selected empirically – typically in the range of 10 to 100 pF.

Though usually effective in achieving stability, empirical values do not necessarily guarantee optimum results[1]. For a detailed analysis, it is necessary to examine the feedback factor, β, which together with the open-loop gain of the error amplifier plays a key role in the stability of the circuit. Redrawing Figure 10.1's feedback network yields the equivalent circuit shown in Figure 10.2. Using this circuit, you can find the feedback factor by applying a test voltage, V_x; determining the resulting inverting-input voltage, V_n; and then taking the ratio: $\beta \approx V_n/V_x$.

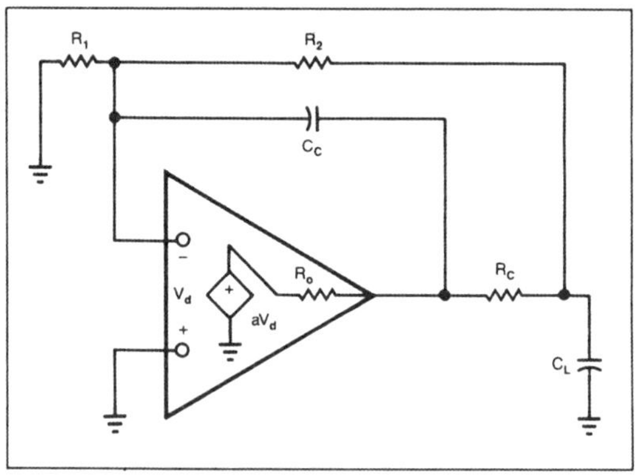

Figure 10.1 This capacitively coupled, resistive-feedback circuit uses compensation network R_CC_C to achieve stability

The equivalent circuit of Figure 10.2 is easily recognizable as a bridged-T network of the type frequently used in audio graphic equalizers[2]. Its distinguishing features are a flat response at the low and high ends of the frequency spectrum

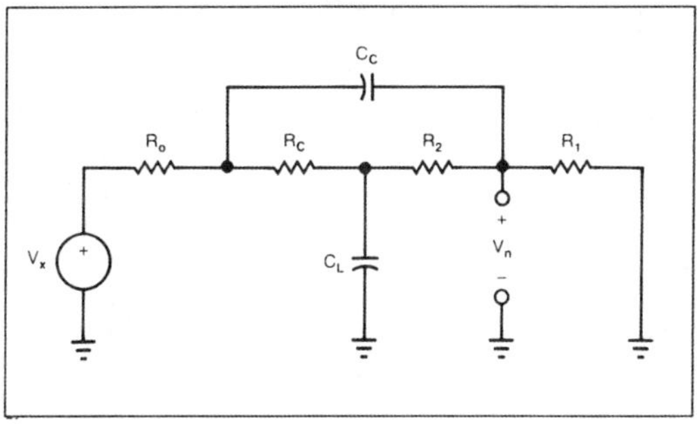

Figure 10.2 You can use this equivalent bridged-T test circuit to determine the feedback factor, β

with a midband bump or dip that is dependent on the relative magnitudes of its components. By selecting R_C and C_C so that β is flat over the entire frequency spectrum, you can achieve neutral compensation – a condition where the phase contribution from the feedback network is zero. In other words, as far as the feedback signal is concerned, the feedback network behaves as if it were purely resistive. This behavior is similar to that obtained in the compensation of an oscilloscope probe.

To achieve neutral compensation, start with the dc value of β, denoted β_0. Exploiting the fact that both capacitors act as open circuits at dc:

$$\beta_0 = R_1/(R_1 + R_2 + R_0 + R_C).$$

The objective is to ensure that $\beta = \beta_0$ over the entire spectrum. To this end, you first impose $\beta_\infty = \beta_0$, where β_∞ represents the high-frequency value of β. The fact that the capacitors now act as short circuits yields the following equation:

$$\beta_\infty = (R_1 \parallel R_2 \parallel R_C)/(R_0 + (R_1 \parallel R_2 \parallel R_C)).$$

By letting $\beta_\infty = \beta_0$ and simplifying, the first design equation becomes

$$R_C = R_o \cdot R_1/R_2.$$

Next, impose the absence of any midband bumps or dips. With the resistances in fixed ratios, these characteristics are controlled by the capacitances. Let f_L be the frequency at which the reactance of C_L equals the resistance represented by the remainder of the circuit. Assuming negligible loading by C_C, R_1, and R_2:

$$f_L = 1/(2\pi (R_o + R_C)C_L).$$

Likewise, let f_H be the frequency at which the reactance of C_C equals the resistance represented by the remainder of the network. Assuming C_L is essentially a short circuit at this frequency and that $(R_1 \parallel R_2) >> (R_o \parallel R_C)$:

$$f_H = 1/(2\pi (R_1 \parallel R_2)C_C).$$

The objective is to position f_H relative to f_L, so that β comes out flat. A laborious analysis, confirmed by SPICE simulation, indicates that flatness is achieved when the ratio f_H/f_L is made equal to the ratio R_o/R_C. Substituting the above expressions for f_H and f_L and using the first design equation yields the second design equation:

$$C_C = (1 + (R_1/R_2))^2 C_L(R_0/R_2).$$

As an example, suppose you configure an op-amp having an R_o of 100 Ω as a gain-of-ten, noninverting amplifier. Further suppose that R_1 = 20 kΩ, R_2 = 180 kΩ, and that the amplifier drives a 0.05-μF load. The previous equations indicate that to achieve neutral compensation you must use the following values for R_C and C_C: R_C = (20/180)100

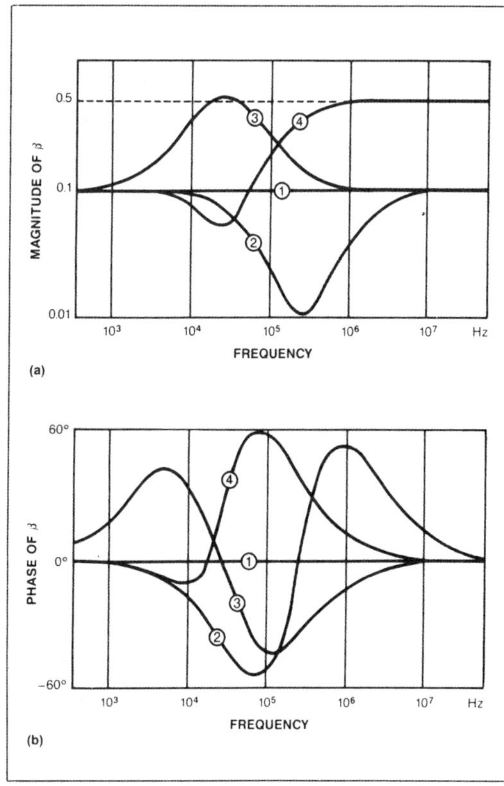

Figure 10.3 These curves show plots of magnitude (a) and phase (b) for different values of R_C and C_C. Curve 1: R_C = 11.1 Ω C_C = 34.3 pF. Curve 2: R_C = 11.1 Ω, C_C = 3.4 pF. Curve 3: R_C = 11.1 Ω, C_C = 340 pF. Curve 4: R_C = R_o = 100 Ω, C_C = 34.3 pF

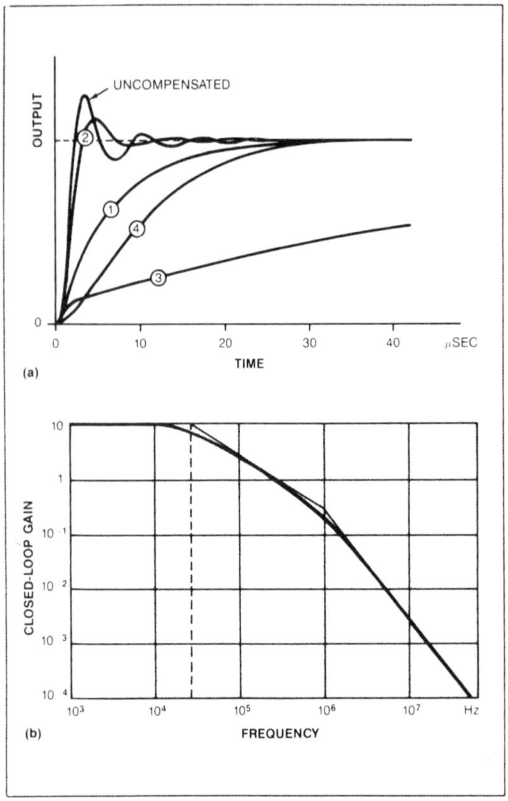

Figure 10.4 These curves show the closed-loop transient response (a) and frequency response (b) for the case of a 10-MHz op-amp having a constant gain-bandwidth product

$= 11.1\,\Omega$ and $C_C = (1 + 20/180)^2(0.1/180)\ (0.05)\ (10^{-6}) = 34.29$ pF.

The curves of Figure 10.3, obtained via SPICE, confirm that neutral compensation (curve 1) yields optimum performance. Making C_C smaller than needed (curve 2) yields a magnitude response with a dip and a phase response with a pole-zero characteristic. The resulting closed-loop transient response (Figure 10.4) exhibits undesirable ringing. Conversely, too large a value of C_C (curve 3) yields a magnitude response with a rise and a zero-pole phase, thus unnecessarily slowing down the closed-loop responses. Shown for comparison is curve 4, which corresponds to component values based on the rule-of-thumb method.

The pole formed by C_C with R_2 establishes the small-signal bandwidth of Figure 10.4's closed-loop response. In the example, this pole is located at $1/(2\ \pi\ R_2C_C = 26$ kHz. The frequency at which the loop gain becomes unity provides an additional pole. This pole is located at (10 MHz)/10 = 1 MHz. Thus, the closed-loop gain rolls off at –20 dB/decade between 26 kHz and 1 MHz and at – 40 dB/decade thereafter.

References

1. Precision Monolithics Inc, 1988 Analog Applications Seminar.

2. Franco, S, *Design with Operational Amplifiers and Analog Integrated Circuits*, McGraw-Hill, New York, NY, 1988.

11 Boost instrument-amp CMR with common-mode driven supplies

Input op-amp rails follow common-mode voltage

This scheme was used in HP Telephone Voltmeters in the 1960s, but note that the acceptable common-mode input range is limited by the supply rails of IC_6 and IC_7. (The CCIR Psophometer spec. specified a CM input at 50 Hz of 200 Vrms, so this scheme was impracticable and a transformer was necessary.)

R Mark Stitt
Burr-Brown Corp

Modern systems are placing ever greater demands on instrumentation amplifiers. When standard instrumentation amplifiers can't deliver the performance you require, consider using enhanced versions. Operating the input amplifiers of a classical 3-op-amp instrumentation amplifier from common-mode-driven subregulated power supplies will dramatically improve their performance.

Instrumentation amplifiers amplify low-level differential signals while rejecting unwanted common-mode signals. Common-mode rejection (CMR) is an important feature of instrumentation op-amps. CMR in ac is especially important because common-mode signals are inevitably dynamic – ranging from the 60 Hz of power-line interference to the hundreds of kilohertz of switching-power-supply noise. By using common-mode-driven subregulated supplies, you can improve an instrumentation amplifier's ac and dc CMR. You'll also get improved ac and dc power-supply-noise rejection as an added bonus.

When you need high gain, input-offset-voltage drift is critical. In some applications, chopper-stabilized op-amps provide the best solution because of their low input-offset-voltage drift. Unfortunately, because many chopper-stabilized op-amps use low-voltage CMOS processes, you can't operate them on standard ± 15 V power supplies. On the other hand, you can operate them from common-mode-driven, subregulated ± 5 V supplies without restriction in ± 15 V systems.

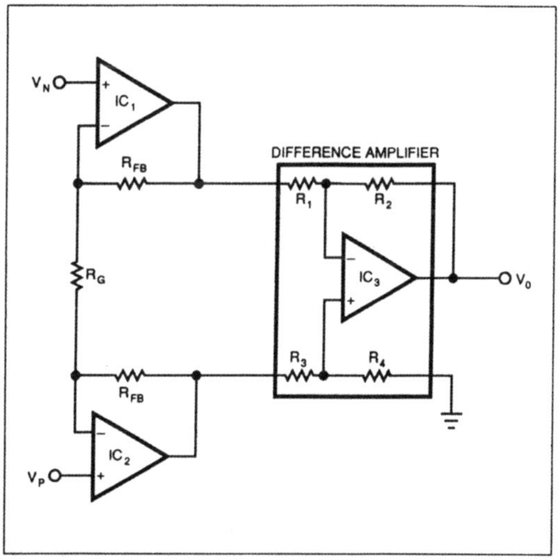

Figure 11.1 A 3-op-amp instrumentation amplifier boosts differential signals by a gain of R_2/R_1, if R_2/R_1 matches R_4/R_3

To understand the technique for maximizing rejection, first consider the 3-op-amp instrumentation amplifier (Figure 11.1). The design comprises an input-gain stage driving a difference amplifier. The difference amplifier consists of op-amp IC_3 and ratio-matched resistors R_1 through R_4. If the resistor ratio R_2/R_1 exactly matches R_4/R_3, the difference amplifier will boost differential signals by a gain of R_2/R_1 while rejecting common-mode signals. Resistor mismatch will almost certainly limit the difference amplifier's CMR if IC_3 is a high -performance op-amp. If the input-stage gain is 1, a unity-gain difference amplifier will require a 0.01% resistor match for CMR of 86 dB.

Add gain ahead of the difference amp

There are significant drawbacks in using a single-op-amp push-pull-to-single-ended converter with gain (difference amplifier) to amplify small signals superimposed on common-mode voltages. First of all, any imbalance in the source resistance will alter the resistor match and degrade the CMR. To avoid this problem, many instrumentation amplifiers precede the difference amplifier with a differential-input, differential-output gain stage consisting of two amplifiers (IC_1 and IC_2 of Figure 11.1) and three resistors. The gain of this stage is $1 + 2 \cdot R_{FB}/R_G$, and the instrumentation amplifier's overall gain is $(1 + 2 \cdot R_{FB}/R_G) \cdot R_2/R_1$.

The CMRR is the ratio of differential gain to common-mode gain. Using buffer amplifiers to add gain ahead of a difference amplifier increases an instrumentation amplifier's CMR (if the buffer amplifiers' CMR is better than that of the difference amplifier). That's why instrumentation amplifier data sheets usually specify one CMR at gain = 1 and a much higher CMR at higher gains.

Most high-performance op-amps have better CMR than difference amplifiers do. However, be careful when selecting an input op-amp. High-grade versions of the venerable 741 op-amp have a minimum dc CMR of 80 dB, and the popular LM324 has a minimum dc CMR of only 70 dB. High-performance bipolar-input op-amps have the best CMR. The OPA177, for example, has a minimum dc CMR of 130 dB. FET-input op-amps usually don't offer quite as much CMR performance. For example, the Burr-Brown OPA627 FET-input op-amp has a minimum dc CMR of only 106 dB.

Driving the input op-amp's power-supply from subregulated power supplies referenced to the instrumentation-amplifier common-mode-input voltage improves the dc CMR of a standard instrumentation amplifier. The device mismatch and thermal feedback that occur as an op-amp's inputs change limit its CMR. (Similar effects also limit the amplifier's power-supply rejection ratio.) Varying the power supply to track the common-mode input signal inhibits changes and reduces errors, which can degrade CMR.

The input amplifier's ac response limits the instrumentation amplifier's ac CMR. The outputs of the input amplifiers in the instrumentation amplifier follow the common-mode input signal. As the frequency of the common-mode signal increases, the input op-amp's loop gain diminishes, which causes differential gain errors to increase and CMR to fall off.

For large common-mode signals, the input op-amp's slew rate can limit the function of the instrumentation amplifier. The instrumentation amplifier will fail to function completely when the maximum rate of change of the common-mode signal exceeds the op-amp's slew-rate limit. For a sine wave, where the maximum rate of change occurs at the zero crossing, the derivation of the slew-rate limit is

$$V = V_P \cdot \sin(2\pi ft),$$

$$dV/dt = 2\pi f \cdot V_P \cdot \cos(2\pi ft).$$

$$\text{At } t = 0,$$

$$dV/dt = 2\pi f V_P,$$

$$\text{therefore the slew-rate limit} = 2\pi f_{MAX} V_P,$$

where

$$V = \text{common-mode voltage vs time } (t),$$

$$V_P = \text{peak common-mode voltage},$$

$$\text{slew-rate limit} = \text{maximum } dV/dt,$$

and f_{MAX} = maximum common-mode frequency at amplitude V_P beyond which a standard instrumentation amplifier will fail to function due to the slew-rate limit of the input op-amp.

Driving the power supply of the input op-amps from common-mode-referenced subregulated supplies improves ac CMR as well as dc CMR. Because neither the amplifier's inputs nor output change relative to the subregulated power-supply rails – at least in the ideal case – nothing within the amplifier moves in response to the common-mode signal. No current flows in the phase-compensation capacitors, which disables the slew-rate-limiting phase compensation for common-mode response.

Figure 11.2 shows a complete circuit for an enhanced instrumentation amplifier. As you can see, this enhanced instrumentation amplifier contains the 3-op-amp instrumentation amplifier from Figure 11.1 plus a buffered common-mode voltage generator and ± 5 V subregulated power supplies.

The 3-op-amp instrumentation amplifier in Figure 11.2 uses an INA106 gain-of-10 difference amplifier, which contains a precision op-amp and ratio-matched resistors (R_1 through R_4) pre-trimmed for 100-dB min CMR. Because the INA106 already contains ratio-matched resistors, you don't have to match critical resistors to build a precision instrumentation amplifier.

The resistor divider network (R_5 and R_6) creates the common-mode signal that drives the subregulated supplies. The instrumentation-amplifier inputs drive the network through unity-gain-connected op-amps IC_4 and IC_5. These buffer amplifiers preserve the instrumentation amplifier's high input impedance. Some applications don't require such impedance. In those applications, the impedance of the R_5, R_6 network may be connected directly to the instrumentation-amplifier inputs. Figure 11.3 shows a circuit without buffer amplifiers IC_4 and IC_5. The signal at the R_5, R_6 connection of the resistor divider is the common-mode or average voltage of the two instrumentation-amplifier inputs.

The negative subregulator consists of IC_6, R_7, C_1, and a 100-μA current source. Since no current flows in the op-amp input, 100 μA flows through the 50-kΩ resistor, R_7, forcing a − 5 V drop from the op-amp input to its output. The op-amp forces the negative input to be at the same potential as the positive input. The result is a − 5 V floating-voltage reference relative to the op-amp non-inverting-input terminal. The positive subregulator is the same as the negative subregulation except for the current-source connection's polarity.

The circuit in Figure 11.3 only connects the positive and negative subregulators' outputs to the power supplies of the input op-amps (IC_1 and IC_2). All other op-amps are connected to ± 15 V power supplies.

The subregulated-supply voltage limits the common-mode input range of the enhanced instrumentation amplifier. The outputs of the subregulator amplifiers (IC_6 and IC_7) must swing

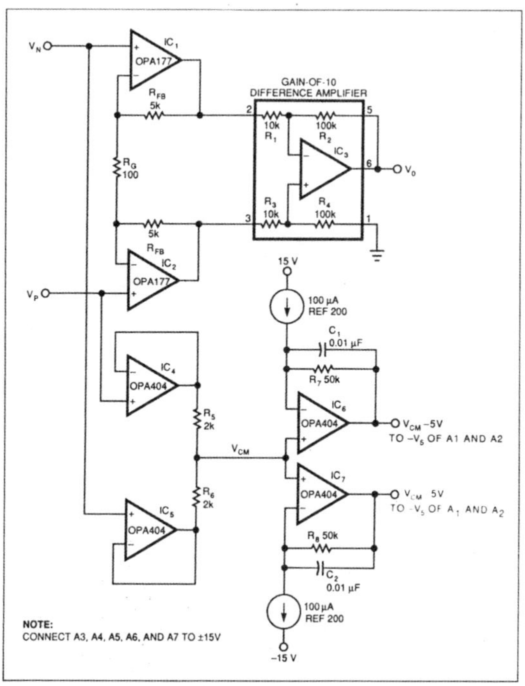

Figure 11.2 A common-mode voltage buffer connected to a resistor divider network drives the subregulated supplies that power the enhanced instrumentation amplifier

the common-mode voltage as well as the subregulator voltage. The larger the subregulator voltage, the smaller the common-mode input range. A subregulator voltage of ± 5 V is low enough to give good input common-mode range and high enough to allow full performance from almost any op-amp. The reduced power-supply voltages lower power dissipation in the input op-amps. They also improve the instrumentation amplifier's performance by reducing thermally induced low-frequency noise.

In all semiconductor packages, thermocouples exist at dissimilar conductor interfaces. Matched-seal metal, side-brazed ceramic, cerdip, and many plastic packages use Kovar leads. Thermocouples exist between the lead plating and the Kovar. Thermocouples also exist between the leads and the solder connections to the printed circuit.

If thermal gradients are properly matched at the amplifier inputs, the thermocouple errors will cancel one another out. In practice, mismatches occur. Even under laboratory conditions, a mismatch may produce several tenths of a microvolt – well above low-noise-amplifier levels. In the output of a high-gain amplifier, errors appear as low-frequency noise or short-term input-offset errors.

In signal op-amps, package leads conduct away much of the heat. The resulting thermal difference between the package and the printed circuit can be a major source of error. Operating the op-amp on ± 5 V supplies instead of ± 15 V supplies decreases quiescent power dissipation and its associated temperature rise by at least 300%. This decrease also provides a commensurate reduction in thermally induced errors.

The common-mode input range of an enhanced instrumentation amplifier is equal to that of most integrated-circuit instrumentation amplifiers. Because an enhanced instrumentation amplifier uses a gain-of-10 difference amplifier rather than a unity-gain difference amplifier, input amplifiers don't limit a difference amplifier's common-mode range. The common-mode input range of both an enhanced instrumentation amplifier and a standard instrumentation amplifier is ± 7 V. With a 10 V output, the common-mode input of a standard instrumentation amplifier is only ± 7 V, not ± 10 V as is commonly believed.

An input amplifier's output swing limits the common-mode swing of a standard instrumentation amplifier. The output swing of subregulator amplifiers limits the common-mode range of an enhanced instrumentation amplifier.

Standard instrumentation amplifiers use unity-gain difference amplifiers for practical reasons. Since standard instrumentation amplifiers are general-purpose devices, they must be adjustable to unity gain. Because maintaining the resistor ratio necessary for good difference-amplifier

CMR is difficult, standard instrumentation amplifiers usually contain a fixed unity-gain difference amplifier. Gain adjustment is made with the input amplifiers, where matching is not critical for good CMR. Also, the more gain placed ahead of the difference amplifier, the better the instrumentation amplifier's CMR.

To investigate the limits on instrumentation amps' input common-mode range, assume the op-amps' inputs can all swing to within 3 V of their power-supply rails (± 12 V when operating on ± 15 V power supplies). In a standard instrumentation amplifier with a unity-gain difference amplifier, the input amplifiers must provide a differential 10 V output to produce a 10 V difference amplifier output. If the input amplifiers have equal gains, each must deliver one-half of the 10 V differential signal. With a common-mode input of 7 V, an input amplifier must deliver 7 V common mode plus 5 V differential mode in order to bring it up to its 12 V-swing limit.

The enhanced instrumentation amplifier also has a ± 7 V common-mode input limit. Its subregulators are set at ± 5 V from input common-mode signal. With a 7 V common-mode input, a subregulator's output will be at its 12 V-swing limit.

In an enhanced instrumentation amplifier using a gain-of-10 difference amplifier, the buffer amplifiers must provide a differential output of only 1 V to produce a 10 V instrumentation-amplifier output. If the input amplifiers have equal gains, each must deliver one-half of the 1 V differential signal. With a common-mode input of 7 V, one input amplifier must deliver 7 V common mode plus 0.5 V differential for a total output of 7.5 V. Obviously, producing the 7.5 V is no problem since the V_S is 12 V (5 V subregulated plus 7 V common mode).

Figure 11.4 offers a performance comparison between a standard instrumentation amplifier and an enhanced-instrumentation amplifier. In Figure 11.2, IC_1 and IC_2 are OPA177 amplifiers; IC_3 is an INA106 gain-of-10 difference amplifier; and IC_4 to IC_7 is an OPA404 quad op-amp in the enhanced circuit. This instrumentation amplifier's overall gain is set at 1000 V/V. The OPA177 is an improved version of the industry-standard OP-07. It offers 10-μV max V_{OS} and 0.1–μV/°C max V_{OS}/dT. The OPA404 provides high speed and low bias current.

The FET inputs of the OPA404 don't add loading at the inputs of the instrumentation amplifier. Their speed is higher than the OPA177's, yielding an improvement in CMR vs frequency. An HP4194A gain-phase analyzer with an input signal to the instrumentation amplifier of 9 dBm made the CMR plots. The enhanced plot shows a dramatic CMR vs frequency boost. At 2 kHz, for example, the CMR of the standard

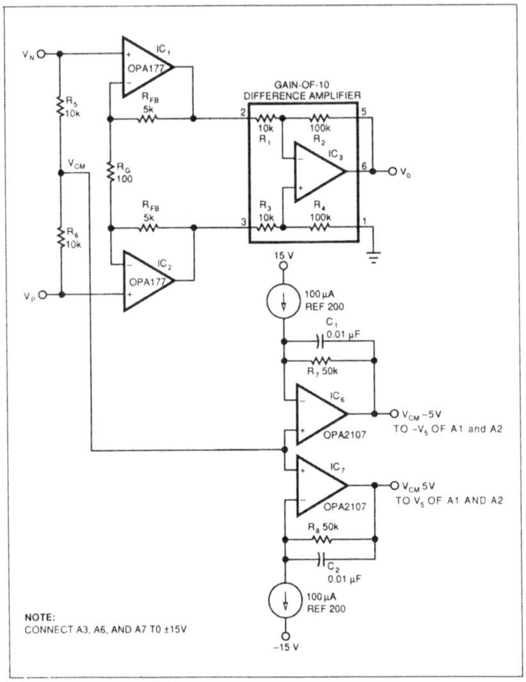

Figure 11.3 If you don't need high input impedance, removing the common-mode voltage buffers will still provide better common-mode rejection than standard instrumentation amplifiers and do it at a lower cost

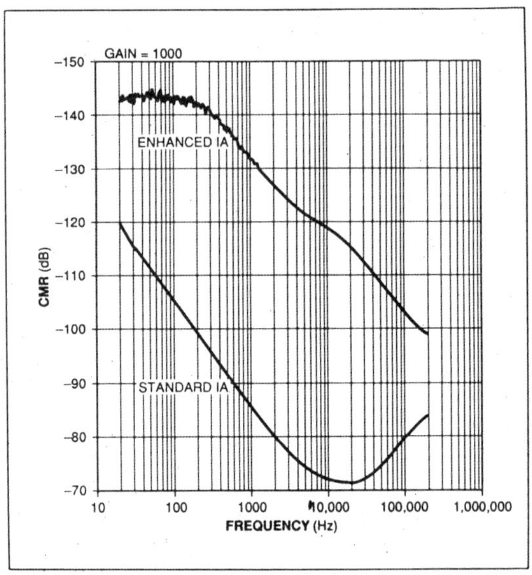

Figure 11.4 With a gain of 1000 V/V, an enhanced instrumentation amplifier offers common-mode rejection of better than 120 dB at frequencies below 8 kHz

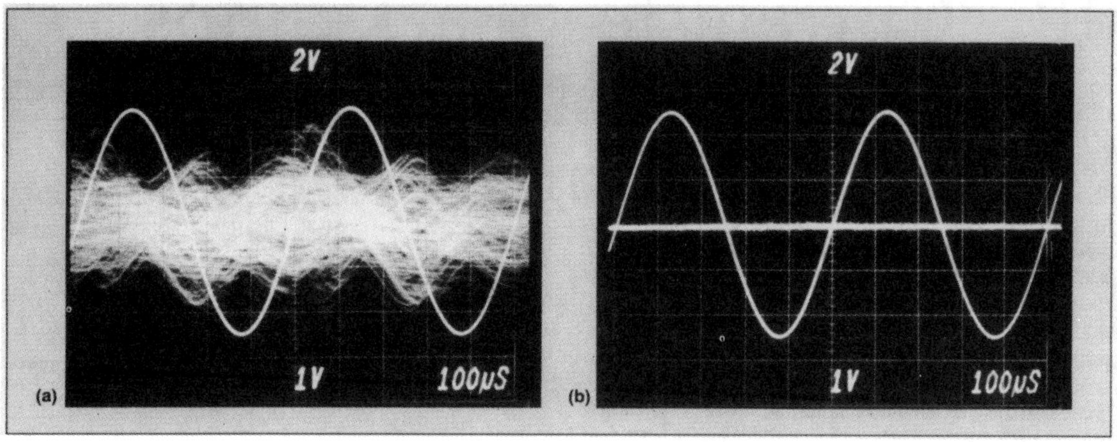

Figure 11.5 These scope photos show a dramatic difference in common-mode response between standard (a) and enhanced instrumentation (b) amplifiers with chopper-stabilized op-amps. In both photos, the sine wave is the common-mode input – a 2-kHz, 5 V p-p signal; the other trace is the instrumentation-amp output with a differential gain of 1000. Without common-mode drive (a), noise limits the usable CMR to 56 dB with the common-mode input shown. With common-mode drive to the input op-amp power supplies (b), noise performance improves so much that the CMR appears essentially infinite under these test conditions. (The CMR actually measures about 82 dB.)

instrumentation amplifier is about 80 dB. At 2 kHz, the CMR of the enhanced instrumentation amplifier is more than 120 dB – an improvement of more than two orders of magnitude.

The scope photos of Figure 11.5 show similar instrumentation amplifiers using LTC1050 chopper-stabilized op-amps for IC_1 and IC_2. When V_{OS}/dT is critical, chopper-stabilized op-amps may be the best choice because they offer 5-μV maximum V_{OS} over temperature. With a ± 2.5 V, 2-kHz input signal, chopper noise limits CMR to about 56 dB. The enhanced circuit improves the usable CMR to about 82 dB with the common-mode input shown in Figure 11.5b.

A difference amplifier will limit CMR performance in enhanced instrumentation amplifiers. The more gain you add ahead of a difference amplifier, the better the potential for improvement. For example, with a gain of 100 V/V ahead of the difference amplifier, an improvement in CMR of as much as 40 dB is possible. The actual performance boost depends on impedance matching and parasitics in the devices you select.

How fast are your amplifiers?

Of course, the way CMR varies with frequency depends on the dynamic performance of all the amplifiers in the circuit. Improvement in dynamic CMR will be most dramatic when the speed of the amplifiers IC_4 to IC_7 is much higher than the speed of IC_1 and IC_2.

You can easily implement high-voltage instrumentation amplifiers using the enhanced instrumentation-amplifier configuration. Use standard-precision signal-level op-amps for the input amplifiers and less critical op-amps for the high-voltage chores. For example, use OPA445 op-amps for IC_6, IC_7, and IC_3 (the difference amplifier). To boost the voltage rating of the current sources in the subregulated supplies, place two REF200 current-source sections in series. If you use 1% resistors for difference resistors R_1 to R_4, you may need a potentiometer to adjust CMR. The resulting instrumentation amplifier will provide outstanding performance on power supplies up to ± 45 V.

12 Current-feedback amps enhance active-filter speed and performance

CF op-amps benefit filter circuits among others

Covers high performance lowpass, bandpass and bandstop types. The lowpass types, though exhibiting a Cauer (elliptic) response, use twin-Tee networks and are therefore not canonic.

Doug Smith
Burr-Brown Corp

Analog designers have historically relied on passive filters for applications with frequencies greater than 1 MHz. Until recently, designing viable active filters with cutoff frequencies at 1 MHz or greater was difficult because voltage-feedback amplifiers with sufficient gain-bandwidth products and short propagation delays were simply too expensive. The emergence of current-feedback, or transimpedance, amplifiers has significantly changed this picture. With these amplifiers and a conscientious design and pc-board layout, you can design active filters that operate at high frequencies. Active-filter applications are no longer restricted to the audio frequency range.

Active RC filters have many advantages over passive filters, and many of these advantages become increasingly important as the frequency increases. For example, there is no insertion-loss penalty, and you can even have power gain if needed. A doubly terminated passive filter would decrease the signal by at least 50%. The elimination of inductors is the biggest advantage offered by active filters. This advantage doesn't involve size considerations alone.

Passive inductors are only linear for low power levels, much like transistors with no negative feedback. As you pump more current through the inductor, the magnetic core material begins to saturate and the inductor generates its own harmonic-distortion terms. The filter's transfer response will not necessarily suppress these signals. In an active RC filter, the amplifier quality and the sophistication of the design set the dynamic range. Theoretically, designers have a good deal of control over both of these parameters.

As a case study, consider three situations – a lowpass antialiasing filter, a bandpass filter, and a high-Q notch filter – in which active filters that incorporate current-feedback amplifiers provide a viable alternative to passive filters. All three filters will be designed around the Burr-Brown OPA603. You could implement the three designs using carefully selected, video-speed conventional op-amps. However, current-feedback amplifiers more readily satisfy the low transit time and large bandwidth at high gain requirements for the example circuits. Let's start with the design of an antialiasing filter to drive the input of an ADC603 – a 12-bit, 10-MHz A/D converter.

When dealing with A/D converters in filter work, the Nyquist theorem states that if any converter input harmonic frequency is greater than half the sampling rate, those frequencies must alias, or fold back, into the passband. Normally, this condition is not desirable. To skirt the issue, you must suppress any input frequencies that exceed the Nyquist rate before the converter sees them. The result of this maneuver is that the required attenuation becomes a function of converter resolution. It's also important for the filter to roll off as fast as possible. An elliptic response is the best choice because the addition of transmission zeros in the stopband creates the sharpest roll-off theoretically possible for a particular number of poles without having to rely on mutual inductance.

The first step in designing the filter is calculating the attenuation requirements. You can do so by estimating the theoretical signal-to-noise ratio (SNR) using the expression

$$\text{SNR} = 6.02N + 1.8\,\text{dB},$$

where N is the number of bits. For the ADC603, the expression yields

$$\mathrm{SNR} = 6.02 \times 12 + 1.8 = 74.04\,\mathrm{dB}.$$

The calculation shows that the guaranteed stopband attenuation must be greater than 74 dB. A search of standard design tables shows that a fifth-order elliptic lowpass response is a reasonable compromise between the transition width and the filter order. The general transfer function for this filter is

$$T(s) = \left(H_a \frac{s^2 + b_{0a}}{s^2 + a_{1a}\, s + a_{0a}}\right) \left(H_a \frac{s^2 + b_{0b}}{s^2 + a_{1b}\, s + a_{0b}}\right) \left(\frac{a_0}{s + a_0}\right).$$

You can now form the filter by cascading two second-order sections and one first-order section Figure 12.1b. The essential equations[1] for the second-order sections are

$$T(s) = H(s^2 + b_0)/s^2 + a_1 s + a_0$$

$$p = 1/\sqrt{b_0}$$

$$q = ((b_0/a_0) - 1)/2\sqrt{b_0}$$

$$K = 2 + (1/2)((b_0/a_0) - 1) - (a_1\sqrt{b_0}).$$

The essential equations for the first-order section are

$$T(s) = a_0/s + a_0$$

$$a_0 = 1/RC.$$

The task is to design a fifth-order elliptic antialiasing filter Figure 12.1a with a guaranteed stopband attenuation of 75 dB and no more than 3 dB of passband ripple. In addition, the maximum attenuation should begin at 5 MHz – half the sampling rate.

The transfer coefficients[1] for this case are

$$a_{1a} = 0.096035$$
$$a_{0a} = -0.945044$$
$$b_{0a} = 10.47185$$
$$a_{1b} = 0.285481$$
$$a_{0b} = 0.413907$$
$$b_{0b} = 4.328514$$
$$a_{0c} = 0.191095.$$

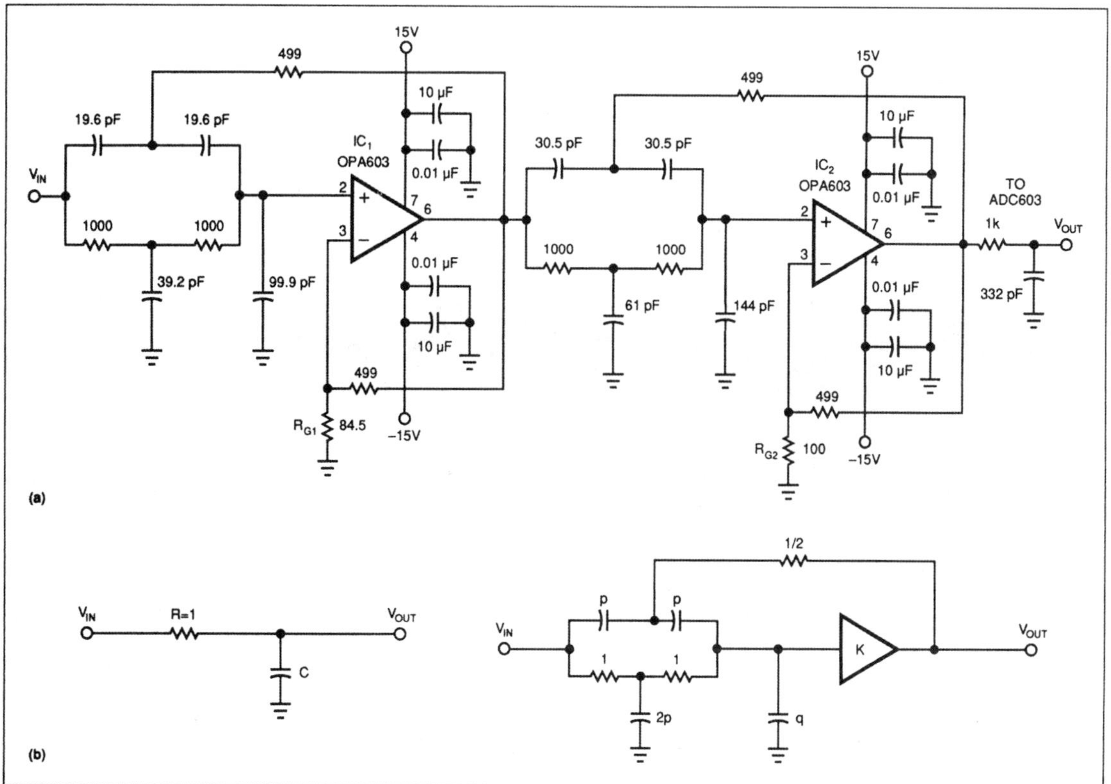

Figure 12.1 You need an antialiasing lowpass filter when you're driving an A/D converter. A fifth-order elliptic design (a) proves to be the best choice in such an application. You can use two second-order sections and one first-order section (b) to form the necessary filter

The corresponding component values are

$$p_a = 0.309021$$
$$q_a = 1.557592$$
$$K_a = 6.875982$$
$$p_b = 0.480652$$
$$q_b = 2.272930$$
$$K_b = 6.011362$$
$$R = 1$$
$$C = 5.232999.$$

This filter prototype has an f_3 bandwidth of 0.15912 (1 rad/sec), and its maximum attenuation begins at $f_{STOPBAND} = 0.3171$. In this case, you have to scale the frequency to $f_{STOPBAND}$ rather than f_3. In addition, you can arbitrarily scale the impedance to 1 kΩ. Multiply each resistor by this impedance value; divide every capacitor value (p, q, and C) by the frequency-impedance scaling factor, K_f:

$$K_f = 1k(5 \times 10^6\,Hz)/0.3171\,Hz = 1.577 \times 10^{10}.$$

The final component values, rounded to three significant figures, are

$$p_a = 19.6\,pF$$
$$q_a = 98.8\,pF$$
$$K_a = 6.88$$
$$p_b = 30.5\,pF$$
$$q_b = 144\,pF$$
$$K_b = 6.01$$
$$C = 332\,pF.$$

Using a feedback resistance of 499 Ω, you can choose the closest 1% values for the gain resistors, or $R_{G1} = 84.5\,\Omega$ and $R_{G2} = 100\,\Omega$.

High-Q bandpass filters have many uses. One is isolating a particular harmonic of a distorted sine wave before amplifying the signal to more easily measure the magnitude. Many common active filter configurations run into problems in such applications because the value of Q is highly sensitive to changes in the gain (and, thus, the frequency response) of the amplifier. One of the best filter topologies in this situation is an extension of the basic Sallen-Key circuit (Figure 12.2a[2]) The addition of a second amplifier can raise the potential value of Q by two orders of magnitude.

For stable operation, K1 should be greater than zero and K2 should be less than zero. The transfer function is

$$T(s) = K1 \times K2s/((1 - K1K2)s^2 + (4 - K1)s + 2).$$

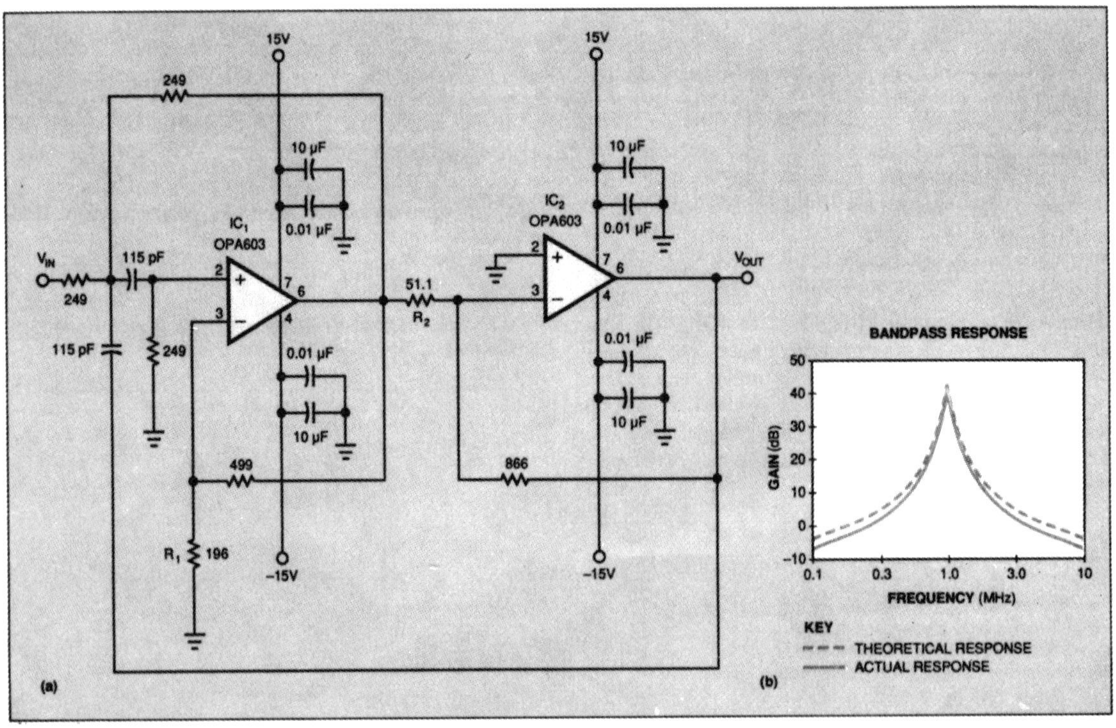

Figure 12.2 To develop high-Q bandpass filters, you can add a second amplifier to the basic KRC circuit (a) to raise the potential Q by orders of magnitude. The actual bandpass response of the filter is very close to the theoretical value (b), although the response shows a slightly lower gain

From this expression, you can determine that

$$Q = \sqrt{2(1 - K1K2)/4 - K1}$$

$$\omega_0 = \sqrt{2/(1 - K1K2)}.$$

The sensitivities of most concern involve the variations of Q when the gain of either amplifier changes. Analysis shows that

$$S_{K1}{}^{Q} = K1(1 - 4K2)/(4 - K1)(1 - K1K2)$$

$$S_{K2}{}^{Q} = K1K2/1 - K1K2.$$

You can neglect $S_{K2}{}^{Q}$ because it is approximately equal to 1 and is not a serious limitation. Although it's probably not obvious, there's a tradeoff between K2 and $S_{K1}{}^{Q}$. The higher the gain of K2, the lower the value of $S_{K1}{}^{Q}$. In a voltage type op-amp, higher gain inherently means lower bandwidth. However, a transimpedance amplifier has the ability to maintain its bandwidth at high gains. This characteristic gives current feedback amplifiers a clear advantage in this situation.

Putting theory into practice

Again, it's time to put theory into practice. Let's say that you have to design a second-order bandpass filter with a center frequency of 1 MHz and a – 3 dB bandwidth of 40 kHz. In addition, the sensitivity to variations in gain should be no greater than 9.
First, the required value of Q is

$$Q - f_0/BW = 1\ \text{MHZ}/40\ \text{kHz} = 25.$$

Next, you can simultaneously solve the equations for Q and $S_{K1}{}^{Q}$ and obtain K1 = 3.556357 and K2 = – 17.01347. The corresponding center frequency for this prototype is then $f_0 = 0.287996$. You have to scale this center frequency to 1 MHz. If you arbitrarily choose a value of 1 kΩ for the resistors, the final value of C becomes C = 115.3 pF. You can realize the required K1 gain by using a 499 Ω resistor for the feedback and a 196 Ω resistor R_1.

K2 gain is a different situation because the feedforward resistor of the second amplifier is the load resistance of the first amplifier. As a result, the feedforward resistor value needs to stay reasonably large. If you limit the second feedforward resistance to 50 Ω, the second stage feedback resistor will be 866 Ω and R_2 will equal 51.1 Ω.

The dynamic range of high-frequency, moderately priced spectrum analyzers is often less than 80 dB. However, you can effectively increase the measurement range by suppressing the fundamental frequency of the input signal by a known amount without affecting the rest of the frequency spectrum. This application doesn't require a high-order, band-reject filter – a low-order, high-Q notch filter will work quite well.

The classic twin-T network (Figure 12.3a) is a promising candidate for the job. The transfer function of this network is

$$T(s) = s^2 + \omega_0{}^2/s^2 + 4\omega_0 s + \omega_0{}^2,$$

and the attenuation at any bandwidth equals

$$A\rangle B + 10 \cdot \log(1 + (4f_0/BW_{\times\ dB})^2).$$

This circuit has two drawbacks – it is somewhat sensitive to passive-component tolerances, and it has an intrinsic Q value of 0.25. The first drawback creates no problem but the second drawback must be overcome. You can substantially increase circuit Q value by adding a second amplifier to the network[3].
The new transfer function is now

$$T(s) + s^2 + \omega_0{}^2/s^2 + 4\omega_0\,(1 - K)s + \omega_0{}^2,$$

and the Q value is now a function of K:

$$Q = 1/4(1 - K).$$

As K approaches 1 from below, Q increases in an unlimited fashion. If K is greater than 1, however, the circuit is unstable. Although wide bandwidth at high gain is not as important here as it was in Figure 12.2's example, the comparatively lower transit time of a current-feedback amplifier should yield superior performance in this application.

A specific example will prove the point. The task is to design a 1.5 MHz notch filter that has a – 3 dB bandwidth of 225 kHz. The first step is to calculate Q using the expression

$$Q = f_0/BW_{-3\,dB} = 1.5\,\text{MHz}/225\,\text{kHz} = 6.66.$$

You can use this value to calculate

$$K = 1 - 1/4Q = 0.9625.$$

If R_1 is set equal to 1 kΩ, then

$$C = 1/2\pi f_0 R_1.$$

If you let R_2 also equal 1 kΩ, then $(1 - K)R_2 = 37.5$ and $KR_2 = 962.5$. Figure 12.3b shows the final notch filter design. Both amplifiers are configured as unity gain buffers, and the feedback resistance is set at 499 Ω. The actual response Figure 12.3c shows a slight excess attenuation beyond the notch frequency, but the performance is still good.

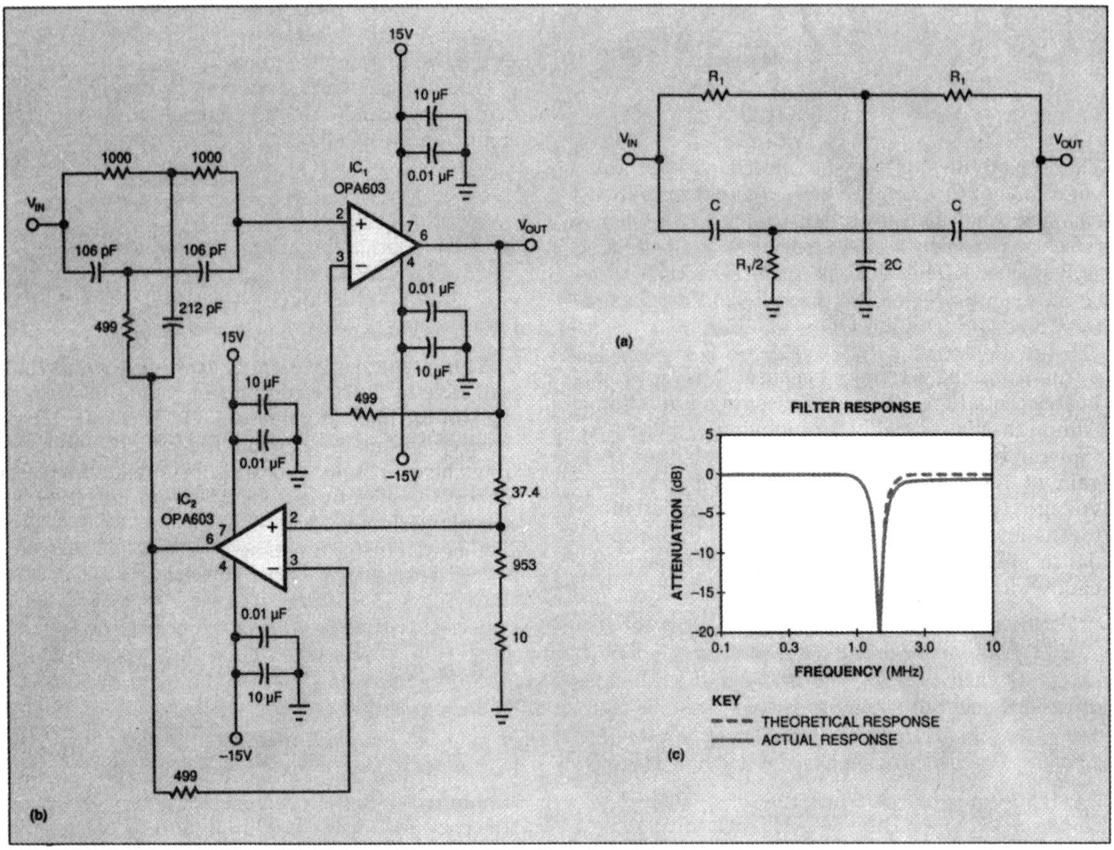

Figure 12.3 When you need a high-Q notch filter, the classic twin-T network (a) is a good starting point. By adding a second amplifier (b), you can substantially raise circuit Q. The actual response of the filter (c) shows a slight excess attenuation beyond the notch frequency.

Making the case for current feedback

Don't get the idea that something is inherently wrong with voltage feedback, even at high speed. In fact, voltage-feedback amplifiers generally have a lower noise-floor specification than do current-feedback amplifiers. However, when comparing voltage- and current-feedback amplifiers, you must take the application into consideration. Current-feedback, or transimpedance, amplifiers have some distinct performance advantages as waveform speed gets higher and higher. These advantages can translate into higher-performance active filters.

The most striking difference between voltage-feedback and transimpedance op-amps is that with a fixed-feedback resistor, the current-feedback amplifier has very low gain-bandwidth tradeoff. A transimpedance amplifier maintains its bandwidth at high gain settings – an advantage in active-filter topologies because a large gain is necessary to minimize sensitivity.

In addition, transimpedance amplifiers have very high slew rates compared with those of conventional voltage op-amps. A typical slew rate for a video-speed voltage-feedback amplifier is in the 200 to 300 V/µs range; a comparable current-feedback amplifier might slew as fast as 2500 V/µs. This slew-rate disparity is simple to explain. In a conventional amplifier, the slew rate is the ratio of the bias current flowing through the slewing node to the capacitance that can be referred back to that node. In a transimpedance amplifier, the feedback current mirrors and adds to the bias current flowing through the slew-rate-limiting node. Because more current is available to charge the capacitance, the slew rate increases. The feedback current is proportional to V^{OUT}, which is proportional to V^{IN}. So the harder you drive a current-feedback amplifier, the faster it slews. In practice, this fact effectively eliminates slew rate as a limiting factor in high-speed, active-filter design.

One final factor favors the transimpedance

amplifier – settling time. Designers often choose a filter's transfer function for best time-domain response. Therefore, ensuring that the amplifier settles to the required level substantially faster than the filter has to settle is crucial.

High-speed amplifiers are complicated devices and acceptable ac response does not necessarily ensure an acceptable settling time. Many conventional voltage-feedback op-amps use internal pole-zero cancellation to increase their bandwidths. Analysis shows that a small mismatch in the pole-zero cancellation has a negligible effect on frequency response, but the scheme can dramatically boost settling time.

Transimpedance amplifiers have settling-time problems, too. Although transimpedance amplifiers settle to 0.1% (10 bits) or 0.02% (12 bits) in as little as 15 ns, the settling time to 0.01% can be relatively long. The same current flow that increases the slew rate of a transimpedance amplifier also upsets the amplifier's bias point slightly, and a finite amount of time is required for the bias point to return to equilibrium. This effect is small, but it can often extend the 0.01% settling time to several microseconds.

References

1. Kerwin, W J, 'An Active RC Elliptic Function Filter', *IEEE Region Six Conference Record*, Volume 2, April 1966, p. 647.
2. Kerwin, W J, *Linear Active Circuit Design*, Part 4, Course Notes for ECE-553, University of Arizona, Spring 1987, p. 51.
3. Williams, Arthur B, *Electronic Filter Design Handbook*, McGraw-Hill, New York, 1981, p. 6.

13 Use Spice to analyze component variations in circuit designs

Take a trip to Monte Carlo to tolerance your circuits

Several proprietary versions of the popular 'Spice' circuit analysis software include a 'Monte Carlo' analysis capability. This permits multiple analysis runs with specified component tolerances chosen at random, giving an insight into likely performance spreads in production. Typically, uniform and Gaussian (normal) distributions are accommodated, and some versions even cope with bi-modal distributions – e.g. the sort where a manufacturer of, say, 10% capacitors has selected out all those within +/– 5% for sale at a higher price, leaving you with the rest.

George Ellis
Kaydon Corp

In the hectic process of design, simulation, prototyping, and testing, it's easy to neglect variations in circuit components. If you do consider variations, you may add some margin to nominal component values and yet be guided by only a rule of thumb or an educated guess. Then, component tolerances can stack up the wrong way and yield unpleasant results.

It's possible, though, to accurately account for component tolerances without expending extraordinary effort. Spice, the most popular circuit simulator today, provides a variety of tools for analyzing component variations. The most powerful of these tools is Monte Carlo analysis.

Circuit performance varies because parameter values vary. The variations either can be deterministic or statistical. Deterministic variations are possible to predict. For example, the resistance of copper is deterministic: It increases about 0.39% for every °C of temperature increase. The effect is the same, for all practical purposes, from one piece of copper to the next.

It's not possible, however, to predict precisely the effects of statistical variations on any given component. Examples of statistical variations include manufacturing tolerance (the variation of values from one unit to the next) and aging (the variation of values over time). Fortunately, you can analyze statistical variations with Spice.

If you're uncomfortable working with statistical variation, it may be because the analysis tools seem mysterious. However, if you already use Spice to simulate your circuits, you may be surprised to learn how easily Spice can simulate variations. In fact, you will probably need to make only a few modifications to your circuit models to investigate the most important sources of variation.

The examples presented here use PSpice[1,2], a version of Spice from Microsim Corp. The implementation of variation analysis differs among Spice versions, but the principles are the same. You should be able to apply this information to any version of Spice that offers statistical-variation analysis.

PSpice lets you add a tolerance value to any parameter by using the DEV directive in the .MODEL command. For example, you can specify a 1%, 20-kΩ resistor with these two lines:

```
R1 1 2 ONE_PERCENT_R 20K.
.MODEL ONE_PERCENT_R RES(R =
  1 DEV = 1%)
```

When you simulate the circuit, the 'ONE_PERCENT_R' in the specification for R1 will point PSpice to a model named ONE_PERCENT_R, which shows that R, the resistance multiplier, has a 1% device tolerance. (Incidentally, the character string 'ONE_PERCENT_R' has no meaning itself; any name would have worked.)

The DEV directive can apply to any parameter in any model by following the parameter with 'DEV = nn%'. For example, a transformer (K) can have a variable-effective air gap if you add the lines

K1 1 2 3 4 VARIABLE__AIR__GAP 100UH.
.MODEL VARIABLE__AIR__GAP CORE
(GAP = 0.1 DEV = 5%)

Here, the air gap is 0.1 cm ± 5%.

Sometimes, parameters vary in groups. PSpice provides the LOT directive for these cases. For example, suppose you're using a resistor pack with eight resistors that you know have a tolerance of ± 1% and are matched to be within 0.2% of each other. The resistor model would be

.MODEL ONE__PERCENT__RPAK RES(R = 1
DEV = 0.2%
LOT = 0.8%)

With this model, you can specify eight resistors in your circuit. (Actually, this same statement can specify any number of resistors in a pack.) The values of the eight will vary ± 1% from one run to the next, but on any given run, they will vary only ± 0.2% from each other. If the circuit has more than one resistor pack, you need a .MODEL directive for each one. These directives let the resistances of one pack vary from the resistances of a second pack by the full ± 1%.

A component's tolerance describes only the limits of variation. For a complete description, you must also know the statistical distribution – how likely a component is to take on a particular value. (Distribution is actually the integral of density. However, it is common to refer to a graph of statistical density as a 'distribution,' and the examples used here follow that conventional usage.)

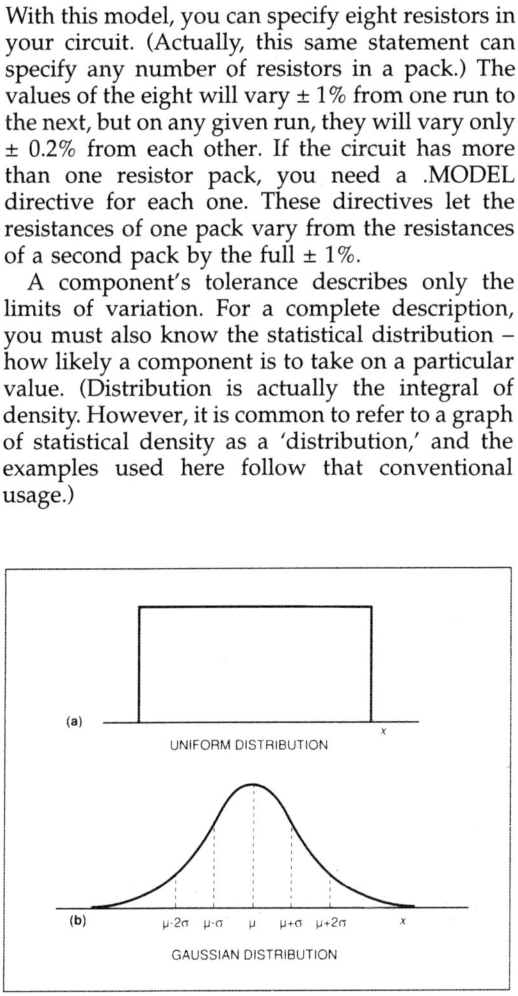

Figure 13.1 If the statistical distribution of components is uniform (a), then component values are equally likely to have any value within their tolerance band. The Gaussian, or normal, distribution (b) is the most common distribution in manufacturing, however

The simplest distribution is uniform (Figure 13.1a), in which components are equally likely to take on any value within the tolerance band. PSpice defaults to uniform distributions. PSpice also supports Gaussian, or normal, distributions (Figure 13.1b), which are the most common distributions in manufacturing.

Gaussian distributions are described by two parameters: the mean, or average (usually represented by $\bar{x}$ or μ), and the standard deviation (σ). Components with normal distributions are much more likely to take on values in the center of the tolerance band rather than at the outside edges.

You can specify a normal distribution for ± 1% resistors with the PSpice command.

.MODEL ONE__PERCENT__GAUSS__R RES(R =
1 DEV/GAUSS = 0.25%)

Here, the resistor values have a normal distribution and $\sigma = 0.25\%$. PSpice limits values of normal distributions to $\pm 4\sigma$, so ONE__PERCENT__GAUSS__R can vary ± 1%.

Manufacturing tolerances usually have normal distributions, so if you're using a component produced in large volume, you would probably specify a normal distribution. However, if you're using components that have been discriminately selected from a low-yield batch, the distribution will be closer to uniform.

For example, manufacturers may select capacitors with ± 20%, ± 10%, and ± 5% tolerances from one manufacturing batch. The selected capacitors have a somewhat uniform distribution (Figure 13.2a); however, the remaining capacitors have a distribution with a 'notch' (Figure 13.2b). To account for a distribution such as this, you have to specify the distribution yourself.

In PSpice, you can divide custom distributions into as many as 100 parts, each part being a section of the distribution curve. You specify the distribution with two *x,y* pairs for each part. For example, you can specify a 5-part distribution for ± 20% capacitors (Figure 13.2c) by using the following statements:.

.MODEL TWENTY__PERCENT__C CAP(C = 1
DEV/BI__MODAL 20%).
.OPTIONS DISTRIBUTION = BI__MODAL.
.DISTRIBUTION BI__MODAL (– 1, 0) (– 0.5, 1)
(– 0.5, 0) (0.5, 0) (0.5, 1) (1, 0)

One problem with statistical analysis is that component manufacturers usually don't provide distributions. One lot of 20% capacitors may contain a substantial number of 10% capacitors because the manufacturer produced more than enough 10% capacitors to fill orders; the next lot may not have any 10% capacitors.

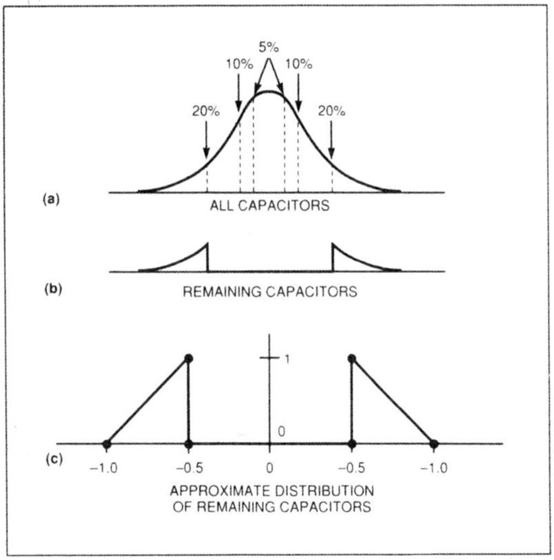

Figure 13.2 Capacitors with a specified tolerance may have a somewhat uniform distribution that doesn't include the Gaussian curve's 'tails' (a). After those capacitors have been removed from a manufacturing batch, the remaining capacitors have a distribution with a'notch' in the middle (b). To account for distributions such as this one in Spice, you have to specify the distributions yourself (c)

If you don't know the distribution, you can take a guess; uniform distribution will usually provide reasonable results. If you're especially concerned about a component, you can contact the manufacturer, but don't expect much help. The manufacturer may provide guidance but probably won't make any commitments.

Spice programs such as PSpice use Monte Carlo analysis to investigate component variations. Monte Carlo analysis, as its name implies, is based on chance. It involves using random numbers in a calculation that has the structure of a statistical, not deterministic, process[3].

Monte Carlo methods often fall into two categories: solution and simulation. Solutions directly solve a problem, such as determining the numerical value of π. Monte Carlo methods for circuit analysis are in the second category – simulations.

The effects of component tolerances on circuit performance are statistical, and those effects can be characterized by simulating a circuit multiple times. Each time, the simulation generates random numbers to create new values for components with tolerances. The combination of many simulations predicts the range of performance you can expect from the circuit.

To compare the many simulations mathematically, Monte Carlo methods use collating functions. Spice programs let you choose a collating function according to the measure of performance you consider most important. For example, if you're simulating a switching circuit, you might want to know when a rising edge occurs. In this case, you can specify that the time of the edge be recorded for each simulation. Different versions of Spice offer different collating functions; for example, PSpice has five to choose from. However, you can choose only one function per Monte Carlo analysis.

The following examples use an intuitive collating function, called YMAX, which compares simulations. The first simulation run uses nominal values for each component; subsequent runs use random numbers to generate values for all components and their tolerances. The value of YMAX results from comparing the output waveform of each run to that of the first, or nominal, run. The YMAX value is the point of greatest difference between the two waveforms.

Analyzing circuits

Two examples illustrate Monte Carlo analysis. The first considers the effects of cascaded gain stages; the second examines the effects of component variation in a high-pole-count analog low-pass filter.

A common question among engineers concerns a cascading op-amp circuit such as the one in Figure 13.3: If each of the six resistors has 1% tolerance, how much should you expect the overall gain to vary? The easy answer is to assume the worst case: ± 6%. This would occur if R1, R3, and R5 were 1% low, and R2, R4, and R6 were 1% high. It is unlikely that this combination would ever occur, but how much variation would be likely to occur? Monte Carlo analysis answers just such questions.

Listing 13.1 contains a model for the Monte Carlo analysis of the circuit in Figure 13.3. Note that high-gain voltage-controlled voltage sources replace the op-amps in order to simplify the PSpice model. Note also that only a few lines make up the Monte Carlo model: an .MC line

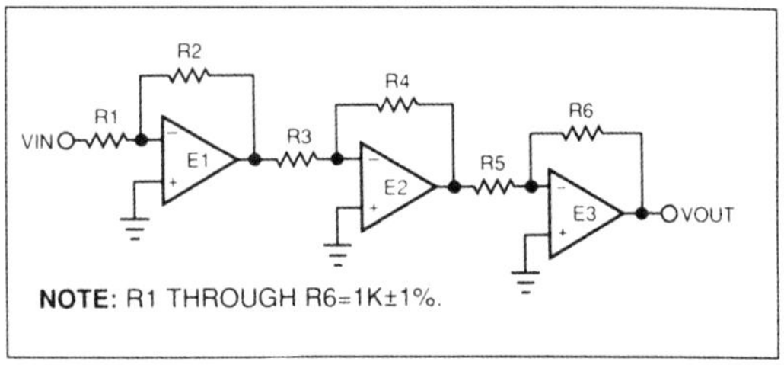

Figure 13.3 In a cascaded op-amp circuit, the tolerances of the resistors combine to affect the circuit's gain. In the worst case, the gain varies by ± 6%; however, it's likely that the actual variation will be less

Listing 13.1 Simulation model for cascaded op amps

```
V1 1 0 1.00                         ;Input signal is 1 volt D.C.

R1 1 2 ONE_PERCENT_R  1K            ;Gain stage #1
R2 2 3 ONE_PERCENT_R  1K
E1 3 0 (0, 2) 1000K

R3 3 4 ONE_PERCENT_R  1K            ;Gain stage #2
R4 4 5 ONE_PERCENT_R  1K
E2 5 0 (0, 4) 1000K

R5 5 6 ONE_PERCENT_R  1K            ;Gain stage #3
R6 6 7 ONE_PERCENT_R  1K
E3 7 0 (0, 6) 1000K

.MODEL    ONE_PERCENT_R  RES(R=1 DEV=1%)   ;1% uniform distribution

.DC V1  0.0 1.0 1.0                 ;D.C. analysis at 0 and 1 volt
.MC 250 DC V(7) YMAX LIST OUTPUT ALL ;250 runs.  Monitor V(7) using
                                    ;YMAX collating function

.END
```

specifying 250 simulations and the YMAX collating function, and a .MODEL line specifying uniformly distributed ± 1% resistors.

Simulations using this model show that circuit gain varies much more than ± 1%. The greatest variation over 250 simulations is + 3.9%, and the gain from seven simulations (of 250) varied more than ± 3%.

The results are very different, however, when the circuit's resistors have a normal, rather than uniform, distribution. To change the distribution, you need only modify the .MODEL line to.

.MODEL ONE__PERCENT__R RES(R = 1
DEV/GAUSS = 0.25%)

With this change, the greatest variation in circuit gain over 250 simulation runs is + 1.8%, and only 21 runs show gains that vary more than ± 1%.

Simulating the effects of component variation in a filter is another use for Monte Carlo analysis. Figure 13.4 shows an 8-pole, lowpass Butterworth filter constructed with four Sallen and Key[4,5] lowpass filters. A common perception is that multiple-pole filters produced in volume will not perform reliably because of component variation. The concern is that component variation will distort a filter's performance; poles will shift so much that the filter will no longer be a true Butterworth filter. Monte Carlo analysis can investigate this concern.

Listing 13.2 Simulation model for an 8-pole Butterworth filter

```
V1 1 0 AC 1.0V                      ;Input signal is 1 volt

R1 1 2 ONE_PERCENT_R  31.6K         ;Sallen and Key Filter #1
R2 2 3 ONE_PERCENT_R  14.3K
C1 2 4 TEN_PERCENT_C 0.082UF
C2 3 0 TEN_PERCENT_C 0.068UF
E1 4 0 (3, 4) 1000K

R3 4 5 ONE_PERCENT_R  22.1K         ;Sallen and Key Filter #2
R4 5 6 ONE_PERCENT_R  16.9K
C3 5 7 TEN_PERCENT_C 0.1UF
C4 6 0 TEN_PERCENT_C 0.068UF
E2 7 0 (6, 7) 1000K

R5 7 8 ONE_PERCENT_R  31.6K         ;Sallen and Key Filter #3
R6 8 9 ONE_PERCENT_R  13.7K
C5 8 10 TEN_PERCENT_C 0.15UF
C6 9 0 TEN_PERCENT_C 0.039UF
E3 10 0 (9, 10) 1000K

R7 10 11 ONE_PERCENT_R  37.4K       ;Sallen and Key Filter #4
R8 11 12 ONE_PERCENT_R  14.7K
C7 11 13 TEN_PERCENT_C 0.39UF
C8 12 0 TEN_PERCENT_C 0.012UF
E4 13 0 (12, 13) 1000K

.MODEL    ONE_PERCENT_R  RES(R=1 DEV=1%)    ;1%  uniform distribution
.MODEL    TEN_PERCENT_C CAP(C=1 DEV=10%)    ;10% uniform distribution

.AC DEC 50 30 300                           ;50 points/decade, 30 to 300 Hz
.MC 50 AC V(13) YMAX LIST OUTPUT ALL        ;50 runs. Monitor V(13) using
                                            ; YMAX collating function

.PROBE V(13)                                ;Specify PSpice "oscilloscope"

.END
```

Listing 13.2 models Figure 13.4's filter circuit using 10% capacitors and 1% resistors, both with uniform distributions. Again, high-gain voltage-controlled voltage sources replace op amps to simplify the PSpice model.

The .AC directive specifies the frequency domain which is well suited for analyzing filters. The .PRO directive specifies PSpice's graphical output.

The filter's simulated gain appears in Figure 13.5. The curves are from three of 25 simulation runs: the run with nominal component values, the run with the lowest gain, and the run with the highest gain. Although the cut-off frequency varies, the filter appears to behave like a Butterworth filter.

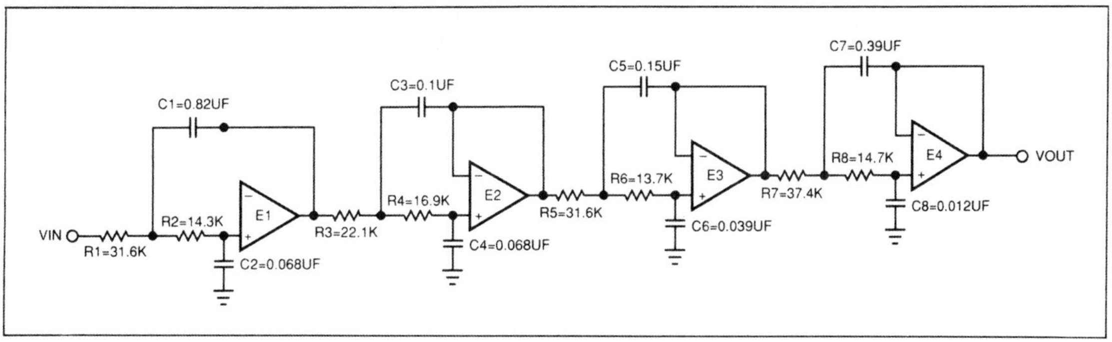

Figure 13.4 For this 8-pole, lowpass Butterworth filter, Monte Carlo analysis can help determine if the filter's poles will shift so much that the circuit will no longer be a true Butterworth filter

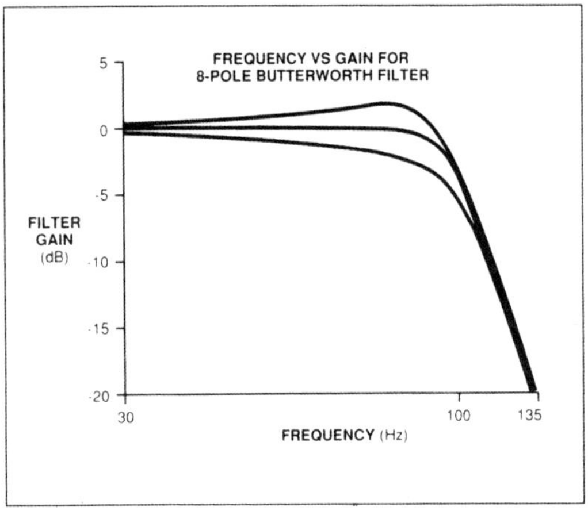

Figure 13.5 The simulated gain for Figure 13.4's circuit shows that the circuit does, as intended, behave like a Butterworth filter. The plot shows the results from three of the 25 simulation runs: the run with nominal component values, the run with the lowest gain, and the run with the highest gain

Monte Carlo analysis helps you understand the effects of component tolerances on circuits without having to rely on rules of thumb or guesses. Because you'll understand these effects before a circuit goes into production, you can correct many problems in the design phase. And depending on the version of Spice you use, your current circuit models should require only a few changes to let you analyze performance using Monte Carlo techniques.

Spice programs with Monte Carlo analysis

Spice programs that include Monte Carlo analysis are available from several vendors, and some programs are available in modestly priced evaluation versions. Evaluation versions let you experiment before buying a complete package. They usually do everything, or almost everything, that production versions do, but they limit the size and complexity of circuits you can simulate.

For information about Spice programs that feature Monte Carlo analysis, contact the following companies:

Intusoft
2525 S Western Ave, Suite 203
San Pedro, CA 90732
(301) 833–0710

Meta-Software Inc
1300 White Oaks Rd
Campbell, CA 95008
(800) 346–5953
Fax (408) 371–5100

Microsim Corp
20 Fairbanks
Irvine, CA 92718
(800) 245–3022
Fax (714) 770–3022

Spectrum Software
1021 S Wolfe Rd
Sunnyvale, CA 94086
(408) 738–4387

Tatum Labs Inc
3917 Research Park Dr, Suite B-1
Ann Arbor, MI 48108
(313) 663–8810

References

1. *The Design Center User's Guide and Reference Manual*, Version 5.1, Microsim Corp, Irvine, CA, 1992.
2. Tuinenga, P W, *Spice: A Guide to Circuit Simulation and Analysis Using PSpice*, Prentice-Hall, Englewood Cliffs, NJ, 1988.
3. Kalos, Malvin H and Paula A Whitlock, *Monte Carlo Methods, Volume 1: Basics*, John Wiley and Sons, New York, NY, 1986.
4. Daryanani, G, *Principles of Active Network Synthesis and Design*, John Wiley and Sons, New York, NY, 1976.
5. Sallen, R P, and E L Key, 'A Practical Method of Designing RC Active Filters', *IRE Transactions on Circuit Theory*, CT-2, May 1955, p. 74–85.
6. Rubinstein, Reuven Y, *Simulation and the Monte Carlo Method*, John Wiley and Sons, New York, NY, 1981.
7. Walpole, Ronald E and Raymond H Myers, *Probability and Statistics for Engineers and Scientists*, Macmillan Publishing Co Inc, New York, NY, 1978.

14 Differential amplifier reduces parts count

Unusual instrumentation amplifier

This circuit shows the properties of an instrumentation amplifier – high input impedance to both differential and common-mode signals at its balanced floating input and high common-mode rejection – but using fewer gain-defining resistors than the usual three-op-amp circuit.

Irwin Cohen
Hewlett-Packard, Waltham, MA

Whereas the most common differential amplifier has six resistors, the differential amplifier in Figure 14.1 has only four, suiting the circuit for designs that must have an absolutely minimal parts count.

In operation, IC_2 drives the noninverting input of IC_3 to a virtual ground. The closed-loop gain of the IC_2/IC_3 loop is less than –1, regardless of potentiometer R_2's setting (use R_2 to null out common-mode signals from IC_1's output). Ideally, R_2's wiper will be exactly in the center, feeding equal currents into the inverting inputs of IC_1 and IC_2 from the output of IC_3. The current flowing to IC_1's inverting input cancels the common-mode signal at its noninverting input. Note that the inverting input of IC_1 is *not* at a virtual ground.

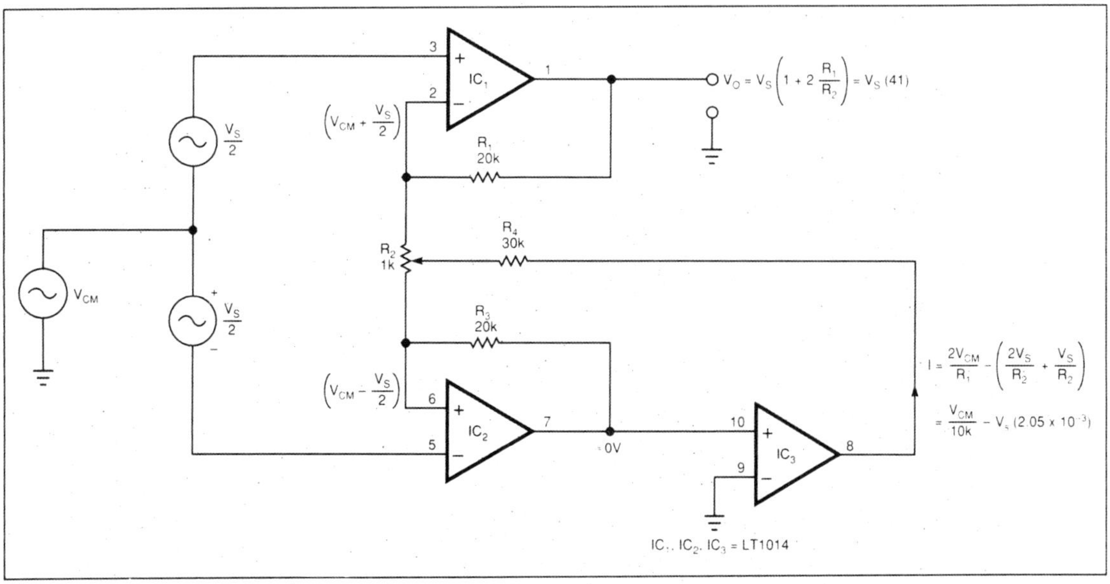

Figure 14.1 This differential amplifier uses two fewer resistors than a conventional design

15 Filters and oscillators

Unusual oscillators and filters

From the series called 'The Jim Williams Papers', this article introduces some unusual AF and RF filters and oscillators, in Jim's distinctive style.

Jim Williams
Linear Technology Corp

Filters and oscillators share a common point of view – they deal with signals in the frequency domain. You can define a filter's function as rejecting frequencies you don't want (the job of a band-reject filter, for example) or including only the frequencies you want (what a bandpass filter does). If you reorient your thinking slightly, though, you realize that all filters reject unwanted frequencies. (The bandpass filter rejects frequencies outside the band of interest.) When you view filters in this way, you see that any filter's function is the inverse of an oscillator's; oscillators synthesize individual frequencies or ranges of frequencies. Although there are more kinds of filters and oscillators than any magazine article of reasonable length can hope to touch on, herein are a few types of circuits that can meet a range of needs.

Figure 15.1a shows a highly selective bandpass filter using a resonant ceramic element and a single amplifier. Except at its resonant frequency, (in this case, 400 kHz) the ceramic element looks like a high impedance. For off-resonance inputs, IC_1 produces no output; it acts as a follower whose input is grounded. At resonance, the ceramic element has a low impedance, and IC_1 behaves as an inverter with gain. The 100 Ω resistor isolates IC_1's summing point from the ceramic element's capacitance. This capacitance is quite substantial and limits the circuit's out-of-band rejection. Figure 15.1b, curve A shows this effect. This plot shows very steep rejection, with IC_1's output down almost 20 dB at 300 kHz and 40 dB at 425 kHz. The device's stray parasitic capacitance causes the gentle rise in the output at higher frequencies and also sets the –20-dB floor at 300 kHz.

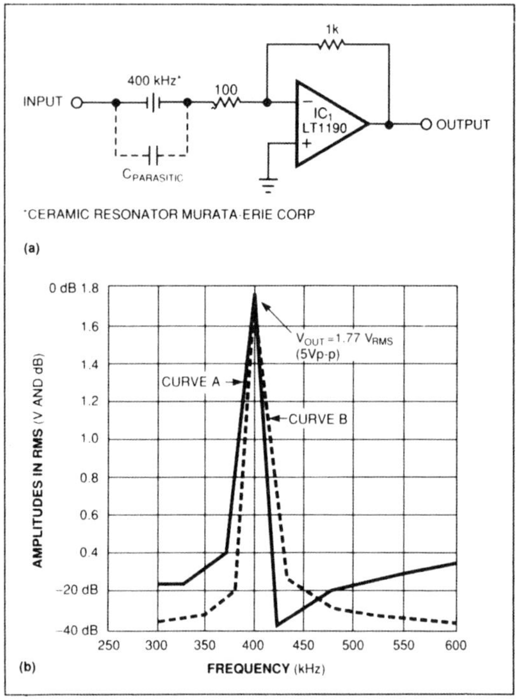

Figure 15.1 One amplifier and a ceramic resonator create a bandpass filter (a). The solid curve of b shows the filter's frequency response. Note the dip to – 40 dB on the high side of resonance. The dip is the result of the resonator's parasitic capacitance

Figure 15.2 shows how to use a nulling technique to partially correct problems caused by the ceramic element's parasitic capacitance. This circuit is similar to the previous one, except that a portion of the input goes to IC_1's positive input. The R-C network at that input has an impedance

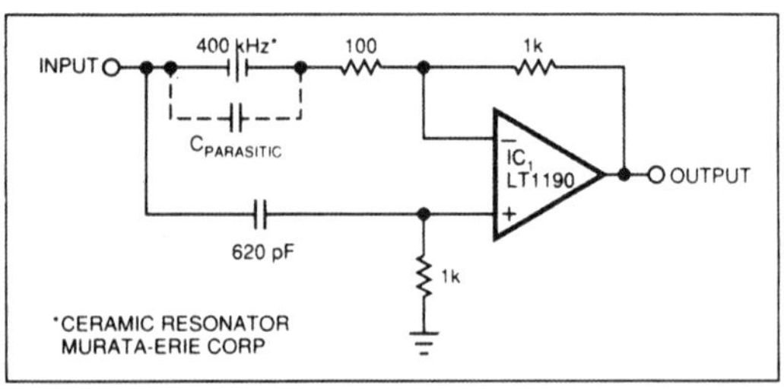

Figure 15.2 A slight modification of the circuit in Figure 15.1a allows you to cancel out the effects of the resonator's parasitic capacitance. The dashed curve of Figure 15.1b shows the effects on the filter response. Below resonance, the modified circuit attenuates by an extra 20 dB. Above approximately 525 kHz, the improvement is even more dramatic

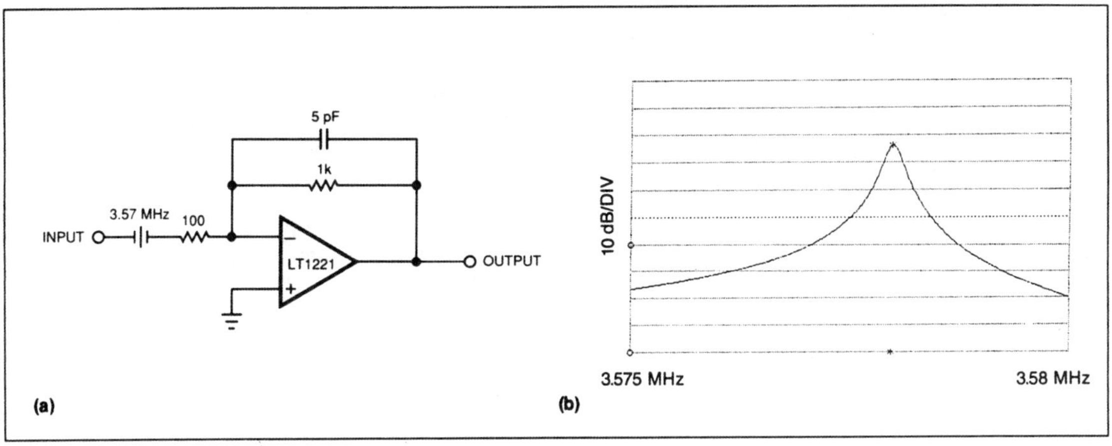

Figure 15.3 Replacing the ceramic resonator of Figure 15.1a with a 3.57-MHz crystal is the most significant change that leads to this crystal filter (a). You can see the crystal filter's response in b

close to the ceramic resonator's off-null impedance. Therefore, out-of-band components produce similar signals at IC_1's inputs, and, because of IC_1's common-mode rejection, produce little output. At resonance, the added R-C network appears as a much higher impedance than does the ceramic element, and the filter response is similar to that of the circuit in Figure 15.1a. Figure 15.1b, curve B shows that this circuit has much better out-of-band rejection than does the earlier circuit. The high-frequency rolloff is smooth, and, at 475 kHz, over 20 dB deeper than that of the circuit in Figure 15.1a. At 375 kHz and below, on the low-frequency side of resonance, the circuits behave similarly.

By using quartz crystals, you can make filters whose high-frequency selectivity is even higher than that of filters based on ceramic resonators. Figure 15.3a replaces Figure 15.1a's ceramic element with a 3.57-MHz quartz crystal. Figure 15.3b shows almost 30 dB of attenuation only a few kHz on either side of resonance! The differential nulling technique used with the ceramic elements is less effective with quartz crystals. Crystals have significantly lower parasitic capacitance, making the cancellation less effective.

Oscillators use crystals and resonators

The circuit in Figure 15.4 places a crystal within the amplifier's feedback path, creating an oscillator. With the crystal removed, the circuit is a familiar noninverting amplifier with a grounded input. The impedance ratio of the elements associated with IC_1's negative input sets the gain. Inserting the crystal closes a positive feedback path at the crystal's resonant frequency, and oscillations commence.

In any oscillator, you must control the gain as well as the phase shift at the frequency of interest. If the gain is too low, oscillation will not occur. Conversely, too much gain produces saturation limiting. In this circuit, gain control comes from the positive temperature coefficient of the lamp at IC_1's negative input. When you first apply power, the lamp's resistance is low, the gain is high, and the oscillation amplitude increases. As the amplitude builds, the lamp current increases and causes heating, which raises the lamp resistance. The increased resistance reduces the amplifier gain and the circuit finds a stable operating point. This circuit's sine-wave output has all of the stability advantages associated with quartz crystals. Although shown with a 10-MHz crystal, the circuit works well with a variety of crystal types from 100 kHz to 20 MHz. Using a lamp to control the amplifier gain is a classic technique, first described by Meacham in 1938. Electronic gain control, though more complex, offers more precise control of amplitude.

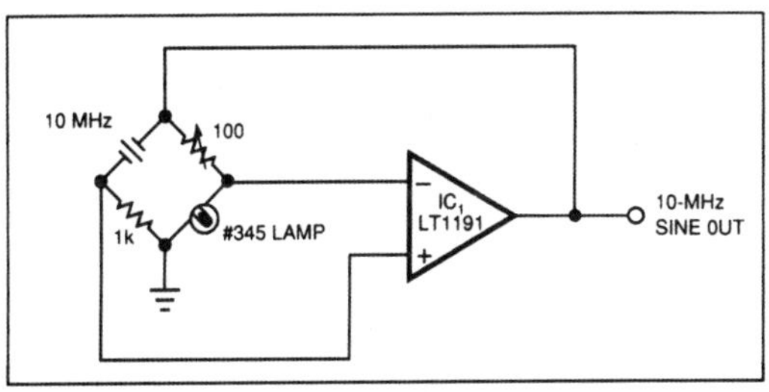

Figure 15.4 An incandescent lamp's current-dependent resistance stabilizes the oscillation amplitude of this 10-MHz crystal oscillator

Figure 15.5a's quartz stabilized oscillator replaces the lamp with an electronic amplitude-stabilization loop. IC_2 compares the IC_1 oscillator's positive output peaks with a dc reference. The diode in the dc-reference path compensates for the rectifier diode's temperature dependence. IC_2 biases Q_1, controlling the FET's channel resistance and influencing the loop gain. The amplitude of the oscillator's output is a reflection of the loop gain. Loop closure around IC_1 stabilizes the amplitude of the oscillator's output; the 1-μF capacitor compensates the gain-control loop.

The dc-reference network provides optimum temperature compensation for the rectifier diode, which sees IC_1's 2 V p-p, 20-MHz output waveform. IC_1's small output swing minimizes the distortion attributable to channel-resistance modulation in Q_1. To use this circuit, adjust the 50 Ω trimmer until 2 V p-p oscillations appear at IC_1's output.

Figure 15.5b is a spectrum analysis of the oscillator's output. The fundamental is at 20 MHz; the second harmonic, at 40 MHz, is 47 dB down. The third harmonic, 50 dB down, occurs at 60 MHz. Resolution bandwidth for the spectrum analysis is 1 kHz.

The circuit in Figure 15.6a replaces the quartz crystal with a Wien network at IC_2's positive input. IC_1 controls Q_1 to stabilize the amplitude of IC_2's oscillations. The operation is identical to that of the circuit in the previous figure. Although the Wien network is not nearly as stable as a quartz crystal, it has the advantage of a variable-frequency output. Normally, you vary the frequency by varying either R or C or both. The use of manually adjustable elements, such as dual potentiometers and 2-section variable capacitors is common. The circuit in Figure 15.6a uses fixed, 360 Ω Wien-network resistors and uses varactor diodes as capacitors. The varactor diodes' voltage-variable capacitance allows dc tuning of the oscillator. Applying 0 to 10 V dc to the varactors shifts the oscillation frequency from 1 to 10 MHz. The 0.1-μF capacitor blocks the dc bias from IC_2's positive input but lets the Wien network function normally. IC_2's 2 V p-p output minimizes the varactors' junction effects and thereby limits distortion.

This 5 V-powered circuit requires a voltage step-up to develop adequate varactor drive. IC_3 and the LT1172 switching regulator form a simple voltage step-up regulator. IC_3 controls the LT1172 to produce whatever output voltage is required to close a loop at IC_3's negative input. The 22-μF output capacitor stores L_1's high-voltage inductive-flyback pulses after they have been rectified by the diode-and-zener-connected Q_2. The 7.5-kΩ/2.5-kΩ divider closes the loop by providing a sample of the output value to IC_3's negative input. The 0.1-μF capacitor stabilizes this feedback action. Q_2's zener drop allows the circuit to produce controlled outputs at voltages as small as zero. This arrangement permits a 0 to 2.5 V input at IC_3 to produce a corresponding 0 to 10 V varactor bias. Figure 15.6b, a spectral plot of the circuit running at 7.6 MHz, shows the second harmonic down 35 dB and the third harmonic down almost 60 dB. The resolution bandwidth is 3 kHz.

Figure 15.7a shows the schematic of an AM

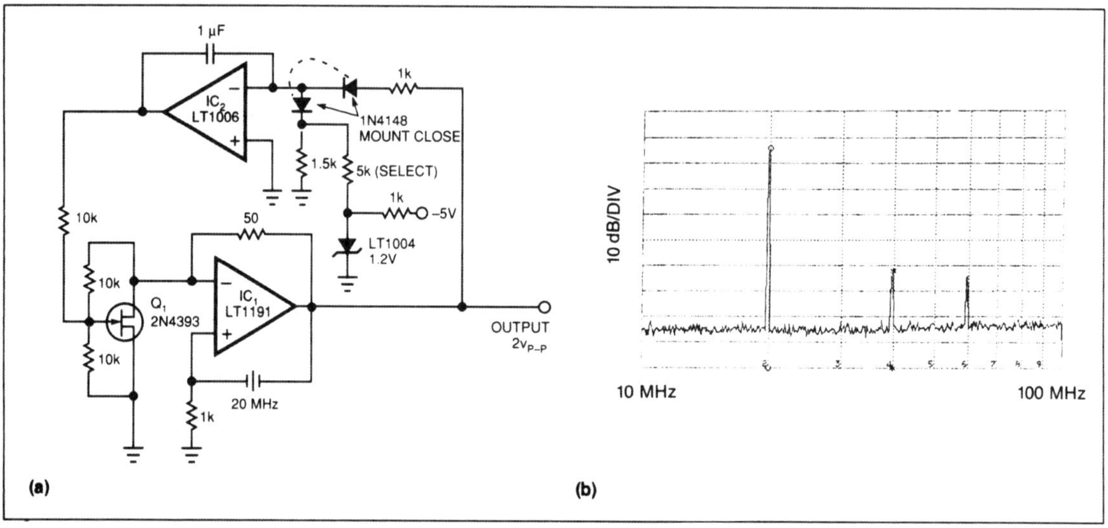

Figure 15.5 An electronic gain-control circuit that uses the voltage-controlled on-resistance of a FET stabilizes the output amplitude of this 20-MHz crystal oscillator (a). In b, you see that the output's harmonics are at least 47 dB below the fundamental

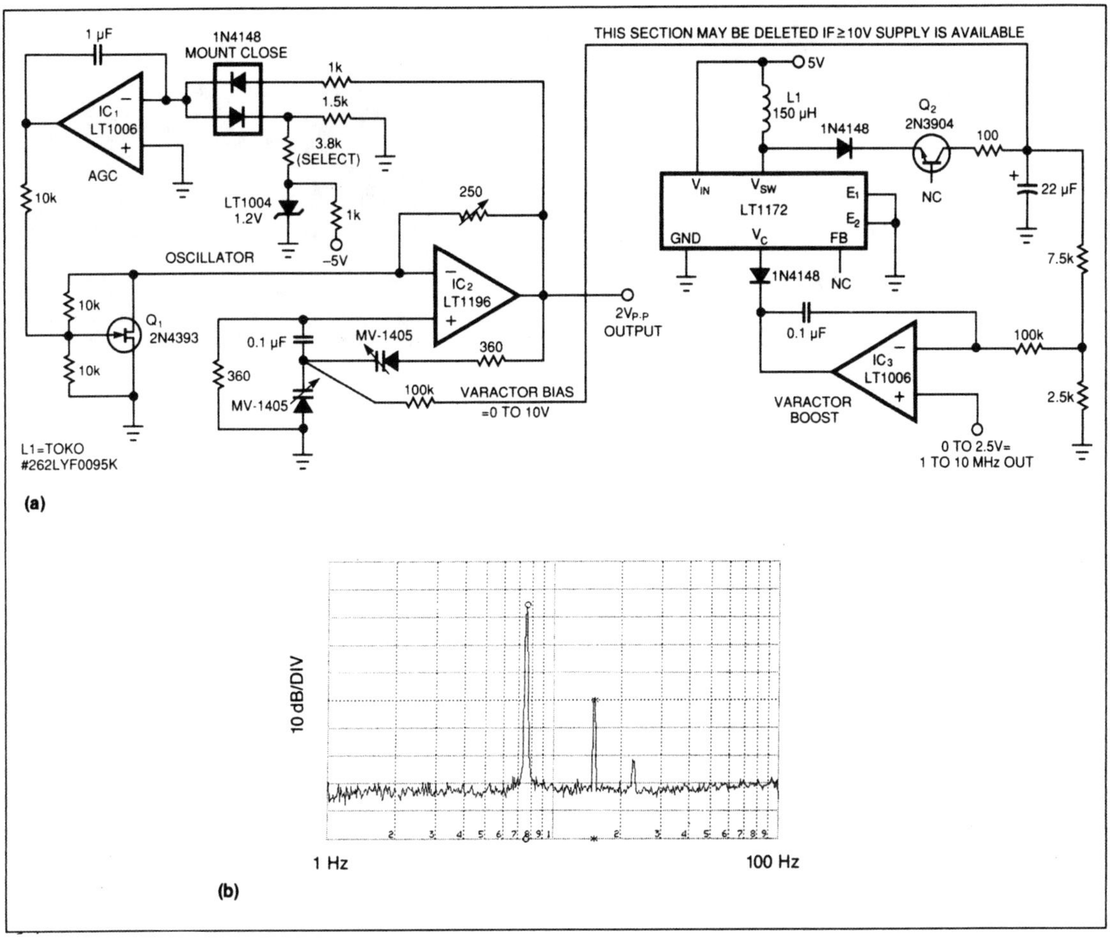

Figure 15.6a A pair of varactor diodes lets you tune this Wien-bridge oscillator (a) from 1 MHz to 10 MHz by applying a 0 to 10 V signal. Adding the components in the right half of the schematic lets you operate the circuit from a 5 V supply and permits controlling the frequency with a 0 to 2.5 V signal. The spectrum analysis in b shows that the sinusoidal output is quite clean

radio station – complete from microphone to antenna, but lacking a Federal Communications Commission license. IC_1, set up as a quartz-stabilized oscillator similar to the one in Figure 15.4, generates the carrier. IC_1's output feeds IC_2, which functions as a modulated RF power-output stage. The bias applied to offset pins 1 and 8 restricts IC_2's input-signal range. (See the LT1194 data sheet for details.) IC_3, a microphone amplifier, supplies bias to the offset pins, resulting in an amplitude-modulated RF carrier at IC_2's output. The dc voltage summed with the microphone output biases IC_3's output to the appropriate level for good quality modulation characteristics. Calibrating this circuit involves trimming the 100 Ω potentiometer in the oscillator for a stable 1 V p-p 1-MHz output from IC_1.

Figure 15.7a does not show on-air personalities – or, in keeping with current trends in AM radio – a means of providing any kind of program other than a talk show. There is no phonograph pickup or connection to the output of a compact-disc player. Nevertheless, you can connect such a music source to the microphone input. Figure 15.7b shows a typical AM carrier output at the antenna. In a throw-back to the days when top-40 formats reigned on the AM band, the modulating signal is Mr Chuck Berry singing the rock-'n'-roll classic 'Johnny B. Goode'.

Start with a triangle; end up with a sine

The oscillators presented to this point have limited tuning-frequency range. Although the circuit in Figure 15.8a is not a true oscillator, it produces a synthesized sine-wave output over a wide dynamic range. Many applications such as audio, shaker-table driving, and automatic test

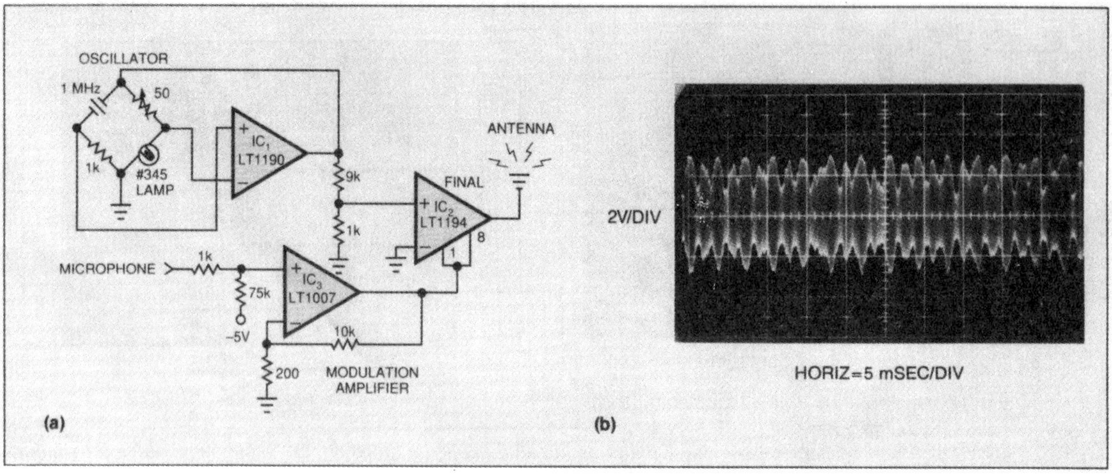

Figure 15.7 Though perhaps not worthy of Wolfman Jack or Dick Biondi, the circuit of a is still a complete AM radio station. When Chuck Berry picks his guitar and belts out 'Johnny B. Goode', the modulated output looks like what you see in b

equipment require voltage-controlled oscillators (VCOs) that have sine-wave outputs. This circuit meets this need, spanning a range of 1 Hz to 1 MHz (equal to 6 decades or 120 dB) for a 0 to 10 V input. The circuit maintains 0.25% frequency linearity and 0.40% distortion.

To understand the circuit, assume Q_5 is on and its collector (Figure 15.8b, trace A,) is at –15 V, cutting off Q_1. IC_3, which inverts the positive input voltage and biases the summing node of integrator IC_1 through the 3.6-kΩ resistor and the self-biased FET's, pulls a current, + I, from the summing point. IC_2, a precision op-amp, provides dc stabilization of IC_1. IC_1's output (trace B) ramps positive until IC_5's input (trace C) crosses zero and causes IC_5's inverting output to go negative. The Q_4/Q_5 level shifter then turns off, and Q_5's collector goes to + 15 V, allowing Q_1 to come on. The values of the resistors in Q_1's path result in a current, + 2I, exactly twice the absolute magnitude of the current,+ I, that flows out of the summing node. As a result, the net current into the junction becomes + I, and IC_1 integrates negatively at the same rate it did during its positive-going excursion.

When IC_1 integrates far enough in the negative direction, IC_5's + input crosses zero and the circuit's two outputs change state. The state change switches the Q_4/Q_5 level shifter's state, causing Q_1 to go off and the entire cycle to repeat. The result is a triangular waveform at IC_1's output. The frequency of this triangle depends on the circuit's input voltage and varies from 1 Hz to 1 MHz with a 0 to 10 V input. The LT1009 diode bridge and the series-parallel diodes provide a stable bipolar reference that always opposes the sign of IC_1's output ramp. The Schottky diodes bound IC_5's + input, ensuring its clean recovery from overdrive.

Sine of the times

The AD639 trigonometric function generator, biased via IC_4, converts IC_1's triangular output into a sine wave, (trace D). To avoid output distortion, you must supply the AD639 with a triangular wave that does not vary in amplitude. At higher frequencies, delays in the IC_1-integrator switching loop result in late turn-on and turn-off of Q_1. Unless you minimize these delays, the triangle amplitude will increase with frequency and cause the distortion level to increase. IC_5, the Q_4/Q_5 level shifter, and Q_1 generate a total delay of 14 ns. This small delay, combined with the 22-pF feed-forward network at IC_5's input, keeps distortion to just 0.40% over the entire 1-MHz range. At 100 kHz, the distortion is typically less than 0.2%. The 8-pF capacitor in Q_1's source line minimizes the effects of gate-source charge transfer, which occurs whenever Q_1 switches. Without this capacitor, a sharp spike would occur at the triangle peaks, increasing distortion. FETs Q_2 and Q_3 compensate for the temperature-dependent on-resistance of Q_1 and keep the + 2I/ –I relationship constant with temperature.

This circuit responds very rapidly to input changes – something most sine-wave generators cannot do. Figure 15.8c shows what happens when the input switches between two levels, (trace A). IC_1's triangle output (trace B), shifts frequency immediately, with no glitches or poor dynamics. The sine output (trace C) reflecting this action, is similarly clean. To adjust this circuit,

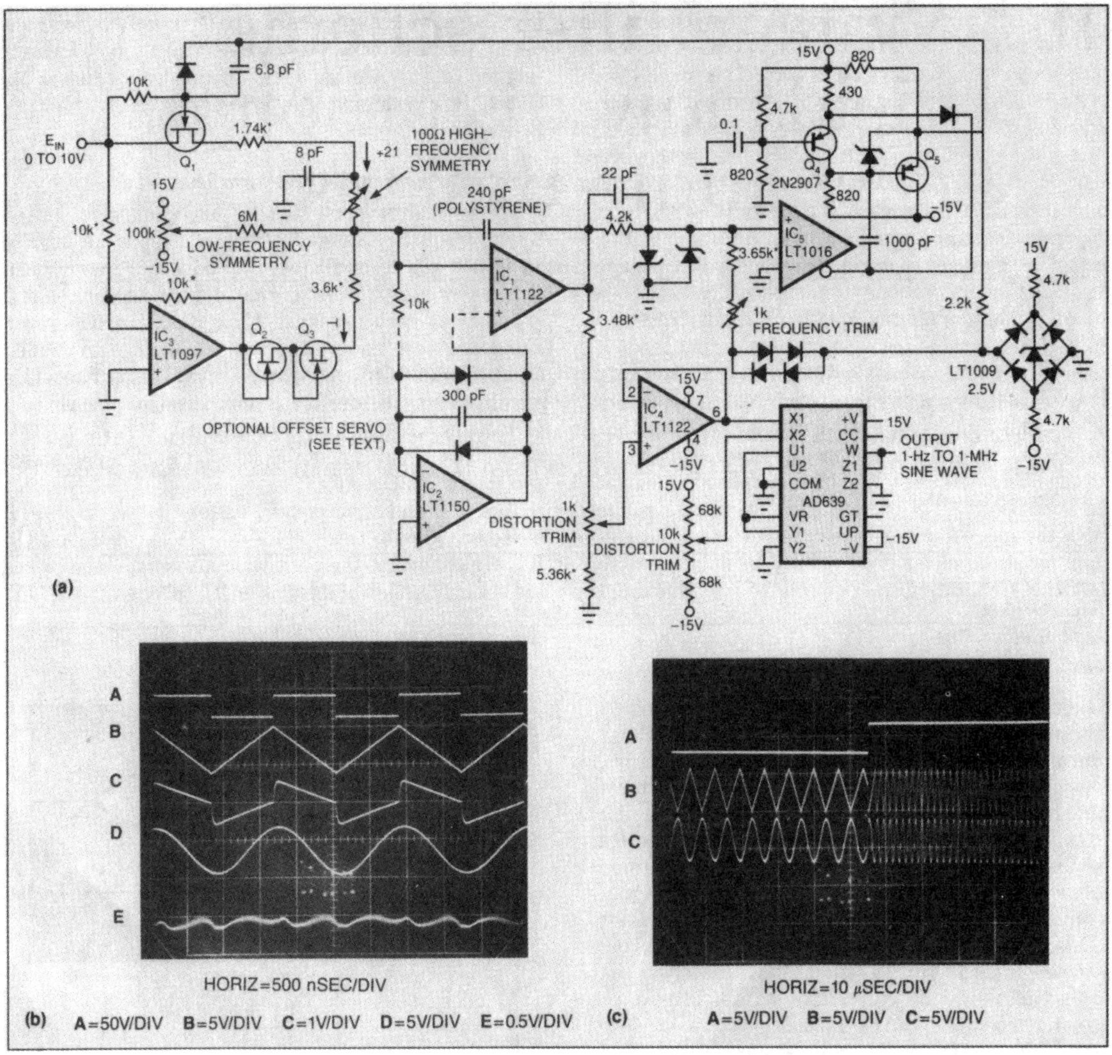

Figure 15.8 A classic function generator, a, creates square and triangular waves whose frequency you can control with a dc voltage. A trigonometric-function generator IC converts the triangle to a sine. The traces in b show waveforms within the circuit. The lowest trace shows the residual distortion after you remove the output's fundamental-frequency component. In c, you see the circuit's quick and clean response to a command to change frequency

apply 10.00 V and trim the 100 Ω potentiometer for a symmetrical triangle output at IC_1. Next, apply 100 μV and trim the 100-kΩ potentiometer for triangle symmetry. Then, apply 10.00 V again and trim the 1-kΩ frequency-trim adjustment for a 1-MHz output frequency. Finally, adjust the distortion-trim potentiometers for minimum distortion as measured on a distortion analyzer (Figure 15.8b, trace E). You may have to readjust the other potentiometers slightly to achieve the lowest possible distortion. If you won't operate the circuit below 100 Hz, you can delete the IC_2-based dc-stabilization stage. If you make this change, you should ground IC_1's positive input.

Many of the filter and oscillator circuits presented here are simple as well as useful. Their simplicity shows that clever circuit designers often take a minimalist approach. When you speak or write, you are more likely to get your point across if you use short words that are familiar to your audience. So it is with circuits. The simplest design that does the job usually costs the least and operates more reliably than complex alternatives.

16 Vintage filter scheme yields low distortion in new audio designs

FDNR filters avoid inductors

This article shows how you can design antialiasing and anti-imaging filters with low sensitivity to component values, using frequency-dependent negative resistors, also called D elements or 'supercapacitors'. These (implemented with generalized immittance converters) take the place of capacitors in an LC filter design, while resistors take the place of the inductors.

Rick Downs
Burr-Brown Corp

Digital-audio designers are constantly concerned about noise, total harmonic distortion, and phase linearity. To prevent aliased noise and distortion, designers, using an antialiasing filter, must limit the bandwidth of the audio signal before the signal reaches the analog-to-digital converter (ADC). Using an anti-imaging filter, designers must also limit the bandwidth of the audio signal coming from the digital-to-analog converter (DAC). However, maintaining an acceptable phase linearity for a filter that matches the dynamic range of converters having 16-bit or greater resolution can be a challenging task. More and more digital audio designers are utilizing an early filter architecture, the Generalized Immittance Converter (GIC), to meet this challenge.

The GIC (Figure 16.1) is a 2-port network whose input impedance is:

$$Z_{in}(s) = \frac{Z_1(s)\ Z_3(s)\ Z_5(s)}{Z_2(s)\ Z_4(s)}$$

where s is the standard Laplacian operator. When impedance Z_{s2} is capacitive and the other impedances are resistive, the GIC's input impedance simulates an inductor.

$$Z_{in}(s) = \frac{s\ R_1\ C_2\ R_3\ R_5}{R_4}$$

You can also simulate an inductor by making Z_4 capacitive and the remaining impedances resistive. When you make any two of the numerator impedances capacitive, such as Z_1 and Z_5, and the other impedances resistive, the GIC simulates a frequency-dependent negative resistor (FDNR).

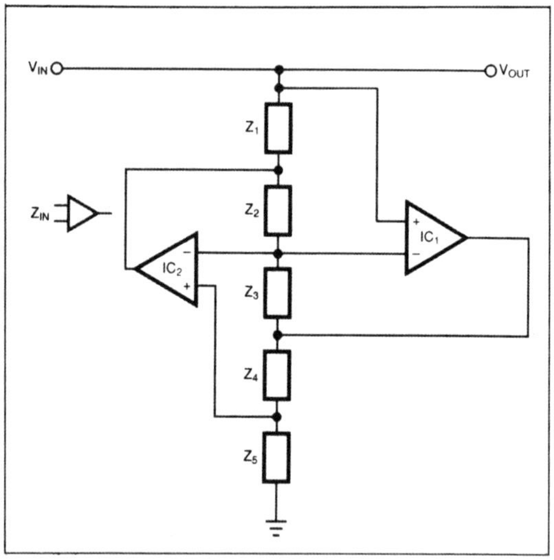

Figure 16.1 The topology of a Generalized Immittance Converter (GIC) can simulate the impedance of any passive component, including a frequency-dependent negative resistor (FDNR)

$$Z_{in}(s) = \frac{R_3}{s^2\ C_1\ R_2\ R_4\ C_5}$$

The impedances you make capacitive have different circuit implications. If you make Z_1 a capacitor, and the filter is ac-coupled, the operational amplifier, IC_1 will not have a bias-current return

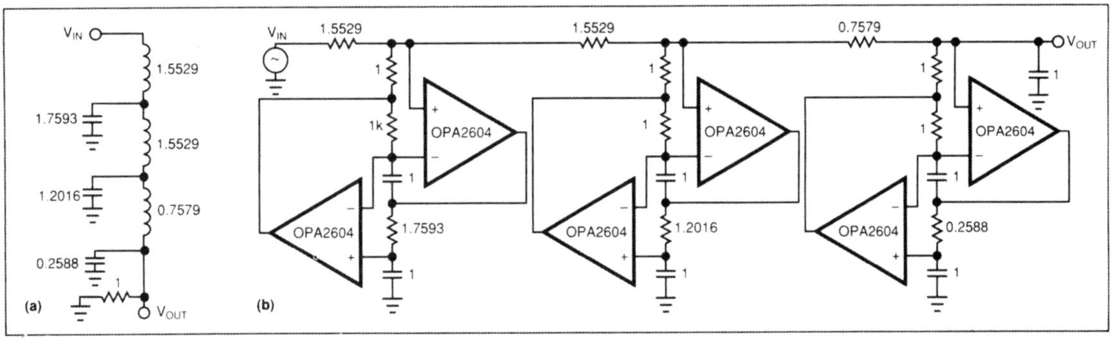

Figure 16.2 An LC ladder network generates any lowpass-filter polynomial (a). You can simulate the filter's transfer function by multiplying each element value by 1/s and realize the filter using GICs (b)

path, which could affect the filter's operation. When you make Z_3 and Z_5 capacitive, the op-amps in the GIC all have a bias-current return path.

Audio designers generally make Z_1 and Z_5 capacitive, letting you set $R_2 = R_3$ ($Z_2 = Z_3$) to minimize the effect of op-amp gain-bandwidth mismatch. If you configure the GIC with capacitors for Z_3 and Z_5, you should employ op-amps with well-matched gain bandwidths.

An active GIC filter can simulate the transfer characteristic of a passive LC ladder network (Figure 16.2a). You transform an LC ladder network into an active GIC filter by multiplying each ladder network element by 1/s. The transformation changes all the inductors to resistors, the capacitors to FDNRs, and the resistors to capacitors (Figure 16.2b). You can set $R_1 = R_2$ and $C_3 = C_5$, allowing you to trim the filter using only one component, R_4.

Because the GIC filter simulates an LC filter, it has a lower sensitivity to component value variations than other RC active filters, such as the familiar Sallen and Key topology[1]. In addition, the GIC filter topology lets you design high-order filters having unity gain, whereas the Sallen and Key filter topology (Figure 16.3) often requires gain greater than unity to derive real resistor values.

To illustrate an active GIC filter design, consider the filter requirements in a practical digital-audio record and playback channel. Modern digital-audio channels oversample the recorded bandwidth using a sampling rate that is greater than 4 × the Nyquist rate. Multiplying the recommended standard 48-kHz audio sampling rate by a factor of 4 produces the standard 192-kHz 4 ×-oversampling rate.

Such a high oversampling rate eliminates the need for 'brick-wall' attenuation slopes and eases limitations on allowable phase distortion, hence allowing you to use lower-order – and lower-cost – antialiasing and anti-imaging filters. In addition, it avoids the manufacturing difficulties and nonlinear group delay associated with brick-wall filters.

Playback devices assist anti-imaging

The anti-imaging filter's attenuation requirements benefit from the attenuation provided by three factors in the playback channel. The first factor is the attenuation characteristics of the digital interpolation filter that precedes the DAC. The filter interpolates the data between samples and removes most of the signal energy above 20 kHz to prevent aliasing by the DAC. The second factor is the frequency response and linearity of the output power amplifier. The power amplifier bandwidth is generally restricted to 20 kHz, which attenuates higher-frequency energy. A caution however: noise-shaping DACs have significant out-of-band energy, which, if the amplifier becomes nonlinear, can create intermodulation products.

The third factor is the frequency response of the DAC. Because a DAC maintains the analog value of each digital sample, the DAC exhibits the

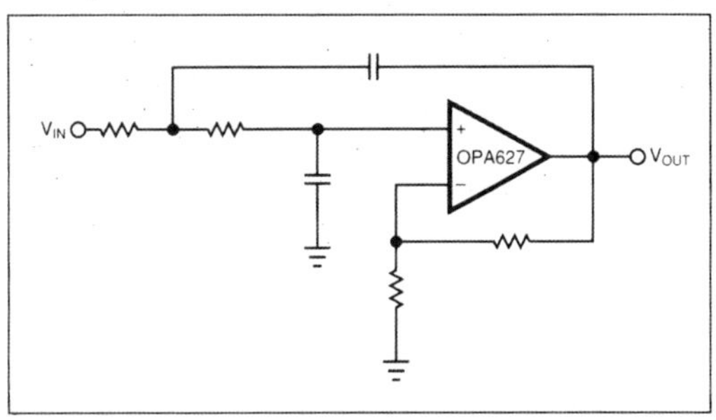

Figure 16.3 Active-filter handbooks contain many 'cookbook' designs for the familiar Sallen and Key lowpass architecture. The topology often requires greater-than-unity gain to realize the filter

frequency response of a zero-order-hold filter given by

$$H(f) = \frac{\sin\left(\pi \frac{f}{f_s}\right)}{\left(\pi \frac{f}{f_s}\right)},$$

where f is the signal frequency and f_s is the sampling frequency. The DAC's sin(x)/x frequency response can provide 20 dB or more of attenuation in the imaging frequency range, which is 20-kHz removed from the sampling rate.

The antialiasing filter's attenuation requirements are more stringent. Because there isn't a digital filter or a power amplifier preceding the ADC, the filter can't benefit from the additional attenuation provided by these devices. In addition, because you mathematically represent the digital samples from the ADC as impulses in discrete-time, an ADC doesn't exhibit the zero-order-hold response of a DAC.

To match the dynamic range of a 16-bit ADC, the antialiasing filter should theoretically provide 96 dB of attenuation to keep alias responses below the quantization noise level. In practice, however, the amplitude of audio signals in the 10- to 20-kHz range is significantly less than the ADC's dynamic range. Therefore, designers often use 65 dB as an adequate rule of thumb for the antialiasing filter's attenuation. Because of the help it gets in the playback channel, the anti-imaging filter can be of a lower order than the antialiasing filter.

Although you have a wide choice of lowpass filter polynomials, many of these polynomials are not suited for audio applications. Chebyshev and elliptic filters have steep cutoff characteristics, but their large pass-band ripple can be troublesome in some audio applications. The Butterworth filter has a maximally flat pass-band response, but its cutoff characteristics are less steep. Because these three types of filters don't exhibit constant group delay, pass-band frequencies experience unequal time delays that can cause excessive overshoot and ringing in the transient response.

The Thompson filter, also known as the Bessel filter, has constant group delay, which provides excellent transient response in audio and DSP applications. However, the Thompson filter's cutoff characteristics are even less steep than the Butterworth filter's. Therefore, you would need a high-order Thompson filter to achieve the same stopband attenuation as a lower-order Butterworth filter. And because oversampling relaxes the attenuation requirements, you can often employ a 40-kHz Butterworth filter, having tolerable group delay from 20 Hz to 20 kHz, in most 4 × oversampled digital audio systems.

Using a 192-kHz 4 ×-oversampling rate, a 40-kHz Butterworth antialiasing filter must attenuate the aliasing components in the 160- to 170-kHz frequency range by the rule-of-thumb 65 dB. You determine the order of a Butterworth filter using the following equation:

$$10^{\frac{K_s}{20}} = \sqrt{1 + \left(\frac{\omega_s}{\omega_c}\right)^{2n}},$$

where K_s, is the stopband attenuation in dB, ω_s is the minimum stopband frequency, ω_c is the 3-dB cutoff frequency, and n is the filter order. Solving for n yields

$$n = \frac{\mathrm{LOG}\left[\sqrt{10^{\frac{K_s}{20}} - 1}\right]}{\mathrm{LOG}\left[\frac{\omega_s}{\omega_c}\right]}.$$

Substituting $K_s = 65$, $\omega_s = 2\pi \times 160$ kHz, and $\omega_c = 2\pi \times 40$ kHz yields n = 5.4. Therefore, you would need a sixth-order antialiasing Butterworth filter. Because the anti-imaging filter benefits from attenuation due to the digital interpolation filter, the DAC's sin(x)/x response, and the restricted power amplifier bandwidth, a third-order Butterworth filter, producing 36 dB of attenuation at 160 kHz, should suffice.

Figure 16.4a shows the component values for a third-order Butterworth LC filter having a normalized cutoff frequency of 1 rad/s. You can extract the component values from standard filter tables such as those found in Reference 1. To realize the filter using a GIC, you first transform each component by multiplying each value by 1/s. Therefore, L_1 becomes R_1, C_2 becomes a

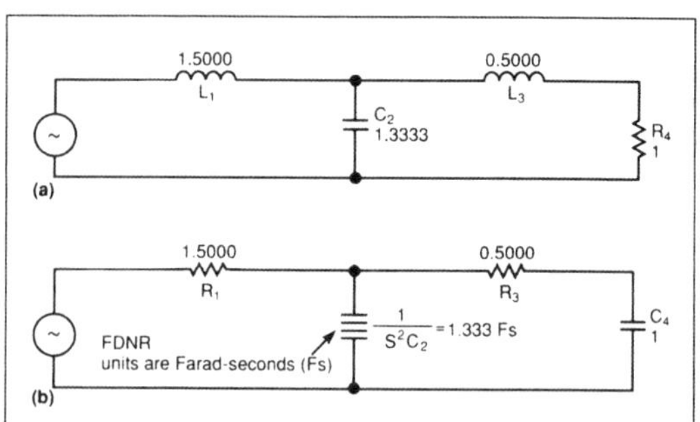

Figure 16.4 A normalized third-order Butterworth filter serves as a building block for the anti-imaging filter (a). You obtain the normalized GIC filter by multiplying each normalized element value by 1/s (b)

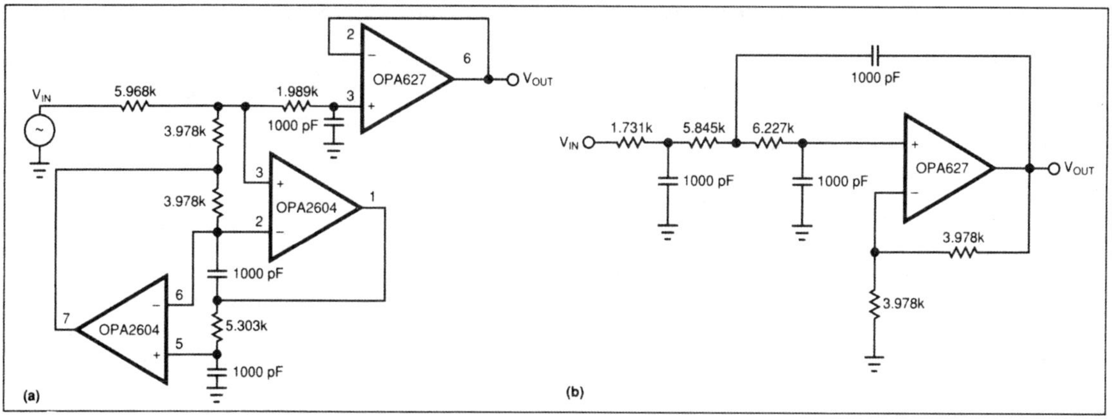

Figure 16.5 This unity-gain third-order Butterworth GIC filter has a 40-kHz cutoff frequency (a). The equivalent Sallen and Key realization requires twice the gain (b)

FDNR having the value $1/(s^2C_2)$, L_3 becomes R_3, and the terminating resistor, R_4, becomes C_4 (Figure 16.4b). By setting the FDNR values, referring to Figure 16.1: $Z_1 = R_1$, $Z_2 = R_2$, $R_1 = R_2 = 1$; $Z_3 = C_3$, $Z_5 = C_5$, $C_3 = C_5 = 1/s$; and $Z_4 = R_4 = 1.333\,\Omega$. Similar to the circuit in Figure 16.2b, a single resistor (R_4) determines the value of the FDNR.

Next you must scale the normalized cutoff frequency to the desired cutoff frequency of 40 kHz. To accomplish frequency scaling, you simply divide all capacitor values by $\Omega_n = 2\pi \times 40$ kHz. Finally, you must scale the large capacitor values and small resistor values by an impedance scale factor to realize practical circuit elements. The impedance scale factor is

$$Z_n = \frac{\text{NORMALIZED C VALUE}}{\text{DESIRED C VALUE}}.$$

Choosing the desired C value to be 1000 pF yields an impedance scale factor of

$$Z_n = 7.23 \times 10^3.$$

Multiplying all resistor values and dividing all frequency-scaled capacitor values by 7.23×10^3

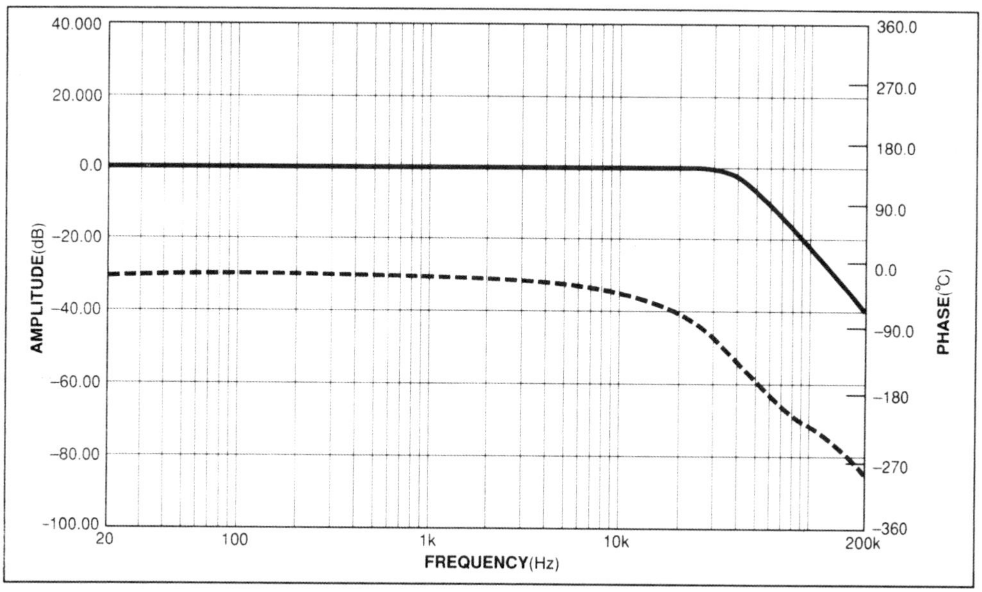

Figure 16.6 The amplitude (solid line) and phase response (dotted line) of the third-order Butterworth GIC filter are nearly constant throughout the audio-frequency band

produces the final filter shown in Figure 16.5a. Because the output impedance of the filter is high, you should buffer the output using an op-amp voltage follower. Figure 16.6 shows the amplitude and phase response of the final filter.

The measured noise-and-distortion and noise levels show a contrast between a third-order Butterworth filter design based on the familiar Sallen and Key architecture (Figure 16.5b) and this GIC realization. Noise-and-distortion measurements were performed using a 1-kHz test signal and a 22-Hz to 80-kHz measurement bandwidth. The noise-only levels were measured in the same bandwidth when the circuit inputs were grounded. The GIC realization achieves a – 96-dB noise-and-distortion level that matches the dynamic range of a 16-bit digital audio system.

The Sallen and Key realization achieves a – 93-dB noise-and-distortion level that's adequate for most consumer audio systems, but is 3 dB higher than the GIC realization. Simple output-noise level was measured in dBu, referenced to 0 dBu, which equals 0.775 V (RMS). The GIC filter, at – 104 dBu, has 7 dB less noise than the Sallen and Key filter. Because the Sallen and Key filter must have greater-than-unity gain to realize the component values, its noise gain is higher than the unity-gain GIC filter.

You can design a sixth-order GIC filter in one of two ways to achieve the 65-dB antialiasing filter requirements. You can extend the previous design procedure using standard element values for a normalized sixth-order Butterworth filter, or you can simply cascade two of the previously designed third-order GIC filters to achieve the sixth-order polynomial.

A design based on a normalized sixth-order Butterworth filter is sensitive to gain-bandwidth mismatches between all the op-amps in the circuit, however. If you use this approach you should employ op-amps with high gain-bandwidth products. Cascading two third-order GIC filters achieves acceptable results and has fewer op-amp matching difficulties.

References

1. Huelsman, L P, and P E Allen, *Introduction to the theory and design of active filters*, McGraw Hill, New York, 1979.

2. R Downs, 'DSP Oversampling to Quiet Noise', *EE Times*, August 8, 1988, p. 68.

3. R Downs, 'High-Speed A/D Converter Lets Users Reap Benefits of Oversampling', *Burr-Brown Update*, Vol XIV, No. 2, May 1988, p. 3.

4. R Downs, 'A Low-Noise, Low-Distortion Design for Anti-Aliasing and Anti-Imaging Filters', Burr-Brown Application Bulletin AB-026, February 1991.

17 Creating phantom circuits simplifies remote monitoring

Extra circuits for free

As a youngster I always found the technique of something for nothing, in the form of extra circuits carried over existing lines as phantoms, a positively fascinating idea. Using super phantoms and super – super phantoms, N pairs can carry 2N-1 signals without crosstalk between them, or even 2N signals if you don't mind using the (usually rather noisy) earth phantom as well. (Note that in Figure 3a, the current in Rld should read i_s, not i_3.)

Jerald Graeme
Burr-Brown Corp

Normally, two wire pairs can transmit two differential signals. A phantom remote-monitoring circuit transmits a third differential signal as the voltage difference between a pair of common-mode signals on the two wire pairs. In effect, the circuit behaves as though a third, or phantom, wire pair were present. In the classic phantom-circuit implementation, transformers let you introduce and then separate common-mode signals. An equivalent circuit that replaces the transformers with instrumentation amplifiers removes the transformers' low-frequency limitation.

Remote monitoring is vulnerable to transmission errors from several sources. These sources are noise pickup, potential differences between grounds, and ground-line voltage drops. Figure 17.1 illustrates the first two of these error sources and how to remove them. Figure 17.1a shows the most basic signal-monitoring configuration in which an output signal, e_O, represents an input signal, e_S. Because of the great distance between the signal source and the monitored output, two other signals, e_N and e_G, interfere with the measurement. The distance itself causes a difference in ground potentials and produces an error signal, e_G. In addition, the long wire that spans the separating distance picks up a noise signal, e_N, from the environment. In Figure 17.1a, both e_G and e_N combine with e_S to degrade e_O.

Switching to differential measurement removes both error sources (Figure 17.1b). Referencing e_S

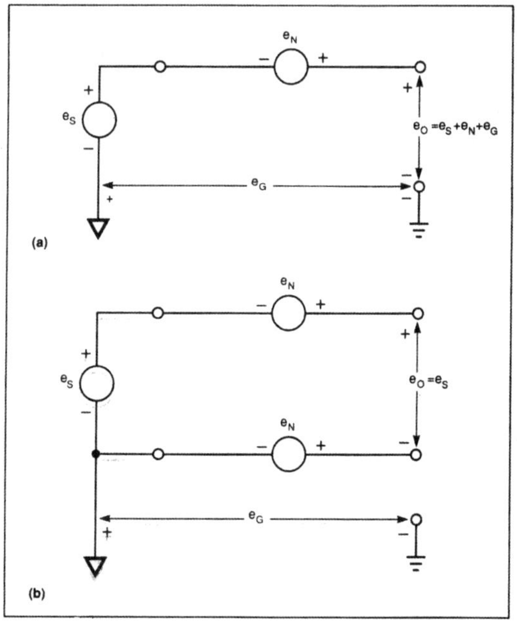

Figure 17.1 Noise and ground-potential differences introduce errors in basic signal monitoring (a). Differential monitoring removes these errors (b)

and e_O to the same common reference potential directly removes the ground-potential difference. The differential measurement's common-mode rejection indirectly removes the noise pickup. Because the same noise environment affects them both, the two wires in Figure 17.1b pick up the same noise signal, e_N. By detecting only the difference in voltage between the wires, the differential measurement of e_O frees the output signal of noise effects.

Even with differential sensing, remote monitoring remains sensitive to line-voltage drops that introduce crosstalk. Crosstalk occurs when multiple differential-signal monitors share a common return line, as in Figure 17.2a. Differentially sensing a pair of signals removes the effects of ground-potential differences and noise pickup, as explained for Figure 17.1b. (For simplicity, Figure 17.2 does not show these canceled ground and noise errors.) When you add output loading as in Figure 17.2, however, the load currents cause line drops. Input signals e_{S1} and e_{S2} drive output load resistors R_{L1} and R_{L2}. The resulting load currents, i_{L1} and i_{L2}, travel through the line resistances and develop line drops e_{L1} and e_{L2}. For remote monitoring, the high line resistances make these line drops significant. The resistance of 22-gauge copper wire, for example, is $16\,\Omega/1000\,\text{ft}$; long runs can cause resistances of hundreds of ohms.

As Figure 17.2 shows, the resulting line drops couple signals between the two monitor circuits. Both e_{S1} and e_{S2} develop line drops in the ground-sense line. Thus, e_{O1} and e_{O2} include signal components that the opposite signal source develops. The resulting signal at the e_{O1} output, for example, is

$$e_{O1} = e_{S1} - 2e_{L1} - e_{L2}.$$

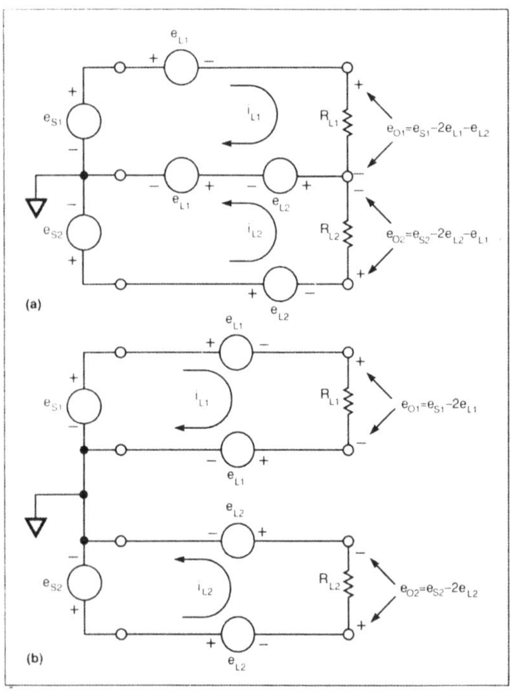

Figure 17.2 A return line common to two differential signals introduces crosstalk (a). Separate return lines remove it (b)

The e_{L2} term is the line drop developed by e_{S2} and coupled into the e_{S1} monitor loop. An analogous crosstalk term occurs in the e_{O2} output as a result of the e_{L1} drop that e_{S1} introduces into the common line.

These output signals also contain error terms that the intended signal sources develop. The second error term of the e_{O1} output equation is $-2e_{L1}$; a $-2e_{L2}$ term reduces the e_{O2} output. These error terms represent only gain errors and are far less serious than the crosstalk components. Subsequent gain adjustments remove the gain errors, but removing the crosstalk signals is difficult.

To avoid crosstalk, each signal source should have a separate sense connection to common, as in Figure 17.2. Each monitor loop contains only the line drops produced by the loop's signal. These drops continue to produce gain errors, but cross coupling does not occur. Thus, accurate remote monitoring requires differential measurement and the use of a separate pair of wires for each signal. The resulting installation and maintenance costs make mundane wire an important consideration in remote monitoring.

Transformers transmit extra signals

This seemingly inevitable need for separate return wires receives dramatic relief from a configuration known as a phantom circuit, which extends the benefits of differential measurement and reduces wire requirements by a third[1]. Conceptually, the phantom approach transmits a third differential signal as a pair of common-mode signals on the two original wire pairs. Each wire pair transmits its own differential signal that is immune to any common-mode signal introduced to the pair. By introducing different common-mode signals on the two pairs, you can transmit a third differential signal. This third signal is simply the difference between the two common-mode signals. In effect, the circuit behaves as though a third, or phantom, wire pair were present.

Historically, phantom transmission used transformers to couple floating signals. In Figure 17.3a, the differential action of the transformer removes any input common-mode signal present in e_1 and e_2. The transformer differential output responds only to the difference between these signals, or e_S. With the 1:1 transformer ratio, the differential signal e_S appears directly at the transformer differential output. This output has no inherent common-mode signal and is free for common-mode control.

The circuit of Figure 17.3a drives the transformer's output center tap to control the common-mode voltage; applying e_3 to the center tap shifts both transformer outputs by e_3. Adding e_3 does not alter the differential output, and the transformer continues to transmit e_S as a differential signal. The reference for e_S becomes e_3, and the two output terminals swing about this level. The resulting single-ended signals at the two output terminals are $e_3 + (e_S/2)$ and $e_3 - (e_S/2)$. This arrangement preserves e_S as a differential output signal and adds e_3 as a common-mode signal.

In practice, load currents flow in impedances that can mix the differential and common-mode signals. To prevent the crosstalk that would result, the two output lines present a pair of balanced impedances to the common-mode signal. Starting at the transformer secondary, the center-tap connection of the common-mode signal ensures equal winding impedances between the tap and the two output lines. Matching the lengths of the output lines results in equal line resistances. Choosing equal common-mode loads, R_{LCM}, balances the output loading. Any impedance beyond R_{LCM} is common to the two lines and introduces no imbalance.

Under these balanced conditions, the common-mode load current, i_3, faces equal impedances along both output lines and divides equally between the two lines. The common-mode signal causes equal line drops. Measuring the output voltage differentially rejects these drops. This operation is analogous to the way the Figure 17.1b circuit rejects the e_N line drops. The common-mode currents do not alter the differential signal voltage that reaches the differential load R_{LD}. As a result, the load current from differential signal e_S is the only current that flows through R_{LD}. The common-mode current drawn by the balanced common-mode load bypasses the differential load.

Center tap also separates signals

The common-mode drive of Figure 17.3a introduces a signal to a wire pair at the input side of the monitor system. At the output side, you must separate the common-mode and differential signals. Once again, a transformer center tap does the job. You can understand the circuit's action by considering what would happen if you reversed the operation of Figure 17.3a's transformer. Interchanging the transformer's input and output windings, as in Figure 17.3b, provides the required load-impedance balance. The signals at the output of Figure 17.3a's transformer – $e_3 + (e_S/2)$ and $e_3 - (e_S/2)$ – become Figure 17.3b's applied input signals. The resulting differential input signal, e_S, appears at Figure 17.3b's differential output. In practice, line drops reduce this signal. These drops do not introduce crosstalk, though, so for simplicity, this discussion omits them.

Imbalances in the common-mode load cause crosstalk between the common-mode signal e_3 and the differential signal e_S. As described for Figure 17.3a, you avoid this crosstalk error by balancing the load impedance to the common-mode signal. In Figure 17.3b, the transformer's center-tapped output provides this balance. The common-mode current, $i_3/2$, faces equal winding impedances on the two halves of the center-tapped output. The resulting voltages on the two impedances are equal and opposite, so they cancel in the differential output. Any impedance following the Figure 17.3b center tap is common mode and produces no impedance imbalance that could introduce crosstalk.

In addition, the center tap of Figure 17.3b supplies the common-mode signal without introducing crosstalk from the differential signal. This center tap follows the midpoint, or average, of the voltages on the two transmission lines. For line voltages of $e_3 + (e_S/2)$ and $e_3 - (e_S/2)$, this midpoint is the common-mode signal e_3. Line drops reduce this signal in addition to the differential signal. Yet equal line drops do not convert the common-mode signal to a differential signal. Line drops from the common-mode $i_3/2$ current components attenuate e_3. Line drops from

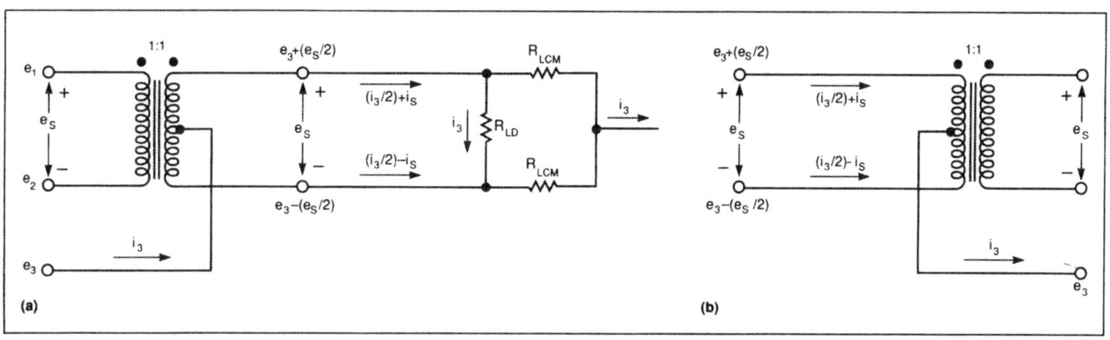

Figure 17.3 Center-tapped transformer windings introduce common-mode signal e_3 to a wire pair (a) and then separate it from the differential signal, e_s (b)

the differential current, i_S, reduce the voltage at each input of the transformer but do not shift the midpoint. For simplicity, Figure 17.3 neglects these line drops.

Phantom circuits cut wire count

Figure 17.4 shows the generalized phantom circuit that combines the roles of the two transformers in Figure 17.3. This discussion of the basic phantom circuit neglects all of the common and ground connections. The upcoming section on converting the basic phantom circuit for use in generalized instrumentation discusses these connections. In Figure 17.4, e_{S1} and e_{S3} drive two pairs of wires differentially. At the circuit's input side, transformers T_1 and T_3 couple these two signals to control the differential signals on the wire pairs. At the circuit's output, T_2 and T_4 separate the differential signals from any common-mode signals on the lines. Thus, dedicated wire pairs transmit signals e_{S1} and e_{S3} from input to output just as though the transformers were not present.

Introducing the transformers provides a free ride for the third signal, e_{S2}. The wire pairs dedicated to e_{S1} and e_{S3} transmit this signal as the difference between the two common-mode signals. The wire pair carrying differential signal e_{S1} couples e_3 from input to output as a common-mode signal. Similarly, the e_{S3} wire pair transmits e_4 as a common-mode signal. Measuring the difference between common-mode signals e_3 and e_4 at the output separates the third differential signal, e_{S2}. The common-mode signals reside at the center taps of output transformers T_2 and T_4. T_5 measures differentially between the center taps to yield e_{S2} at the output.

Thus, Figure 17.4 shows how two pairs of wires transmit three differential signals without errors from ground potential differences, noise pickup, or crosstalk. All three signal transmissions are differential and remove ground potential differences and noise pickup, as described for Figure 17.1. The balanced impedances provided by the transformer center-tap connections prevent crosstalk, as explained for Figure 17.3. For simplicity, Figure 17.4 excludes these rejected signals. In practice, some crosstalk remains as a result of the mismatch between the two halves of each center-tapped winding. However, this crosstalk is normally 60 dB below the level of the original signal[1]. Also, line drops reduce the transmission gains for all three differential signals, as described for Figure 17.2. For simplicity, Figure 17.4 also omits those drops.

Figure 17.4 includes common and ground connections for generalized instrumentation. Historically, engineers have used the phantom circuit with floating, unpowered sources and loads that

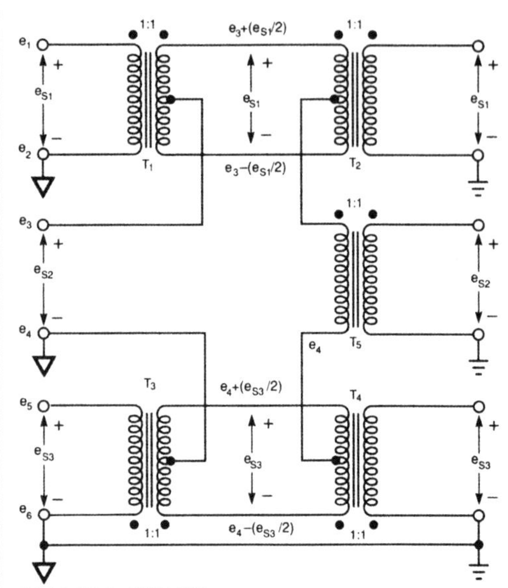

Figure 17.4 Combined transmission of differential and common-mode signals adds a third, or phantom, signal to the differential transmissions of two wire pairs

require no ground reference. As originally used in telephone transmission, the phantom circuit most often had a microphone as its signal source and a speaker as its load. In more general applications – for example, transducer bridges – the source circuit must conduct current from a power supply, and the conducted current must have a ground-return path. Also, a monitoring system that has several sources normally has a common reference. Similar ground-return and common-reference requirements also hold for the output sides of monitoring systems.

To avoid reintroducing errors from ground-potential differences and crosstalk, take care in adding common and ground connections to the phantom circuit. In Figure 17.4, all the differential inputs, e_S, have a common reference connection. You can connect either terminal of any of the differential inputs to this common. Similarly, you can ground either terminal of any differential output. Finally, a separate return line ties the input common to the output ground. By carrying the power-supply currents that bias the source circuits, this line keeps the supply current out of the differential-signal paths. As Figure 17.3 illustrates, the signal currents the wire pairs conduct do not introduce crosstalk.

The grounding scheme of Figure 17.4 also avoids errors from ground-potential differences. Normally in remote instrumentation, the output side of the system supplies power for the input circuit. In Figure 17.4, the most reliable system ground exists on the output side. The common reference on the input side is not connected to earth ground at the input. Instead, a separate return line connects the input common to the output ground. Because the differential inputs of the circuit are insensitive to the ground-line drop, this connection reduces the effect of the ground-potential difference caused by the voltage drop across the return line.

Instrumentation amps replace transformers

Transformers offer a simple implementation of the phantom circuit for most signal frequencies. However, for low-frequency signals, transformers that retain adequate signal coupling become unacceptably large. In remote instrumentation that monitors physical parameters, very-low-frequency signals are common, and transformer coupling is impractical.

For such cases, instrumentation amplifiers provide transformer-like action extending to dc. These amplifiers also remove the gain error caused by line drops and further reduce crosstalk. Instrumentation amplifiers can replace transformers by transmitting a differential input signal to a differential output while permitting independent control of the output common-mode signal. This amplifier action parallels the transformer operation of Figure 17.3. Figure 17.5 demonstrates this similarity using a difference amplifier (Figure 17.5a), a generalized instrumentation amplifier (Figure 17.5b), and the transformer equivalent (Figure 17.5c). Note that with a given input, all three circuits produce the same output signal. In Figure 17.5, a differential output reproduces differential signal e_S. This output rides on common-mode signal e_3.

Most people recognize that an instrumentation amplifier has a differential input, but they usually think of the amplifier's output as being single ended. As Figure 17.5b shows, however, most instrumentation amplifiers have a reference pin (REF) that is the output signal's reference return. Normally, you ground this pin to obtain a single-ended ground-referenced output. However, you can shift the output reference by returning the REF pin to another voltage. You can also modulate the reference. Even when you drive an instrumentation amplifier's REF pin, the amp supplies a differential output voltage referred to the voltage at the REF pin. Thus, like a transformer, an instrumentation amplifier provides a differential output with a terminal free for common-mode control.

The transformer equivalent shown in Figure 17.5c differs in two respects from the transformers of Figure 17.4. First, Figure 17.5c's transformer inverts the differential signal's polarity. This change is merely for drawing convenience to more closely parallel the operation of instrumentation amplifiers. The second difference in the transformer configuration is in the e_3 drive at the secondary. Figure 17.5c shows e_3 connected to one side of the secondary rather than to a center tap.

With a real transformer, this last difference is significant because the transformer inserts a winding impedance between e_3 and only one of the output leads in Figure 17.5c. In Figure 17.4, driving the center tap equalizes the associated voltage drops because equal winding impedances exist between e_3 and both transformer outputs. For the instrumentation-amplifier case, however, the equivalent winding impedance is zero. The instrumentation amplifier accurately transmits e_3 to the second line of the transmission pair without a loss equivalent to that in a winding impedance.

Instrumentation amps extend phantom to dc

If you substitute instrumentation amplifiers for transformers, you extend the low-frequency response of phantom circuits to dc. The type of

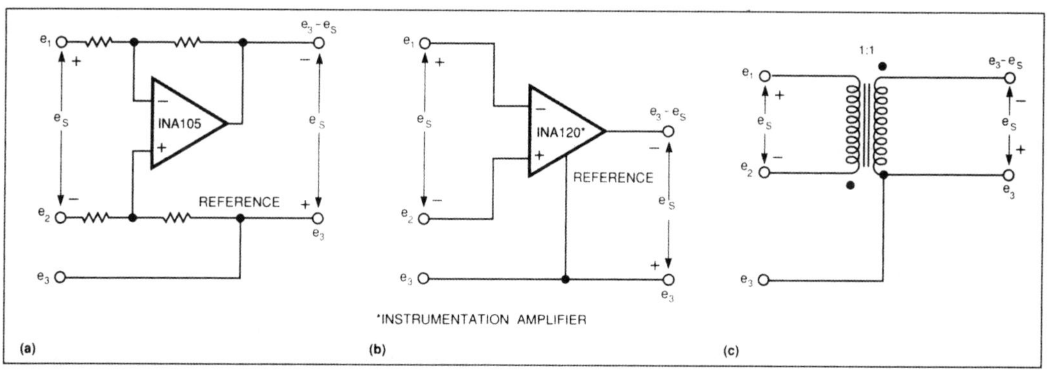

Figure 17.5 Difference- (a) and instrumentation-amplifier (b) circuits parallel the behavior of a transformer circuit (c) but without the need for the impedance balancing of a center tap

instrumentation amplifier you use for each transformer function depends on input-impedance characteristics. The simplest instrumentation amplifier is the difference amplifier, such as the INA105 of Figure 17.5a. However, difference amplifiers have a low input resistance. This resistance is usually acceptable on the input side of the phantom circuit because the resulting input currents do not flow through the resistances of long lines.

On the output side, however, a difference amplifier's input resistance would draw significant input currents through the line resistances. This case is similar to that of the transformer because transformers also have a low input impedance. Instrumentation amplifiers offer the option of virtually removing this line current and avoiding the gain error of Figure 17.2. Modern instrumentation amplifiers have a very high input resistance. The INA120 of Figure 17.5b has a differential input resistance of $5 \times 10^{12}\Omega$. This resistance draws essentially zero current from a transmission-line pair. In addition to reducing line-drop errors, the tiny line currents permit smaller-diameter wires for even greater economy.

To optimize low-frequency performance and gain accuracy, the phantom circuit of Figure 17.6 includes instrumentation amplifiers in place of the transformers of Figure 17.4. In Figure 17.6, instrumentation amplifiers IC_1 through IC_5 replace the similarly numbered transformers of Figure 17.4. For simplicity in comparing Figure 17.6 with Figure 17.4, Figure 17.6 shows the instrumentation amplifiers with unity voltage gain. For greatest economy, INA105 difference amplifiers replace input transformers T_1 and T_3. At the output, high-input-impedance INA120s replace transformers T_2, T_4, and T_5. Figure 17.6 also includes four input-return resistors that provide ground return paths for the amplifier-input currents. The four 100-kΩ resistors provide short return paths for the input bias currents of IC_2, IC_4, and IC_5. Because these resistors shunt the input impedances of the INA120s, the circuit uses matched high resistances.

The circuit operation of Figure 17.6 parallels that of Figure 17.4. Amplifiers IC_1 and IC_3 transfer differential input signals e_{S1} and e_{S3} to their outputs and remove the associated input common-mode signals. At the amplifier outputs, driving the REF pins with e_3 and e_4 adds different common-mode signals to the two differential signals. The combined common-mode and differential signals then pass down the wire pairs to the output amplifiers. Amplifiers IC_2 and IC_4 remove the common-mode components from e_{S1} and e_{S3} and present these signals as ground-referenced outputs. The third, or phantom, signal, e_{S2}, is the difference between common-mode signals e_3 and e_4. Amplifier IC_5 detects this difference and supplies e_{S2} as a third ground-referenced output signal.

Amplifiers decrease phantom's errors

The common-mode characteristics of Figure 17.6's amplifiers determine the circuit's rejection of crosstalk, ground-potential differences, and noise pickup. Each of these error sources reacts with the amplifiers' common-mode rejection ratios to develop differential error signals. Because the circuit greatly reduces line drops, they no longer cause crosstalk. However, the common-mode signals that e_3 and e_4 introduce develop differential errors from the two wire pairs. Signal e_3 first exercises the common-mode swing of IC_1 then adds an error of $e_3/CMRR_1$ to IC_1's differential output. Next, e_3 reacts with the common-mode rejection of IC_2 to develop an additional differential error of $e_3/CMRR_2$. Together, these common-mode error signals produce an output crosstalk error of $e_3(1/CMRR_1 + 1/CMRR_2)$ at the IC_2 output. An analogous crosstalk signal of $e_4(1/CMRR_3 + 1/CMRR_4)$ develops at the output of IC_4.

Still, the instrumentation-amplifier version of the phantom circuit reduces crosstalk. For the amplifiers shown, $CMRR_1 = CMRR_3 = 100{,}000$, and $CMRR_2 = CMRR_4 = 30{,}000$; the net crosstalk attenuation is 23,000:1, or 87 dB. This attenuation is far greater than the 60 dB that is common to

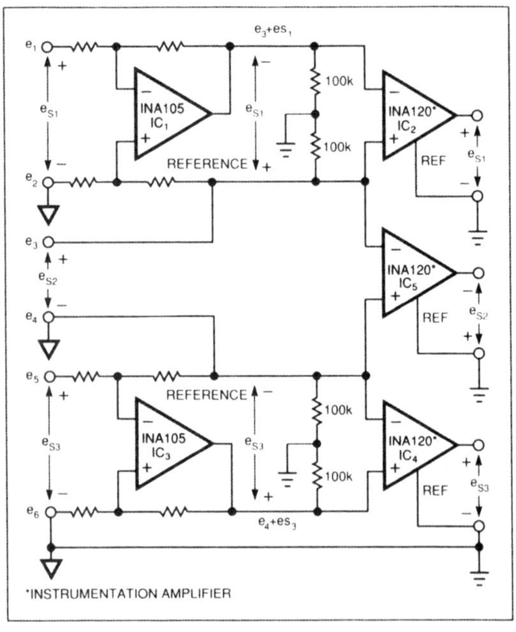

Figure 17.6 Substituting instrumentation amplifiers for transformers converts Fig 4 to a phantom circuit with a frequency response that includes dc

transformer-coupled phantom circuits. In addition, the crosstalk of the instrumentation-amplifier circuit is far less sensitive to line-impedance imbalances. The low currents drawn through these impedances maintain 60-dB attenuation for line-impedance imbalances as large as 100 Ω.

The Figure 17.6 circuit rejects other common-mode signals caused by ground-potential differences and noise pickup. However, these error signals impose a new range limit. The common-mode signal range of instrumentation amplifiers is more restricted than that of transformers. Transformers readily accept common-mode signals of hundreds of volts, whereas the INA120 instrumentation amplifier accepts only ± 12.5 V.

Although the reduced common-mode range seldom restricts noise rejection, it is not sufficient for all ground-potential differences. To limit such differences to values within the amplifier's common-mode range, you should add a direct connection between the input and output commons. For this reason, you should leave the input circuit floating with respect to 'earth' potential. Establish a common return point for the input signals and return this point to the output-circuit ground. This connection returns the current that the input circuit conducts from the power supply. As described for Figure 17.4, the power supply usually resides on the output side of the monitor system.

Another option available in the Figure 17.6 circuit is greater voltage gain. The INA120 output amplifiers can easily supply voltage gains of several thousand at low frequencies. Gains for the INA105 input amplifiers are fixed at unity. However, you can replace these amplifiers with INA106s to achieve voltage gains of 10 at two of the circuit inputs.

Reference

1. Tucker, D, 'A technical history of phantom circuits', *Proceedings of the IEEE*, Vol 126, No. 9, September 1979.

18 Phase compensation optimizes photodiode bandwidth

Speeding up photodetectors

With so much data being transferred over fiber-optic links, the subject of high-data-rate photo detector circuits has assumed a high profile.

Jerald Graeme
Burr-Brown Corp

Photodiodes' large capacitance severely restricts the bandwidth of basic photodiode circuits. An op-amp connected as a current-to-voltage converter greatly improves the bandwidth by isolating the capacitance from the signal voltage. Removing the signal voltage from the photodiode prevents the diode's capacitance from shunting the signal away from the load. However, the current-to-voltage converter's interaction with the photodiode capacitance complicates calculating the stability conditions, the phase compensation, and the resulting bandwidth. Even so, by examining the circuit behavior, you can develop a simple, intuitive approach to each of these calculations.

When operated with a direct resistor load, as in Figure 18.1a, a photodiode exhibits a bandwidth limited mainly by its internal capacitance. In Figure 18.1b, which models the bandwidth limit, the photodiode acts primarily as a current source. A large resistance, R_D, and the capacitance of the diode junction, C_D, shunt this source. The capacitance ranges from 2 to 20,000 pF depending for the most part on the diode area. In parallel with the shunt is the monitor amplifier's input capacitance, C_{IA}. With the monitor amplifier shown, $C_{IA} = C_{ICM}$, the common-mode input capacitance of the op-amp.

In practice, load resistances are small compared with R_D, so you can usually ignore the diode resistance. Similarly, the input resistance of the op-amp is so high that the amplifier exhibits little shunting effect on R_L. The net input-circuit capacitance and R_L then determine the input circuit's response rolloff. The resulting input circuit response has a break frequency, f_I. For Figure 18.1 the response is

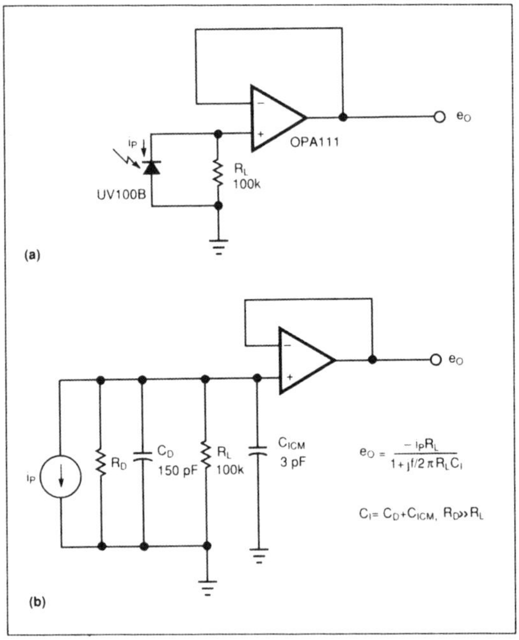

Figure 18.1 Load-voltage swing across the diode capacitance limits the basic photodiode bandwidth

$$\frac{e_O}{i_P} = \frac{-R_L}{(1 + jf/f_I)},$$

where, $f_I = 1/2\pi R_L C_I$,
and $C_I = C_D + C_{ICM}$.

For this single-pole response, the circuit's – 3-dB bandwidth equals the pole frequency and the typical components of the Figure 18.1 circuit set BW = f_I = 10 kHz.

The above expression reflects a typical gain-vs-bandwidth compromise. Increasing R_L gives greater gain but reduces f_I. From a circuit perspective, this compromise results from impressing the signal voltage on the circuit capacitances. The signal voltage in Figure 18.1b appears across C_D and C_{ICM}. The resulting capacitive currents shunt a portion of i_P, the signal current, away from the load resistor. Increasing R_L to raise the gain also increases the signal voltage on the capacitances and increases the portion of the signal current that the capacitances shunt away from the load. Such changes move the – 3-dB response point of the circuit to a lower frequency.

To avoid the gain-bandwidth compromise, you would like to develop the signal voltage across the load resistor but not across the capacitances. The current-to-voltage converter approximates this ideal, providing a dramatic improvement in bandwidth.

I to V isolates signal voltage from C_D

The op-amp current-to-voltage converter of Figure 18.2a removes the signal voltage from the photodiode capacitance. The op-amp and its feedback resistor translate the diode current to a buffered output voltage with excellent linearity. Added to the figure is a feedback capacitance, C_L, that provides phase compensation as described later. An ideal amplifier holds its two inputs at the same voltage. In Figure 18.2, such an amplifier would hold the signal voltage across the photodiode (and across the diode capacitance) to zero. The op-amp transfers the signal voltage to its output and isolates the signal voltage from the diode. Zero signal across the photodiode also improves the response linearity because it keeps the diode's voltage-dependent sensitivity from varying.

In practice, the amplifier's high, but finite, open-loop gain limits the isolation of Figure 18.2a's circuit. Part of the circuit's output voltage remains on the photodiode and produces a new bandwidth limit. Determining this new bandwidth limit is more difficult than determining the bandwidth of Figure 18.1's circuit. Despite Figure 18.2's simplicity, the current-to-voltage converter exhibits complex ac performance as analyzed below. As a result of an input circuit that appears as an inductance and capacitance in parallel, this circuit has a two-pole – rather than a single-pole – response. Feedback resistances above some maximum cause the circuit to resonate and oscillate. A direct mathematical analysis of this ac behavior is complex, but a more intuitive analysis results in simple design equations.

To ensure that a current-to-voltage converter is stable, you must usually supply phase compensation. Because phase compensation and bandwidth are related, you must consider them

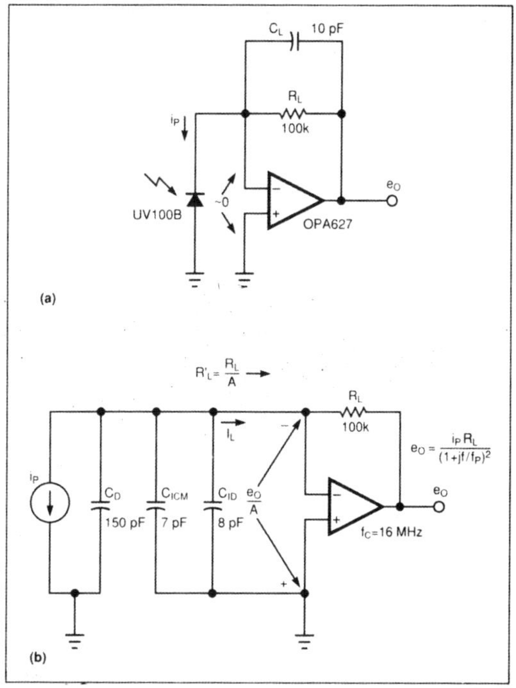

Figure 18.2 The simple current-to-voltage converter isolates the load voltage swing from the photodiode capacitance

together. This discussion develops a bandwidth and phase-compensation background that extends to other photodiode amplifiers. This background also applies to other op-amp applications that present source capacitance to the amplifier. Also, this background applies to any op-amp circuit in which high feedback resistance reacts with the amplifier's input capacitance.

To find the bandwidth of the current-to-voltage converter, you first determine the locations of the circuit's response poles. Then, you design the phase compensation, which defines the overall bandwidth. Figure 18.2b models the circuit for these analyses. Here, a current source and a capacitance, C_D, replace the photodiode. Also, the op-amp input capacitance is separate from the amplifier. The remainder of the amplifier replaces Figure 18.1's R_L with an effective load resistance R_L. For the first step of locating the poles, Figure 18.2b excludes Figure 18.2a's phase-compensation capacitor, C_L, as well as the negligible, high resistances of the reverse-biased diode and the op-amp input.

The input break frequency controls the response of Figure 18.2b's circuit. At the op-amp summing junction, this circuit faces the impedance R_L'. By definition, R_L' equals the voltage across this impedance divided by the current, i_L, supplied to the impedance. The relevant voltage is that from the op-amp's inverting input to ground – simply the amplifier's gain-error signal, e_O/A. Because of the amplifier's finite open-loop gain, A, this signal must exist between the amplifier inputs to support the output voltage, e_O.

The output voltage is $e_O = i_L R_L$, so the voltage across R_L' becomes $i_L R_L'/A$. Dividing this voltage by the current, i_L, defines $R_L' = R_L/A$. This resistance breaks with the capacitance of Figure 18.2b's input circuit at

$$f_P = \frac{A}{2\pi R_L' C_I} \approx A f_I,$$

where $C_I = C_D + C_{ID} + C_{ICM}$.

Above, the break frequency of the input circuit increases from f_I of Figure 18.1 by a factor approximating the open-loop gain, A.

This factor is approximate because the input capacitance C_I is actually smaller for Figure 18.1. There, C_I is the diode capacitance plus the C_{ICM} presented by the voltage follower. In Figure 18.2, however, the amplifier adds its differential input capacitance, C_{ID}, to the total input-circuit capacitance, so $C_I = C_D + C_{ICM} + C_{ID}$. For most photodiodes $C_D >> C_{ID}$, and the C_{ID} difference between the two circuits is not significant. Therefore, the current-to-voltage converter increases the response-pole frequency by a factor essentially equal to the gain, A.

However, this gain varies with frequency, so the actual improvement factor isn't immediately obvious. To calculate the actual pole location, you must determine the relevant ac value of A. This value is the open-loop gain at f_P. To find this gain, consider an approximation to the op-amp's open-loop response. In all practical cases, f_P occurs where the gain of the amplifier exhibits a single-pole roll-off. There, you can approximate the amplifier's gain magnitude as $|A| = f_C/f$, where f_C is the amplifier's unity-gain crossover frequency. At f_P, $|A| = f_C/f_P$. For Figure 18.2's circuit, substituting this expression for A in the f_P equation yields a pole location of

$$f_P = \sqrt{(f_I f_C)},$$

where $f_I = 1/2\pi R_L C_I$,
and $C_I = C_D + C_{ID} + C_{ICM}$.

In this new f_P expression, the pole location is the geometric mean of the old pole frequency, f_I, and the op-amp crossover frequency, f_C. Thus, as long as $f_C > f_I$, the current-to-voltage converter increases the response speed. A typical increase is a factor of 10 to 100, as seen from evaluating $f_P/f_I = \sqrt{(f_C/f_I)}$. With the high-speed OPA627 and the other components of Figure 18.2, the improvement factor is 38:1, and the pole is at 380 kHz. In the rare cases where $f_C < f_I$, the current-to-voltage converter reduces the bandwidth. Even then, however, the current-to-voltage converter provides the improved response linearity mentioned before.

Input circuit forms an L-C tank

Once you have found f_P, you can determine the required phase compensation. Further analysis shows f_P to result from a double – rather than a single pole. Consequently, you must pay careful attention to bandwidth and stability. With the simple, resistive load of Figure 18.1, a single pole controls the response, and the – 3-dB frequency, f_I, coincides with the pole location. Capacitive shunting of a resistive load defines this simple pole. Figure 18.2 exhibits similar shunting, but of a frequency-dependent load rather than a purely resistive one. As shown above, R_L varies with frequency and is an impedance, Z_L, not a resistance.

In Figure 18.2, as the frequency increases and the gain, A, declines, the load of $Z_L = R_L/A$ rises. A load impedance that rises with frequency is inductive. Confirming the inductive character of R_L/A is the phase shift of the gain, A. Over most of the amplifier's useful frequency range, A has a phase lag of 90°. The 180° phase inversion of the basic amplifier gain converts this lag to a 90° phase lead. You can see this effect by including phase information in the previous approximation for A, where $|A| = f_C/f$ for most of the amplifier frequency range. If you include phase in this approximation, $A = 2\pi f_C/s$. Then, the load impedance is $R_L/A = R_L s/2\pi f_C$. With s in the numerator, this impedance appears inductive.

This inductive load resonates with the capacitance of the input circuit at a frequency equal to f_P above. If the resonance occurs at a low enough frequency, it produces oscillation in the current-to-voltage converter. Oscillation occurs if the amplifier's open-loop gain is above unity at the resonant frequency, f_P. Above the unity-gain crossover frequency, the amplifier lacks the gain needed to sustain oscillation. In most cases, $f_P < f_C$, which meets the condition for oscillation.

In L-C tank circuits that can oscillate, you can introduce degeneration by adding resistance in series with either the capacitor or the inductor. For Figure 18.2, this solution would add resistance in the input path of the photodiode-signal current. Signal voltage developed on this added resistance would appear across the photodiode and would degrade the response bandwidth and linearity. In Figure 18.2a, capacitor C_L degenerates the inductive $Z_L = R_L s/2\pi f_C$. Adding C_L in parallel with R_L converts the resistive load to $R_L/$

$(1 + R_LC_Ls)$. Then, $Z_L = R_Ls/2\pi f_C(1 + R_LC_Ls)$, which adds an s term to the denominator of the impedance. This denominator s term counteracts the numerator's s term to degenerate the L-C tank circuit.

Feedback analysis quantifies stability

The feedback analysis that guides the selection of the degeneration capacitor, C_L, quantifies the component's effect. Plotted comparisons of the amplifier and feedback characteristics illustrate how this phase compensation controls the frequency stability. A plot of both the op-amp open-loop gain and the feedback demand for that gain indicates the net conditions for a stable feedback loop. Figure 18.3 shows this graphical analysis for the uncompensated current-to-voltage converter of Figure 18.2b. This figure combines the amplifier's open-loop gain response with the reciprocal of the feedback factor, $1/\beta$. Superimposed on the plot is the resulting current-to-voltage frequency response. As expected from the previous discussion, this response reveals a resonant peak at f_P.

Figure 18.3's $1/\beta$ curve represents the feedback demand, which arises from the feedback factor, β – the fraction of the output fed back to the amplifier input. The voltage-divider action of the feedback network determines β. In Figure 18.3, the voltage divider formed by R_L and C_I produces $\beta = 1/(1 + R_LC_Is) = 1/(1 + s/2\pi f_I)$. Here, C_I is the total input-circuit capacitance or $C_I = C_D + C_{ID} + C_{ICM}$. The feedback factor reflects the pole at f_I introduced into the feedback path by the input circuit. The pole attenuates the feedback signal supplied to the amplifier input. The attenuated input signal requires the amplifier gain to increase at higher frequencies to sustain the amplifier output. The rise in the $1/\beta$ curve, which begins with the response zero of the expression $1/\beta = (1 + s/2\pi f_I)$, reflects this greater gain demand.

Within the limit of its open-loop gain response, the amplifier meets the feedback demand. At low frequencies, the $1/\beta$ curve is flat at unity as expected from simple resistor feedback around an op-amp. At f_I, in response to the attenuated feedback signal, the $1/\beta$ curve begins its rise. Initially, the curve remains below the open-loop gain curve and the vertical distance between the two curves represents the excess gain available, that is, the loop gain. As the frequency increases, the rising $1/\beta$ increases the gain demand. Moreover, the op-amp's gain curve drops simultaneously.

Where the two curves meet, the required gain equals the total available gain so that there can be no further frequency increase. This meeting point is the critical intercept fundamental to feedback stability analysis. At frequencies beyond the

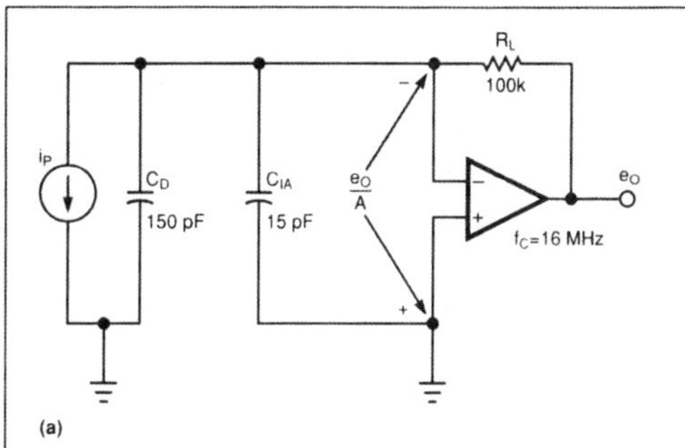

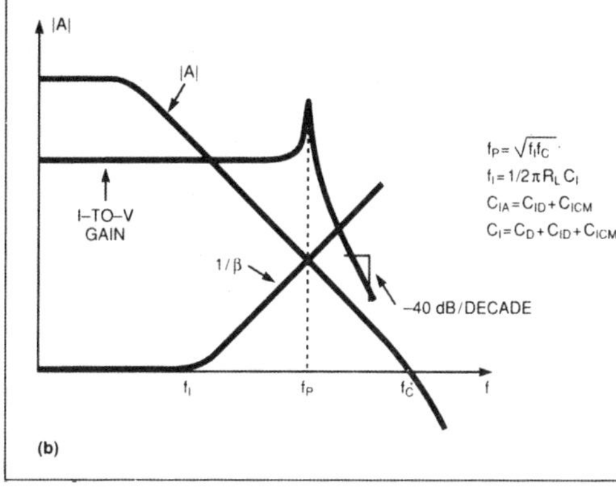

Figure 18.3 Capacitance at the input of the current-to-voltage converter causes the $1/\beta$ curve to rise and results in a resonant response peak

intercept, the amplifier gain is insufficient to support the feedback demand and the response rolls off. Thus, the intercept defines the frequency of the response poles. Graphical analysis of the curves[2] confirms that $f_P = \sqrt{(f_I f_C)}$, as indicated above.

The slopes of the $1/\beta$ and gain curves predict the frequency stability conditions through the two curves' rate of closure. For stability analysis, the rate of closure is the difference between the slopes of the two curves at their intercept. Oscillation can occur where the rate of closure is 40 dB/decade. Each 20 dB/decade of slope corresponds to 90° of phase shift, so the 40 dB/decade of the criterion corresponds to 180°[2]. Added to this is the 180° phase shift of the op-amp gain inversion, producing a net feedback phase shift of 360°. At the intercept, the loop gain is unity. If the phase shift is 360° at the unity-gain frequency, the feedback signal becomes self-sustaining; that is, the circuit oscillates.

For Figure 18.3, both the $1/\beta$ rise and the op-amp roll-off are the result of a single zero or pole, so each has a 20-dB/decade slope. The difference in slopes at the intercept is the critical 40 dB/decade, as anticipated from the earlier resonance discussion. The current-to-voltage response curve of the figure reflects this resonance with a high, sharp peak at f_P, where oscillation will probably occur. Even if oscillation doesn't actually occur, the stability will be poor, with excessive overshoot and ringing. Such stability problems are familiar to everyone who has used high feedback resistances with op-amps. With large feedback resistors, the phase shift introduced by the input capacitance alone disturbs the circuit response.

Phase compensation levels 1/β

In Figure 18.4, to restore stability, place phase-compensation capacitor C_L across feedback resistor R_L. This compensation was added in Figure 18.2a and removed in Figure 18.2b for determining the phase-compensation requirements. Capacitor C_L bypasses R_L at high frequencies to boost the feedback signal at the amplifier input. C_L produces a response zero in the feedback factor and counteracts the pole created by capacitance of the input circuit. Then, for Figure 18.4,

$$\beta = \frac{1 + s/2\pi f_L}{1 + s/2\pi f_I},$$

where $f_L = 1/2\pi R_L C_L$,
and $f_I = 1/2\pi R_L(C_D + C_{ID} + C_{ICM} + C_L)$.

The response zero added to β at f_L is a pole of the inverse function, 1/β. In Figure 18.4, the pole levels off 1/β and reduces the rate of closure for improved stability. To increase the bandwidth, you must sacrifice some stability. Choosing a large C_L could easily make the rate of closure a simple 20 dB/decade, which would yield uncompromised stability. However, this choice would unnecessarily limit the bandwidth. Although the bypass action of C_L counteracts a feedback pole, it degrades the circuit's ability to convert current to voltage at high frequencies. To produce an output signal, the current-to-voltage converter depends on the voltage developed across R_L. Bypassing that resistor to re-establish frequency stability also shunts the output signal and limits the bandwidth.

To optimize the ± 3-dB bandwidth, use a simple guideline to choose a compromise that provides 45° of phase margin. This guideline holds for all practical circuit cases. The phase margin is the difference between the critical 360°, which produces oscillation, and the actual phase shift of the feedback loop. This phase difference is important only at the intercept of the 1/β and gain-magnitude curves[2]. For basic feedback stability, the op-amp, because of its gain inversion, starts by injecting 180° of phase shift. The phase margin is thus 180° minus the added phase shifts through the op-amp and the feedback network.

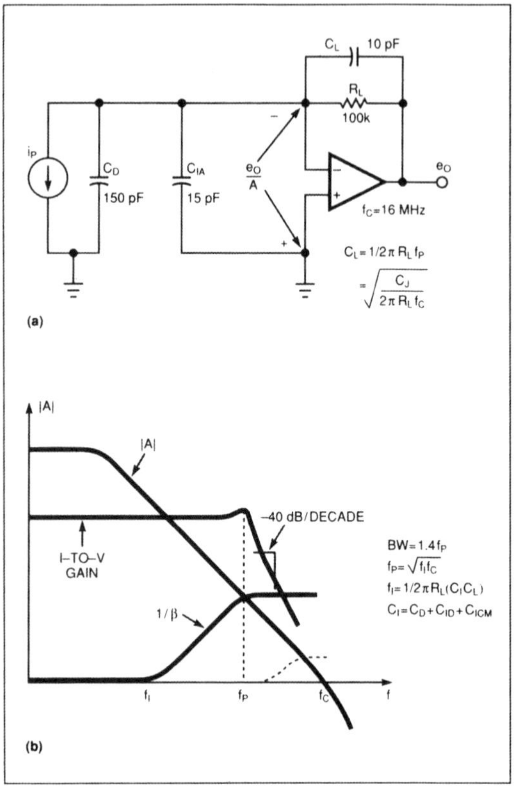

Figure 18.4 A simple design guideline establishes the phase compensation provided by feedback capacitor C_L

The following analyses determine the phase margin of the current-to-voltage converter under two conditions. The relative proximity of f_P to the other circuit-response singularities differentiates these cases. In the simpler case, f_P is more than a decade away from any of the circuit's other response poles or zeros. In that case, any pole or zero at a frequency lower than f_P develops essentially a full 90° of phase difference at the intercept frequency. Similarly, any pole or zero at a frequency higher than f_P contributes essentially zero phase shift at the intercept. For Figure 18.3, this simple case results in 90° of added phase shift from both the first op-amp pole and the 1/β zero. For this case, the far-removed poles around f_C cause no phase shift at f_P. Therefore, the amplifier and feedback loop add a net phase shift of 180°, leaving a phase margin of zero and ensuring oscillation.

To restore the phase margin in Figure 18.4, add C_L to reduce the phase shift from the 1/β curve. For 45° of phase margin, choose C_L to break with R_L right at the intercept frequency, f_P. At its break frequency, a response singularity's phase effect is exactly 45°. Therefore, placing the f_L break frequency at f_P reduces the 1/β phase shift at the intercept from 90° to 45° and boosts the phase margin from zero to 45°.

Fortunately, this simple guideline remains accurate even as f_P approaches f_C, as often occurs in practice. In this second case, the frequency difference between f_P and the other singularities is less than a decade. Hence the phase contribution of these singularities differs from the simple 90° of the first analysis. This condition occurs with smaller photodiode capacitances, which move the input break frequency f_I to the right in Figure 18.4. A dashed curve in Figure 18.4 represents $1/\beta$ for this second condition, in which f_P moves down the open-loop gain curve toward f_C, and the phase shift at the new f_P enters the higher-frequency poles' range of influence around f_C. However, frequency f_I simultaneously moves to the right (the dashed curve in Figure 18.4). This movement compresses the distance between the new f_I and the corresponding new f_P. This compression reduces the phase effect of the lower-frequency singularity at f_I.

For first-order analyses, these two phase adjustments cancel, leaving the choice of C_L unchanged. To demonstrate this effect, consider the op-amp response to be essentially two-pole in nature with the second pole occurring at the unity-gain crossover frequency, f_C. Although this situation is not the actual one, it accurately portrays the op-amp phase response at frequencies as high as f_C. This simple model shows that the amplifier phase shift increases with frequency and produces 135° of phase shift at f_C. Such phase shift is a conservative model of the performance of most op-amps. Beyond f_C, the exact phase response of the amplifier is not usually important. At these frequencies, the loop gain is below unity and will not support oscillation.

With the two-pole amplifier model, four response singularities determine the net phase margin. Two of these singularities follow from the first case: the first amplifier pole and the break frequency of C_L. As before, this amplifier pole decreases the phase margin from 180 to 90° and the $1/\beta$ leveling provided by C_L restores 45°. In the second analysis, the closer proximity of f_P to f_I and f_C alters the phase from the initial 135°. No longer does f_I introduce a complete 90° of phase shift nor is the influence of f_C zero.

To find the actual effects on the phase margin in Figure 18.4, the following equations express the influences of f_I and f_C with higher resolution:

$$\phi_M = 135° - \arctan(f_P/f_I) - \arctan(f_P/f_C).$$

From before, $f_P = \sqrt{(f_I f_C)}$. Substituting this expression in the above equation produces:

$$\phi_M = 135° - \arctan\sqrt{(f_L/f_I)} - \arctan\sqrt{(f_I/f_L)}.$$

The variable terms of this equation are of the form arctan(a/b) + arctan(b/a). Trigonometric analysis shows that this combination always equals 90°. Thus, independent of the location of f_I, for Figure 18.4

$$\phi_M = 45°,$$

for $C_L = 1/2\pi R_L f_P$,

where $f_P = \sqrt{(f_I f_C)}$,

and $f_I = 1/2\pi R_L(C_D + C_{ID} + C_{ICM} + C_L)$.

Note that the above equations interact in the determination of C_L. C_L depends on f_P, which depends on f_I, which in turn depends on C_L. This situation occurs because C_L adds to the capacitance that causes the break frequency at f_I. The added phase compensation moves the target of the compensation. To select C_L, you can remove the interaction either by approximating or by simultaneously solving the three equations above. In the simpler case, large-area photodiodes make $C_D >> C_L$. In Figure 18.4, the above three equations then combine directly for a phase compensation of

$$C_L = \sqrt{(C_I/2\pi R_L f_C)},$$

where $C_D >> C_L$
and $C_I = C_D + C_{ID} + C_{ICM}$.

This result simplifies to an easily memorized relationship in which C_L is the geometric mean of two capacitances. Defining an artificial capacitance, $C_C = 1/2\pi R_L f_C$, relates f_C to R_L just as the resistance relates to f_I and f_L. The above result simplifies to the geometric mean $C_L = \sqrt{(C_I C_C)}$. The phase compensation capacitor C_L equals the geometric mean of the input circuit's total capacitance and the capacitance that represents f_C. This result parallels the expression $f_P = \sqrt{(f_I f_C)}$ in which f_P is the geometric mean of the analogous frequencies. For C_L, one of these capacitances, C_I, is real and the other simply represents the op-amp bandwidth, f_C. For the typical current-to-voltage photodiode amplifier, set the phase compensation at

$$C_L = \sqrt{(C_I C_C)},$$

for $C_D >> C_L$

where $C_I = C_D + C_{ID} + C_{ICM}$. and $C_C = 1/2\pi R_L f_C$, and where f_C is the unity-gain crossover frequency of the op-amp.

In the more comprehensive case, select C_L to accommodate even small photodiode capacitances. Don't use the previous approximation.

Instead, solve the preceding simultaneous equations for $\Phi_M = 45°$. For Figure 18.4, this approach yields

$$C_L = (C_C/2)(1 + \sqrt{(1 + 4C_I/C_C)},$$

where $C_C = 1/2\pi R_L f_C$,
and $C_I = C_D + C_{ID} + C_{ICM}$.

You need this more exact expression where there are lower circuit capacitances that are more sensitive to parasitic capacitances. Depending on where they occur, these parasitics can alter the value of C_L in either direction. Some board parasitics add to the C_I term but others supplement C_L. A final tuning adjusts for these unknowns empirically.

Two features benefit bandwidth

When you set the phase compensation by choosing C_L, you determine the bandwidth of the current-to-voltage converter. This circuit's two-pole response is actually advantageous because gain peaking extends the bandwidth. Excessive damping is inherent in the single-pole response of Figure 18.1; this damping fixes the 3-dB bandwidth at the pole location. The two-pole case of Figure 18.4 permits an underdamped response and extends the bandwidth beyond the pole frequency. Just how much the bandwidth increases depends on the required response accuracy. Where you can accept the traditional ± 3-dB deviation, the damping factors and the resulting responses[3] show a factor of 1.4 increase for a 45° phase margin. Gain peaking is then just + 3 dB followed by the final bandwidth limit at the − 3-dB point. Thus, for the current-to-voltage converter of Figure 18.4, with C_L breaking at f_P,

$$BW = 1.4f_P = 1.4\sqrt{(f_I f_C)},$$

where $f_I = 1/2\pi R_L(C_D + C_{ID} + C_{ICM} + C_L)$.

For the components shown in Figure 18.4, f_I = 9.1 kHz and f_C = 16 MHz for BW = 534 kHz. This represents a 53:1 bandwidth improvement over the 10-kHz limit of the basic circuit in Figure 18.1.

The above expression displays an advantageous gain-bandwidth relationship because of the square-root function. The R_L in the expression for f_I is the element that sets the current-to-voltage converter's transresistance or gain. Increasing R_L for greater gain reduces the bandwidth, but by less than you might expect. Normally, in voltage-amplifier applications, an increase in gain causes an equal reduction in bandwidth. For the current-to-voltage converter, the gain-bandwidth product is R_L(BW). Substituting BW from its equation above shows this product to be $1.4\sqrt{(R_L f_C/2\pi C_I)}$ – proportional to $\sqrt{(R_L)}$. Thus, the maximum practical value of R_L yields the maximum gain-bandwidth product. Above a certain R_L value, parasitic capacitance rolls off the gain that this resistor provides.

The Bode plots of Figure 18.4 explain this reduced gain-bandwidth sensitivity. Consider what happens when you start with the dashed curve and move back to the solid $1/\beta$ curve. Increasing R_L moves f_I down in frequency and shifts the $1/\beta$ curve in direct proportion, lowering the bandwidth-defining intercept of $1/\beta$ with the amplifier gain-magnitude curve – but not in direct proportion. Because the gain-magnitude curve rises as the frequency decreases, the intercept recedes more slowly. The equal slopes of the gain and $1/\beta$ curves make this bandwidth decrease one-half that of $\log(f_I)$, and the log scale converts this fraction to a square root.

An alternate approach to increasing the gain-bandwidth product enjoys the same square-root benefit. By using larger area photodiodes, you increase the overall circuit response to the light source at a rate greater than the accompanying bandwidth decline. Both the photodiode's capacitance and its responsiveness to light are directly proportional to the diode area, A_D. Increasing A_D produces a directly proportional increase in the light-to-voltage gain of the circuit. However, the bandwidth, described by the previous equation, declines only by the square-root of C_D. Thus, gain-bandwidth product for the current-to-voltage converter is proportional to $\sqrt{(A_D)}$. The maximum gain-bandwidth product results from a photodiode area that covers as much of the area illuminated by the light source as is practical.

To maximize the bandwidth instead of the gain-bandwidth product, choose R_L to take advantage of the full amplifier bandwidth. In Figure 18.4, making R_L smaller moves f_I to the right – to the limit imposed by f_C. Beyond f_C, the amplifier lacks the bandwidth required for further extension of the current-to-voltage converter response. To maximize the bandwidth, select R_L to place the intercept frequency, f_P, at the amplifier's unity-gain crossover frequency, f_C. This choice moves the $1/\beta$ curve to the right, compressing its rise to zero and making the three response-defining frequencies coincide; $f_I = f_P = f_C$. Given this condition, the expression for f_I sets the feedback resistor in Figure 18.4 to

$$R_L = 1/2\pi f_C(C_D + C_{ID} + C_{ICM}), \text{ for maximum bandwidth.}$$

Any further increase in bandwidth must come from using a higher speed op-amp that moves the f_C limit to a higher frequency. Once again, a square-root relationship determines the improvement, because f_P is proportional to $\sqrt{(f_C)}$. Figure 18.4 shows the wideband OPA627 instead of Figure 18.1's slower OPA111. This change increases f_C from 2 to 16 MHz for a $\sqrt{(8)}$ increase in current-to-voltage-converter bandwidth. In Figure 18.1, changing amplifiers would offer no benefit because the photodiode in front of the op-amp limits the bandwidth.

References

1. J Graeme, 'FET op amps convert photodiode outputs to usable signals', *EDN*, October 29, 1987, p. 205.
2. J Graeme, 'Feedback plots offer insight into operational amplifiers', *EDN*, January 19, 1989, p. 131.
3. J Graeme, 'Feedback models reduce op-amp circuits to voltage dividers', *EDN*, June 20, 1991, p. 139.

Part Three

Analog/Digital Conversion

Introduction to Part 3

The first two parts dealt with digital and analog topics; in this part, we see the merging of the two worlds. For a computer to deal with analog information, you must first convert the information into bits and bytes. The first three articles in this part discuss techniques and products used to take samples of analog signals, and to convert those samples into digital words.

I've always been a little mystified by some aspects of the Nyquist theorem (that's right, if you sample a signal at exactly twice its frequency at the signal's zero-crossing points, you'll obtain exactly zero output). The second article in this part demystifies the Nyquist criterion to a large extent, and dispels some common misconceptions and myths surrounding it. The last article shows a clever technique for configuring a high-resolution D/A converter by using two low-resolution converters.

19 Undersampling reduces data-acquisition costs for select applications

Sub-Nyquist rate sampling saves money

If all the signal of interest is limited to a narrow-percentage bandwidth, e.g. an IF signal, it can be effectively transferred to baseband by sampling at a frequency which is a subharmonic of the IF center-frequency (giving a complex baseband signal, requiring I Q processing) or a subharmonic of a frequency which is just to one side of the IF band (giving a real baseband signal). This article describes the considerable economies that can result from the reduced sampling rate, compared with the alternative of a sampling rate higher than twice the frequency of the upper edge of the passband.

Jeff Kirsten and Tarlton Fleming
Maxim Integrated Products

In its familiar form, the Nyquist sampling theorem requires that an A/D converter sample the highest significant input frequency at least twice per cycle. Undersampling – a deliberate violation of this rule – is a useful option in certain carefully chosen applications such as synchronous demodulation, pulse-height measurement, and waveform digitization. The technique allows you to use an A/D converter that is slower and therefore less expensive than the converters necessary to perform Nyquist sampling.

Undersampling does comply with an extended or generalized version of the sampling theorem. This extended theorem states that you can reconstruct an input from sampled data if the input's frequency components lie entirely between adjacent integer multiples of one half the f_S sample rate (see 'Spectral limiting enables undersampling'). An input spectrum violates this generalized sampling theorem if it includes or straddles one or more boundaries of the $f_S/2$-wide frequency bands.

The resulting alias frequencies usually distort the sampled output, making it virtually impossible to interpret, except in special cases where the input possesses a certain spectral order. Two examples of such order, to be discussed later, are the double sidebands that result from linear modulation and the extended spectra that result from undersampling a periodic waveform. To avoid unwanted alias frequencies in all cases, you must first evaluate the input spectral characteristics with respect to the converter's sampling rate.

Track and hold is crucial

The burden of speed and accuracy falls on the T/H (track-and-hold) amplifier rather than the A/D converter when undersampling wideband analog signals. The T/H amp must acquire fast-changing signals at uniform and precisely timed intervals, and it must present each sample to the A/D converter as a fixed input voltage. To do so, the T/H amp must have sufficient bandwidth and slewing to pass the input signal without introducing significant distortion. The A/D converter, on the other hand, has an easier task during undersampling operations. For applications in this article, the converter samples at a rate less than twice the highest input frequency of interest.

Undersampling allows a system to acquire narrowband signals whose frequencies are higher than one half the A/D converter's sample rate. Such applications arise in systems that digitize the intermediate frequency produced by a radio receiver (Figure 19.1).

Sampling the signal at f_S shifts the input frequency components in groups or bands to the sampled-data range of 0 to $f_S/2$. These $f_S/2$-Hz-wide bands are contiguous and extend to infinity. Each has a location $Nf_S/2$ to $(N+1)f_S/2$, where N equals zero or a positive integer. In Figure 19.1, f_S = 100 k samples/s, f_{IN} = 510 kHz, and N = 10. To exclude unwanted input frequencies, a bandpass

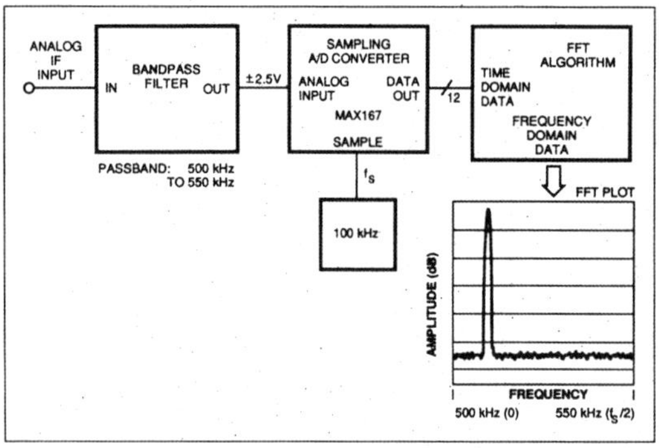

Figure 19.1 The undersampling process translates – you might say intentionally aliases – the desired 500- to 550-kHz passband of this narrowband, high-frequency digitizing system to the 0 to $f_S/2$ FFT baseband. The input bandpass filter excludes unwanted alias frequencies

filter restricts the converter's input to just one band: 500 to 550 kHz. Regardless of the converter's relatively low conversion rate, the resulting output data includes unwanted frequencies only to the extent of spectral leakage outside this filter's $5f_S$ to $5.5f_S$ passband.

You can measure the performance of this system by applying high-purity sine waves at the input and by performing an FFT on the output. A 510-kHz input, which equals $5f_S$ + 10 kHz, produces a 10-kHz response in the FFT. Inputs at 490 kHz ($5f_S$ – 10 kHz) or 590 kHz ($6f_S$ – 10 kHz) could also produce a 10-kHz response, but the filter excludes those frequencies. Barring spectral leakage, the filter assures that each response in the FFT output derives from one and only one input frequency.

Figure 19.2a illustrates the FFT response for the MAX167 12-bit sampling converter operating on a pure 510-kHz input signal. The converter's internal T/H amp has a 6-MHz, predominantly single-pole roll-off, which reduces signal amplitudes in the 500- to 550-kHz band by no more than 0.04 dB. Frequency-dependent non-linearities of the T/H amp produce second and third harmonics of 510 kHz that represent distortion levels of –60 dB (0.1%) and –70 dB (0.03%).

The device's A/D converter alone contributes no more error when processing the 500- to 550-kHz band than when processing the band from 0 to 50 kHz. The harmonics, which originate in the T/H amplifier's upper passband, disappear into the noise floor if you remove the bandpass filter and apply a pure 10-kHz input signal.

Besides harmonic distortion, T/H amps contribute phase noise as a result of aperture jitter that occurs in the sampling process. This phase noise originates both in the T/H amp and in the control signal applied to it. Wideband components of phase noise raise the FFT's overall noise floor; 1/f components cause a widening of the frequency lines, especially near their bases. Phase noise ultimately limits the attainable FFT resolution for larger sets of sampled data.

Figure 19.2b displays the frequency resolution available in Figure 19.1's system by taking a 16,384-point FFT and expanding it around 510 kHz. Note that the 510-kHz input component is spread 67-Hz wide at a level 90 dB below the peak. About 4 Hz of this spreading is caused by the converter's 25 ps of aperture jitter, 14 Hz is

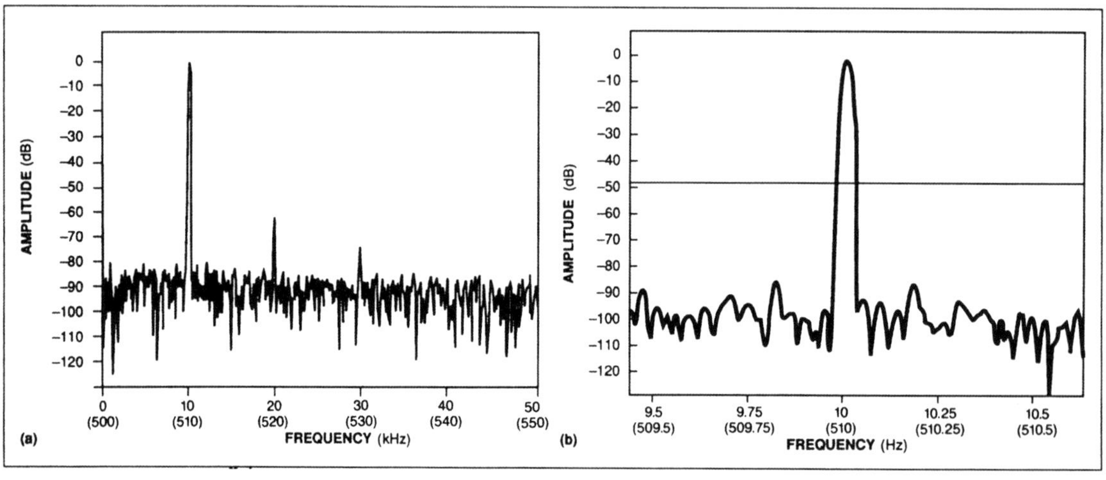

Figure 19.2 A closer view of Figure 19.1's FFT output reveals that the second and third harmonics associated with a 510-kHz input (a) are down more than 60 and 70 dB, respectively. Note that the second harmonic (1020 kHz) appears in the FFT output as 20 kHz. Increasing the sample record from 2048 to 16,384 points (b) illustrates the frequency resolution – the width of the main signal in the FFT – for the DP7 sampling window

caused by phase noise in the test setup's sine wave and clock-signal sources, and 49 Hz is due to the window function applied to the data prior to the FFT.

Except in specialized applications in which the sample record is synchronous with the input signal, window functions must precede an FFT. The window suppresses severe distortion due to spectral leakage, which is caused by the discontinuity between the end and the starting points of consecutive waveform records. Different window types impress different filter responses on the spectral components that pass through them.

The Blackman-Harris four-term window, for example, is frequently used in digitizer testing. It widens spectral components in the 16,384-point FFT by 37 Hz at −90 dB. By comparison, the model DP7 window (Data Physics, San Jose, CA) used to generate the data in Figures 19.2a and b widens the components by 49 Hz at −90 dB. Below that level, however, the DP7 window exhibits a lower widening rate. The DP7 window, suitable for use in testing high-resolution A/D converters, remains free of side lobes to −110 dB and is usable to −150 dB.

A sampling A/D converter can perform synchronous demodulation if the modulation sidebands are no wider than half the sample rate. To ensure this condition, you need to limit the modulating signal's bandwidth or add a suitable filter ahead of the converter. Take the case of a 500- to 550-kHz upper sideband of a linearly modulated 500-kHz carrier. If the sample rate's fifth harmonic (500 kHz) closely matches the carrier in phase and frequency, then the converter's output will return a synchronously demodulated version of the signal. Applying the lower sideband of 450 to 500 kHz to the converter will produce the identical demodulated signal.

Applying both sidebands together – a double-sideband signal – produces the same demodulated output at twice the amplitude. Although this input violates the extended sampling theorem (because it straddles an integer multiple of $f_S/2$ at 500 kHz), the aliasing problems you might expect do not arise. As mirror images, the sidebands make identical contributions to the demodulated output. This property enables the sampling process to return the modulating signal to the baseband just as it does for single-sideband signals.

Measure pulse heights

Undersampling techniques also apply to the measurement of pulse heights, in the sense that only one sample of many is of interest – the one corresponding to maximum amplitude. Such measurements apply to the narrow, flat-topped pulses found in pulsed-RF generators and receivers, and in instruments that employ optical detectors. Pulse-height measurements also drive the leveling loops and output-power indicators of radar transmitters and other pulsed emitters. An 8-bit flash converter can do these jobs, but the combination of a fast T/H amp and a moderate-speed A/D converter can provide 12-bit-accurate measurements at comparable cost for pulses as narrow as 750 ns (Figure 19.3a).

The circuit in Figure 19.3a lets you sample a single point that characterizes the pulse's amplitude. Applying the flat-topped waveshapes to the PULSE INPUT, you drive the pulse-measurement section with two externally generated signals: PULSE SYNC, which marks the rising edge of the pulse to be measured (Figure 19.3b); and an asynchronous, 1.5-MHz clock signal, which drives the converter.

The PULSE-SYNC signal triggers IC_4's one-shot, which in turn generates an acquisition-time interval sufficient for IC_1's T/H amp to acquire its input signal with specified accuracy. In this case, a 600-ns interval allows IC_1's on-chip hold capacitor to charge within 1/2 LSB of the pulse amplitude. The T/H amp then switches to hold mode and the conversion begins. Because pulse inputs are asynchronous with the processor bus, the system includes a bus interface that allows the processor to detect and read current data before a new acquisition cycle begins. This interface, which consists of flip-flop IC_5 and bus drivers IC_6 and IC_7, also requires two external control signals. READ STATUS enables the output buffers, and READ DATA loads new data onto the bus. By clearing the internal CONVERSION-COMPLETE signal, READ DATA also arms the circuit for the next PULSE-SYNC trigger.

The external 1.5-MHz clock is asynchronous with both the bus and the arriving pulses. Consequently, when clock edges coincide closely with the track-to-hold command, some noise energy usually couples to the analog channel via parasitic paths in the IC and in the external printed circuit. The resulting offset error, whose amplitude varies with clock phase over a range of several LSBs, appears as noise. IC_2 minimizes this offset noise by blocking the clock signal during the track-to-hold transition. IC_3 then re-enables the clock during the next full cycle, following a delay of at least half a clock cycle.

Ideally, an output sample represents pulse height at the instant the converter switches from track to hold. Finite resistance in the sampling switch, however, contributes an RC delay that causes hold-capacitor voltage to lag the input signal. The sampling converters of Figure 19.1 and Figure 19.3a thus exhibit a step response that closely resembles a single RC-network exponential. This time constant, about 24 ns for the

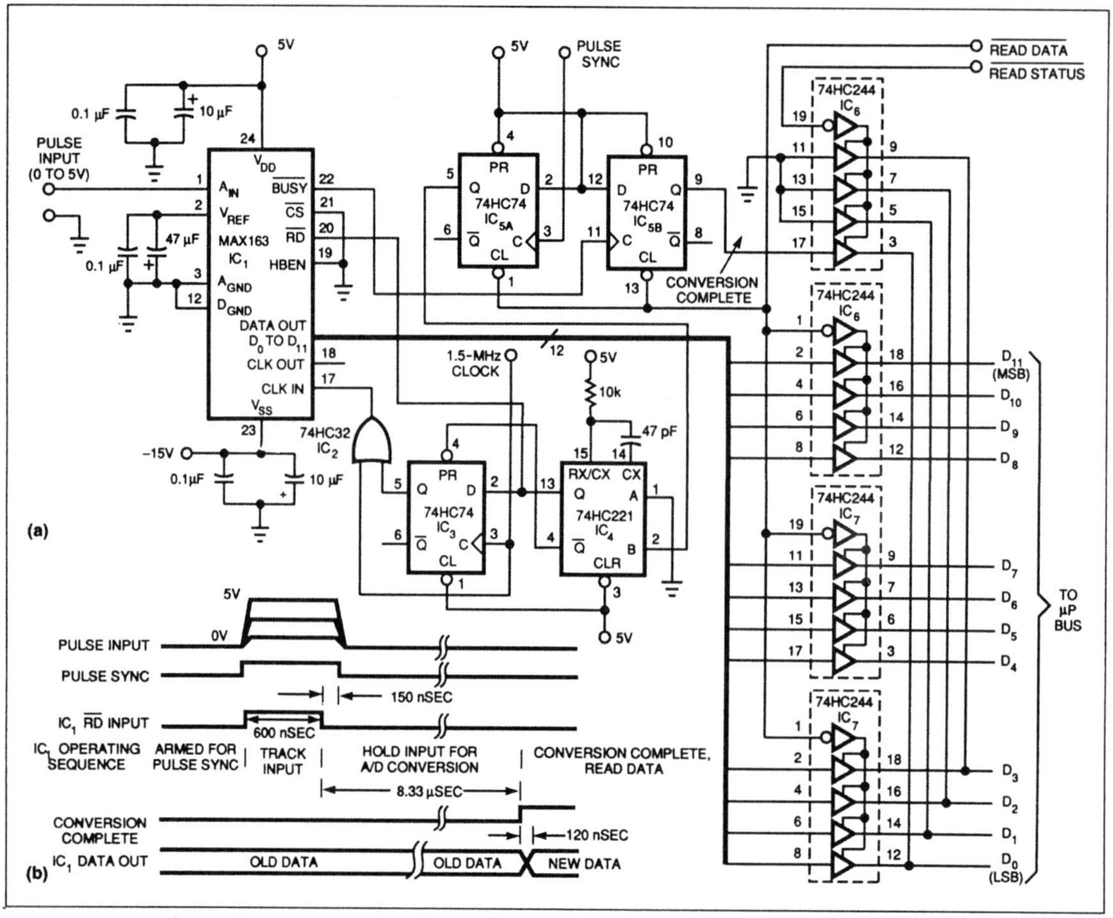

Figure 19.3 This pulse-height measurement circuit (a) adds delay to the externally generated PULSE SYNC signal to position the sampling point on the waveform (b)

MAX163, largely determines the converter's –3-dB bandwidth: $f_{-3\,dB} = 1/2\pi\tau$, where $\tau = RC$. The settling time depends on additional and larger time constants. The minimum acquisition time (settling time) necessary to achieve 8-, 10-, or 12-bit accuracy is approximately 200, 350, or 500 ns, respectively.

In the pulse-height application, a selected feature of the input waveshape – the rising edge of the pulse for example – triggers a fixed delay preceding the sampling instant. However, if you add a small increment to this delay after each conversion, you can digitize a periodic signal whose bandwidth exceeds the converter's sample rate. Again, the benefit is the cost savings realized by using a slower A/D converter. The converter's action is similar to that in a sampling oscilloscope, in which a fast S/H amplifier 'walks' its sampling point along a wide-bandwidth waveshape.

A test circuit (Figure 19.4a) demonstrates this undersampling technique by applying the recurring stimulus f_{STIM} to a network and digitizing the response. By controlling f_{STIM}, the circuit is able to synchronize the signal to be measured to the 20-MHz oscillator. The frequency of f_{STIM} must be low enough to allow each network response to decay nearly to zero before the next stimulus arrives. IC_1, a combination A/D converter and high-speed, 6-MHz T/H amp, then samples the network response at apparent 50-ns intervals.

The circuit's 20-MHz oscillator drives a 256-count divider, which in turn drives the sampling converter's SAMPLE input. The oscillator also drives a 255-count divider, whose output triggers the STIMULUS signal. The sample command and the stimulus square wave occur nearly at the same rate, but the sample command advances one 50-ns count with each repetition of the stimulus (Figure 19.4b).

The converter output represents signal amplitudes at 50-ns intervals along the input wave-

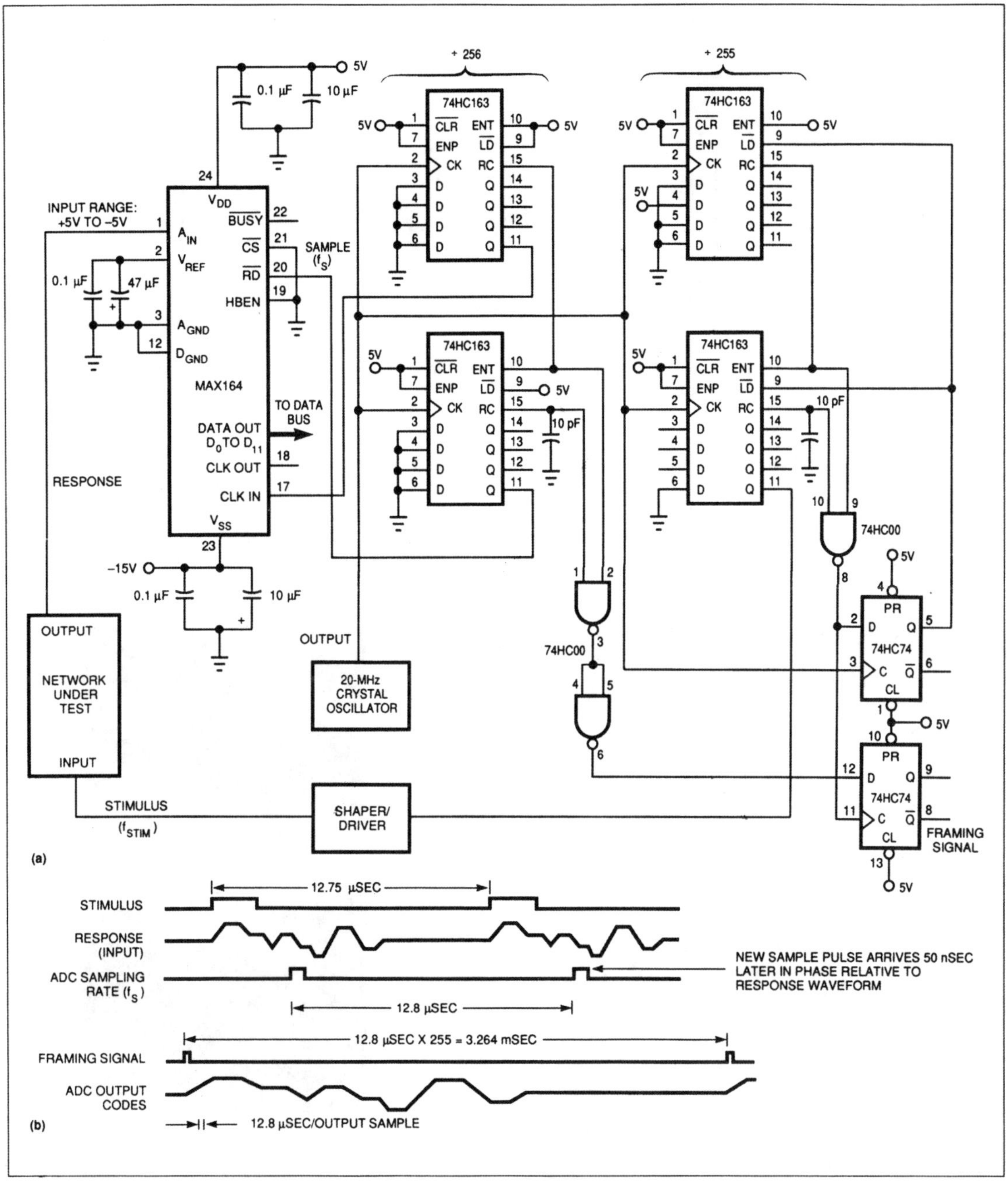

Figure 19.4 By synchronizing the sampling time with the stimulus signal, the system in a employs undersampling techniques to digitize wideband periodic signals. The sampling rate, slightly less than the input signal's repetition rate (b), provides an effective sampling interval of 50 ns

form. Each successive group of 255 consecutive points is identical and represents a complete, one-cycle record of the input signal. Each such record has a duration of 12.75 μs in real time. By producing only one sample per input cycle, the A/D converter requires only 3.264 ms to complete a record. The FRAMING SIGNAL denotes each record with a pulse that occurs each time the SAMPLE and STIMULUS signals are in phase. Thus, the converter's output data defines an expanded time scale with respect to that of the input samples.

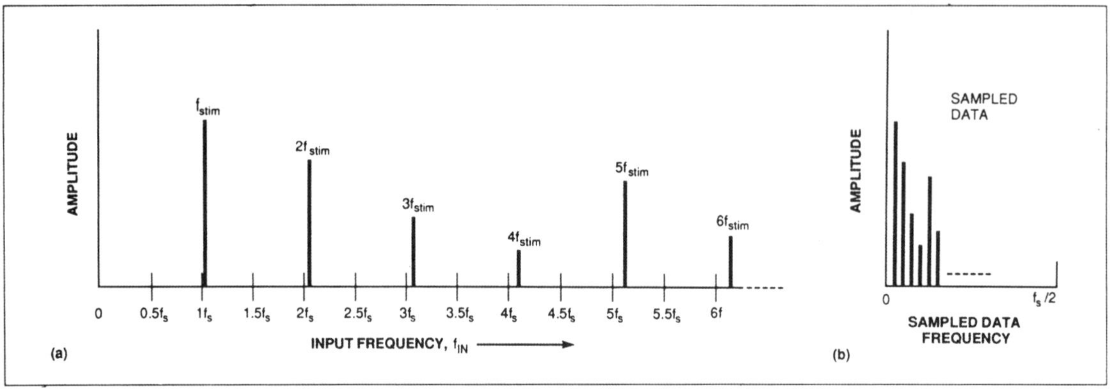

Figure 19.5 The periodic signal f_{STIM} recurs at a slightly higher rate than f_S (a). The sampling process shifts all the signals to nonoverlapping lines in the FFT baseband (b)

Figure 19.5 illustrates this circuit's action in the frequency (b) domain. The input spectrum consists of a dc component and harmonic multiples of f_{STIM}. By setting f_{STIM} slightly higher than the sample rate, you assure for each successive f_S-wide band an additional frequency difference – offset – between the harmonics of f_{STIM} and corresponding multiples of the sample rate (Figure 19.5a). The sampling process shifts each of the f_{STIM} harmonics to unique, nonoverlapping locations in the sampled-data baseband (Figure 19.5b), where the harmonics comprise a frequency-scaled replica of the input spectrum.

Figures 19.6a and b illustrate Figure 19.4's output using a function generator as the network under test. The generator produces a single sine-wave or triangle-wave cycle for each stimulus trigger. In each photo, the top trace displays the input signal and the lower trace is the reconstructed output codes from the A/D converter. Twenty-one samples from the waveform digitizer represent one cycle of the input signal, spanning approximately 1 μs in real time.

The 50-ns equivalent-sampling interval in this application provides an effective sampling rate of 20 M samples/s for periodic signals. According to the sampling theorem, the circuit can process input bandwidths to 10 MHz without aliasing. Moreover, the circuit response is virtually free of slew-rate effects, thanks to small- and large-signal bandwidths that match within 0.3 dB over a range of dc to 20 MHz.

As a second example of wideband waveform digitizing, consider the use of undersampling in capturing a still frame of video data. Assuming a picture bandwidth of 6 MHz, the Nyquist sampling rate must equal at least 12 M samples/s. The 100 k-sample/s converter must therefore sample and convert every 120th pixel on each pass. Each video frame, which occurs at a rate of 30/s in the US and Japan, will begin one pixel later if you adjust the sample rate slightly below 100 k samples/s. The result will be a conversion of pixels 1, 121, 241, and so on in the first frame; 2, 122, 242, and so on in the second frame; and so on for four seconds – 120 frames – until the entire image of

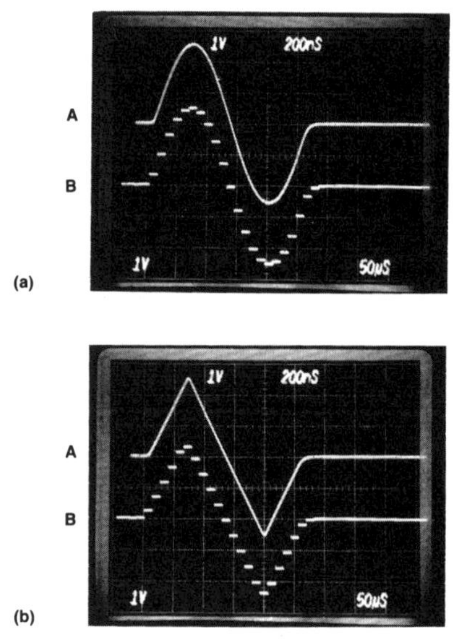

TRACE	HORIZONTAL	VERTICAL
A	200 ηSEC/DIV	IV/DIV
B	50 μSEC/DIV	IV/BIV

Figure 19.6 These photos illustrate the results of Figure 19.4's equivalent-time-sampling technique. The sine-wave generator in a and the triangle-wave generator in b produce one input cycle per stimulus. The bottom traces display a DAC's reconstruction of the sampled signal. Note that each reconstruction step represents a complete pass of the input signal, resulting in different time scales for traces A and B

400 k pixels is completed. Obviously the picture must not move, just as the undersampled signal must be periodic.

In each preceding example, knowledge of the input spectrum enabled an A/D converter operating at a sub-Nyquist sampling rate to derive useful information from the input signal. Inputs confined to a single band between integer half-multiples of the sample rate – as in the narrow-band-IF system and the synchronous demodulator with a single-sideband input – always satisfy the generalized sampling theorem.

Inputs that occupy more than one of the $f_S/2$-wide bands normally produce unwanted alias frequencies that result in output distortion. In a few special cases, however, the spectral components of such an input possess order that assures an undistorted output. Such order is present, for instance, in double-sideband signals.

Spectral ordering is also present in the waveform-digitizing application, in which as many as 128 input-frequency components, each in a different band, contribute to a single, undistorted output. In this case, the system provides the order by using the stimulus rate to make the input signal periodic.

Spectral limiting enables undersampling

Sampling converters measure a signal's amplitude at precise intervals and provide a digital output word corresponding to the amplitude of each sample. The digital word's sequence is only a partial description of the input signal; the sequence includes no information about variations between the samples.

You can often find a complete description, however, by applying suitable constraints to the input spectrum. The sampling theorem states that you can exactly recover a signal from sampled

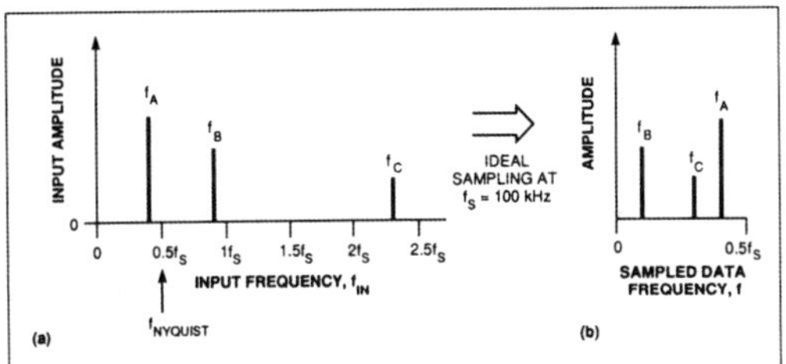

LABEL	INPUT FREQUENCY	M	SIGN	SAMPLED DATA FREQUENCY
f_A	40 kHz ($0.4 f_S$)	0	+	$\lvert 40\text{ kHz} - 0\text{ kHz}\rvert = 40\text{ kHz}$ ($0.4 f_S$)
f_B	90 kHz ($0.9 f_S$)	1	–	$\lvert 90\text{ kHz} - 100\text{ kHz}\rvert = 10\text{ kHz}$ ($0.1 f_S$)
f_C	230 kHz ($2.3 f_S$)	2	+	$\lvert 230\text{ kHz} - 200\text{ kHz}\rvert = 30\text{ kHz}$ ($0.3 f_S$)

(c)

Figure 19.7 Sampling three input frequencies, f_A, f_B, and f_C, at a rate of f_S (a) creates a sampled-data spectrum (b) that repeats every $f_S/2$ Hz. The 0 to $f_S/2$ spectrum that results from the FFT thus includes an alias for every input frequency above $f_S/2$ (c)

data if the input signal contains no components above half the sampling rate. In other words, spectral limiting allows sampled data to describe the input completely.

Consider the effect of sampling three distinct input frequencies – f_A, f_B, and f_C – at 100 k samples/s (Figure 19.7a). At 40 kHz, f_A is less than the 50-kHz Nyquist frequency. Components f_B and f_C at 90 kHz and 230 kHz, respectively, are greater than the Nyquist frequency. Figure 19.7b shows the spectrum after sampling. The location of component f_A is unchanged, but the sampling process has shifted f_B and f_C by multiples of the sample rate. Now, f_B's and f_C's sampled-data versions reside at 10 and 30 kHz, respectively. All three inputs now appear below ½ f_S in the sampled-data spectrum.

Sampling maps these input-signal components (Figure 19.7c) into the range below ½ f_S using the formula

$$f_{SAMPLED} = |f_{IN} - Mf_S|,$$

under the condition that $f_{SAMPLED} < ½ f_S$.

This equation applies to an ideal sampling process – one with zero aperture time. The equation finds the multiple of f_S that is closest to f_{IN}, and subtracts it from f_{IN}. The absolute value of this difference is the sampled-data frequency, $f_{SAMPLED}$.

Using the equation above, you can't determine the input spectrum from sampled data alone because a given sampled-data spectrum can result from one of several input frequencies. Component f_C in Figure 19.7b, for example, could be the result of inputs at 30, 70, or 130 kHz as readily as from 230 kHz. Thus, to reconstruct an input signal, you must have some prior knowledge of the input spectrum, in addition to the sampled data.

Figure 19.8, a plot of frequency ($f_{SAMPLED}$) vs frequency (f_{IN}), shows that the sampled-data frequency is cyclic in the range 0 to $f_S/2$. The spectrum of $f_{SAMPLED}$ actually extends to infinity, but repeats around multiples of f_S. The spectrum also exhibits mirror images on either side of each multiple. So all the information in $f_{SAMPLED}$ is available in any of the intervals $Nf_S/2$ to $(N + 1)f_S/2$.

The input-frequency scale has an M value and a sign term associated with each $f_S/2$-wide interval, as Figure 19.8 displays. Input signals residing in any single interval, or band, will shift directly to the 0 to $f_S/2$ sampled-data band without loss of information or distortion caused by overlap. (Note that input spectral components that lie within any '+' interval will translate to the FFT output in the same sequential order; components within a '–' interval will translate in reverse order.)

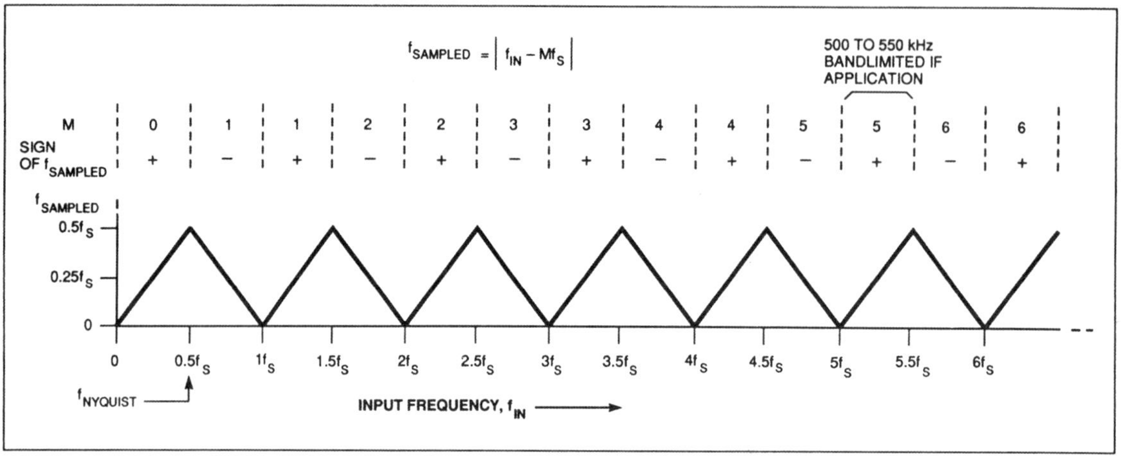

Figure 19.8 An input of arbitrary frequency maps to a specific position within the 0 to $f_S/2$ sampled-data baseband. Note that this plot uses the convention that all information in $f_{SAMPLED}$ is contained in the interval 0 to $f_S/2$

In the bandlimited-IF application discussed in the main text, the 500- to 550-kHz IF signal fits neatly within the '+' band above $5f_S$ (Figure 19.8). The IF signal therefore preserves its spectral order as it shifts – aliases – to the sampled-data band of 0 to 50 kHz. Knowing that the input originated in the band marked M = 5 and sign = '+' allows you to reconstruct the input exactly.

For the bandlimited-IF case, you simply translate the sampled-data spectrum upwards by 500 kHz. You can similarly reconstruct the spectrum of any '–' band by reversing the order of the sampled-data spectrum and making the appropriate frequency translation. Inputs confined to the 0 to $f_S/2$ band satisfy the requirements for Nyquist sampling as stated earlier. The bands above $f_S/2$ are called undersampled bands, although the same sampling principles apply for them as for Nyquist sampling. In either case a faster converter – one with a higher sampling rate – widens the bands and allows accurate sampling with less stringent bandlimiting of the input signal.

Track/hold, sample/hold, or sampling converter?

The terms track/hold and sample/hold appear almost interchangeably in data-acquisition literature. Both amplifier types store a signal's amplitude on a capacitor, and both have an analog switch that isolates the capacitor. The difference between them lies in the closed-switch interval. Corresponding to the sample or the track mode, this interval is short and fixed in a S/H amp but of arbitrary length in a T/H amp.

Drive circuits for the switch provide a practical distinction between the devices. S/H amps, for instance, are usually high-speed signal blocks used primarily in RF applications. The amp's edge-sensitive control input triggers an acquisition interval that is followed rapidly by an automatic return to the hold mode. T/H amps, on the other hand, have a level-sensitive input that allows a control bit to shift the device between track and hold modes as the application requires. Thus, T/H amps can sample but S/H amps can't track. The term track/hold is correct for most of the devices used to deglitch D/A converters or condition the input signals for A/D converters.

An A/D converter can, for a given level of accuracy and resolution, digitize much higher frequencies when preceded by a T/H amp or a S/H amp. Some manufacturers combine the convert and hold functions as a sampling converter, and fabricate it in a single IC or as a multichip device in a single package. These products simplify design by eliminating the need for an explicit accounting of track/hold errors. As a further bonus, the data sheets usually specify ac performance in addition to the dc parameters normally associated with an A/D converter.

Acknowledgement

The authors wish to recognize the contributions and recommendations of many individuals at Maxim, especially Charlie Allen, Bruce Furman, Greg Schaffer, and Dave Watson.

20 Apply fundamentals to avoid surprises with sampled systems

Sense and nonsense about sampled systems

This primer outlines some of the fundamental aspects of sampled analog data. The author points out that the amplitude error in an 11 kHz sinewave sampled at around 44 kHz can be almost 30%. Note however that this worst-case error relates only to the measured (*peak*) value; the rms value of the samples over any one cycle of the 11 kHz sine wave is independent of the phasing of the samples. By the same token, a single sample taken at the peak and lasting a single clock period is subject to a 45 degree delay, whilst two successive samples each of 0.707 times the peak amplitude are in fact also centred on a point 45 degrees after the peak.

Gerard C A Fonte
The PAK Engineers

It is a mystery to me how basic concepts can get so twisted around that they become nearly meaningless. In this new world of digital sampling systems, you would think that the Nyquist theorem would be as familiar as Ohm's law. Nevertheless, as time passes, I see more and more people and publications state the theorem incorrectly and misunderstand its implications. The results of these misunderstandings are circuits that don't work well and designs that are inappropriate.

Simply stated, the Nyquist theorem tells you the minimum sampling rate needed to reconstruct a signal's frequency without aliasing. The important phrase here is 'reconstruct the frequency without aliasing'. (I will define aliasing later. Right now, think of it as the generation of unwanted frequencies.) Note that the theorem does *not* make any reference to the signal's phase or amplitude. In addition, the signal is, by implication, a sine wave – the only kind of signal that contains just one frequency. You can apply the Nyquist theorem to nonsinusoidal waves; however, doing so requires additional effort. Last, but far from least, the Nyquist theorem specifies what *must not* be included in a signal for proper reconstruction. In truth, this is probably the most important concept to understand.

First, to avoid any confusion, let's define what a sampled system is. For our purposes, a sampled system is one that breaks up a continuous (analog) signal into discrete samples and operates on the samples individually. The samples represent the signal's value at specific points in time. The definition includes things like analog-to-digital converters, switched-capacitor filters, and sample/hold circuits, as well as operations like Fourier analysis. Sampled systems *don't* include voltage-to-frequency converters, analog multipliers, phase-locked loops, or digital-to-analog converters.

Secondly, unless noted differently, we will assume that any input to our sampled system is in the form of a sine wave of a specified frequency.

Now let's analyze a simple A/D sampled system, ignoring the Nyquist theorem for the time being. We will use an ADC to cover the audio spectrum (20 Hz to 16 kHz). Suppose we first choose an A/D sampling rate of 10,000 samples/s. Figure 20.1 shows a 200-Hz signal sampled at 10,000 samples/s. Simple algebra tells us that there will be 50 samples in each cycle of a 200-Hz signal (10,000 samples/s divided by 200 cycles/s). The result looks pretty much like a sine wave.

At this point, we should pause for an important note about sampling. By definition, the value of a sample doesn't change until the next sample, whereupon the value changes instantly. The resulting waveform is a stepped wave consisting of horizontal lines and vertical lines. The horizontal lines are the result of the constancy of the sample. The vertical lines represent the instantaneous changes of sample values.

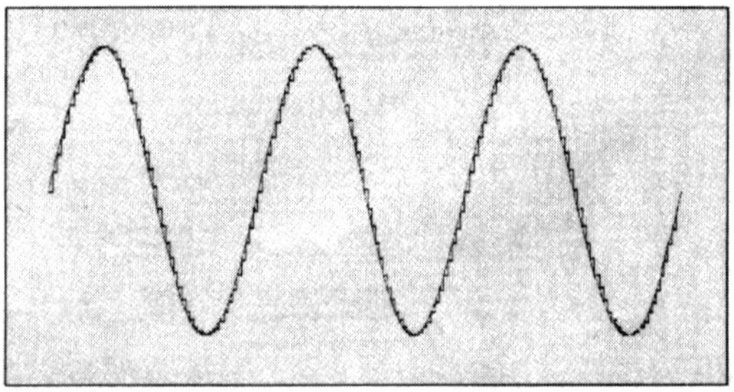

Figure 20.1 When you sample a 200-Hz signal at 10 k samples/s, you take 50 samples/cycle. The smooth curve is the input signal; the stepped curve is the sampled signal. Note that the sampled wave has nothing but vertical and horizontal segments

The sample values must initially be obtained from the waveform. (The Basic program used to create the figures is included as Listing 20.1. It can be very useful.) The sampled waveform must start at a point on the input sine wave, move horizontally until the next sample time, and then move vertically to intersect with the input sine wave again. This means that one right angle of the stepped wave lies on the input wave. The other right angle *must* lie to the right of the input wave (that is, later in time). You'll see the importance of this offset later.

Take a look at any of the figures. They all are step-function waveforms. Whenever you see 'reconstructed waveforms' that have diagonal lines, curved lines or lines that cross the input wave, you are looking at an error. I have seen more articles that are in error in this respect than ones that are correct. (Note: You *can* have diagonal and curved lines if you further process the sampled data, for example, in an analog filter.)

Listing 20.1 Program that created figures

```
10 REM   ***************************************************************
20 REM   *                                                             *
30 REM   *                 THE PAK ENGINEERS                           *
35 REM   *              SIMPLE-SAMPLING PROGRAM                        *
40 REM   *                                                             *
50 REM   *    3985 TONAWANDA CREEK ROAD,  E. AMHERST NY 14051          *
60 REM   *                                                             *
70 REM   *    GERARD C. A. FONTE       PRINCIPAL SCIENTIST             *
80 REM   *    DECEMBER 15, 1991        PROG. NAME "SINSAMP"            *
90 REM   *                                                             *
100 REM  *    NOTE: DIVIDE 360 BY NUMBER OF SAMPLES/CYCLE              *
110 REM  *          TO OBTAIN DEGREES PER STEP.                        *
120 REM  *          OFFSET IS IN DEGREES FROM START OF CYCLE.          *
130 REM  *                                                             *
140 REM  ***************************************************************
150 PRINT "ENTER THE NUMBER OF SINE WAVE CYCLES TO GENERATE.  USUALLY 1-10.";
160 INPUT CYCL  'REM HOLDS THE NUMBER OF SINE WAVES TO GENERATE
170 CLS         'REM CLEAR THE SCREEN OF JUNK
180 OVLY=0      'REM INITIALIZE OVERLAY COUNTER
190 KEY OFF     'REM GET RID OF THE KEY-DISPLAY AT BOTTOM OF SCREEN
200 SCREEN 2,1  'REM INITIALIZE FOR COLOR-GRAPHICS
210  OPTION BASE 1 'REM START ALL TABLES AT 1 INSTEAD OF 0
220 VIEW (1,1)-(635,130),0,1  'REM TOP PART OF DISPLAY SET FOR GRAPHICS
230 WINDOW (1,1.1)-(360*CYCL,-1.1) ' REM DEFINE GRAPHICS AREAS USER UNITS
240 REM USER UNITS ARE DEGREES AND TRIG VALUES OF +1.1 TO -1.1
250 CLS
260 CNTU$=""  'REM INITIALIZE SWITCH FOR DISCRETE OVERLAYS
270 LOCATE 20,1   'REM PLACE CURSOR FIRST COLUMN OF 20th LINE
280 PRINT "ENTER DEGREES PER SAMPLE AND OFFSET DEGREES                    "
290 INPUT DEGSTP, OFSET 'REM DEGREES-PER-STEP AND STARTING POINT
300 IF DEGSTP>0 THEN GOTO 330 'REM CHECK FOR REASONABLE VALUES
310 PRINT "DEGREES PER SAMPLE CANNOT BE 0 OR NEGATIVE"
320 GOTO 270
330 IF CNTU$="C" THEN GOTO 370 'REM ONLY NEED TO DRAW SINE WAVES ONCE
340 FOR X=1 TO (360*CYCL)-1 'REM LOOP THAT DRAWS SINE WAVES
350  LINE (X,SIN(X/57.29579))-(X+1,SIN((X+1)/57.29579)),11
360 NEXT X 'REM 57.29579 CONVERTS RADIANS TO DEGREES
370 FOR X=OFSET TO CYCL*360 STEP DEGSTP 'REM DRAWS THE SAMPLED WAVE
380 Z=INT(X)  'REM CAN'T USE FRACTIONS FOR GRAPHICS POINTS
390 LINE (Z,SIN(Z/57.29579))-(Z+DEGSTP,SIN(Z/57.29579)),15 'REM HORIZ. STEP
400 LINE (Z+DEGSTP,SIN(Z/57.29579))-(Z+DEGSTP,SIN((Z+DEGSTP)/57.29579)),15
410 NEXT X   'REM ABOVE LINE DRAWS VERTICAL PART OF STEP
420 REM  LINE (0,0)-(360*CYCL,0),15 'REM REMOVE REM AT START TO DRAW X AXIS
430 LOCATE 20,1 'REM CURSOR MOVED TO LINE 20 COL 1 (SCROLLING RUINS DISPLAY)
440 IF CNTU$="X" THEN GOTO 490 'REM IF MULTIPLE OVERLAY NO NEED TO ASK BELOW
450 PRINT "ENTER 'C' TO OVERLAY NEW SAMPLING WAVE-FORM OR 'X' FOR AUTO-OVERLAY";
460 INPUT CNTU$
470 IF CNTU$="C" THEN GOTO 270 'REM SINGLE DISCRETE OVERLAY
480 IF CNTU$<>"X" THEN GOTO 550 'REM MULT. AUTOMATIC OVERLAY WITH STEPPED OFFSET
490 OVLY=OVLY+9  'REM OFFSET INCREMENT =9 DEGREES.  ARBITRARY CHOICE. LOOKS OK
500 OFSET=OVLY   ' REM CHANGE OFFSET FOR MULTIPLE OVERLAY
510 IF OFSET <DEGSTP  THEN GOTO 370  'REM ONLY DO MULT. OVERLAY FOR 1 FULL STEP
520 CNTU$="" 'REM RESET VARIABLES WHEN COMPLETE
530 OVLY=0
540 GOTO 430 'REM GO BACK TO DO MORE IF DESIRED
550 END
```

First sample location is arbitrary

In addition, there is a subtle, yet significant, factor that hasn't yet been directly addressed – the position of the first sample point. The location of this point is completely arbitrary (Figure 20.2). It could lie anywhere on the wave because the input wave is not (and cannot be) synchronized with the sampling rate. (Synchronizing the sampling rate to the input signal is impossible in this case because the sampling rate is fixed, whereas the input signal frequency is variable and arbitrary.)

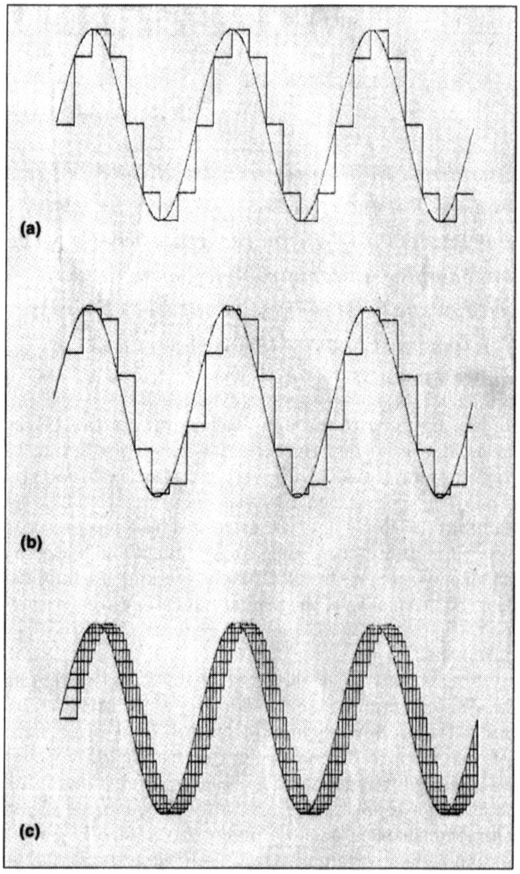

Figure 20.2 The stepped nature of the sampled wave is more apparent when there are only 8 samples/cycle. Note that you cannot control the location of the first sample point (and hence the location of any subsequent sample point). As the first sample point moves along the input wave, the results change significantly (compare (a) and (b)). In (c) the results of multiple sample positions are overlayed to indicate the range of possible results

The effect of this asynchronous nature is that there is an uncertainty in relating any particular point on the wave to a sample period. Examining an arbitrary point on the wave shows that its distance from a sample point cannot exceed ½ of the sample interval, but could be much less. This variation is a significant source of error that we will examine in more detail later.

Now, let's examine a 15-kHz signal sampled at 10,000 samples/s (Figure 20.3). Clearly the sampled result doesn't even come close to the input signal. But we really can't expect it to do so. To measure the length of an event (cycle), you have to have a clock rate faster than the event. But how much faster?

Take a closer look at Figure 20.3. What you see is the mixing of two frequencies (in this case, the input frequency of 15 kHz and the sampling rate of 10,000 samples/s). When you mix two frequencies, they heterodyne (or beat) and produce the sum and difference of the two signals. In this case, the sum is a 25-kHz signal and the difference is 5 kHz. We cannot reproduce the 25-kHz signal, so the resultant signal must be 5 kHz.

This is *'aliasing'* and it is definitely not good – we put in 15 kHz and get out 5 kHz. However, there is a relationship that's worth noting: If we sample at 11,000 samples/s, the heterodyne signals are 26 and 4 kHz, respectively. The faster we sample, the farther apart in frequency the heterodyne signals are. As we sample faster and faster, the aliasing frequency gets lower and lower until it hits 0 Hz. After that, it increases. (Work it out. Start with an input signal of 10,000 Hz and a sampling rate of 5000 samples/s and increase the sampling rate.) For the difference signal not to be a problem, it must lie above the frequency band of interest (which, in our case, extends to 16 kHz). Therefore, the sampling rate must be at least 32 kHz. Only in this way will every lower heterodyne be higher than the highest frequency of interest. This is the Nyquist limit.

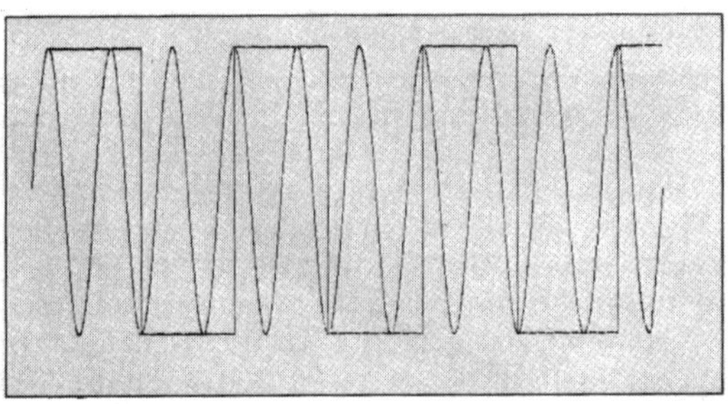

Figure 20.3 When you sample at a rate lower than the signal frequency, the result is aliasing – the sampled wave has a frequency different from that of the input. In this special case, the signal is at 15 kHz and the sampling rate is 10k samples/s. The (aliased) sampled wave is a 5-kHz square wave. Its frequency is ⅓ that of the input sine wave

In the simplified algebra that follows, the highest frequency of interest is represented by HF, the lower heterodyne is LH, and the sampling rate is SR. We want LH ≥ HF, where LH = SR – HF. So by substitution, SR – HF ≥ HF. Isolating SR gives SR ≥ HF + HF or SR ≥ 2HF.

The Nyquist theorem: not complex

In English: To prevent the lower heterodyne from aliasing, the sampling rate must be equal to or greater than twice the highest frequency of interest. (Actually, the sampling rate must be equal to or greater than twice the highest frequency present in the input signal with an amplitude that you consider significant, whether that frequency is of interest or not.) This is the Nyquist theorem.

Note what the theorem *doesn't* say. It doesn't say that there is no distortion in the reconstructed signal. It doesn't say that the reconstructed signal's amplitude doesn't change. It doesn't say that the reconstructed signal has no phase error. All it says is that the original sine-wave signal can be reproduced without adding aliasing frequencies.

Let's now look at the errors introduced by a perfect sampling system. (That is, we will not consider the practical realities of actual components and how they apply.) We will use a compact-disc system for illustration. The sampling rate is approximately 44,000 samples/s. Our signal will be an 11,000-Hz sine wave, which is clearly within the audio band. (Please note that 44 kHz is close to the 'playback' sampling rate. The 'recording' sampling rate is the rate of interest. Although 44k samples/s may have been the original playback rate when CDs were introduced, manufacturers nowadays often use a multiple of this rate to improve fidelity. Therefore, this error analysis may not apply to current CD recording systems.)

Let's first look at the possible amplitude error in an 11-kHz signal sampled at 44k samples/s. But, remember the asynchronous nature of the signals. We touched that topic earlier when we referred to Figure 20.2. The first sample can be at any point on the sine wave. The worst case occurs when the peak of the sine wave lies exactly midway between two sample points as shown in Figure 20.4. In this case, there is a very significant error in amplitude.

The easiest way to work out the amount of error is to convert the sampling rate to relative degrees of the sine wave. In this case, there are four samples per cycle. Therefore, the samples take place on the signal at 90° intervals. The worst case occurs when the peak (with a sine wave, the peak is defined as 90°) is exactly between two sample points. To learn the worst-case sample

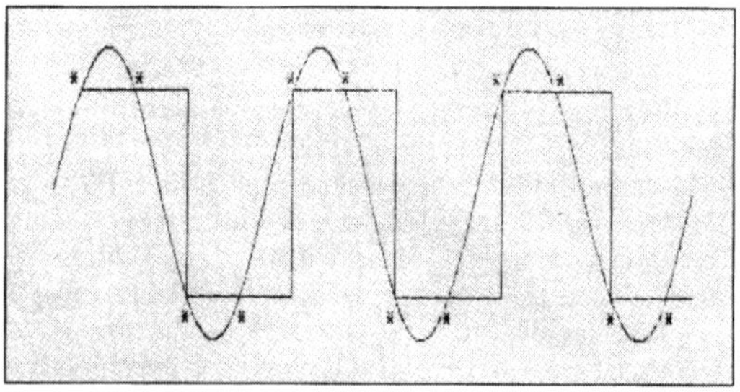

Figure 20.4 The sampled signal has an amplitude different from that of the input. When you take four samples/cycle, the worst-case amplitude error, shown here, is nearly 30%. (The asterisks represent the sample points)

locations, divide the degrees/sample (90) by 2 and subtract the result from the location of the peak in the cycle. The result is 45° (90 – (90/2) = 45), which means that the worst-case amplitude can be defined as the signal value at the 45° point on the sine wave. To figure out what this value is, take the sine of 45° – 0.7071. In this case, the amplitude of the signal could be in error by nearly 30% (1 – 0.7071). Note that the amplitude of the sampled signal must always be less than or equal to that of the input signal.

You have probably noted that the value of a sine wave near its peak varies less as a function of time than does the value at the point where the wave crosses the X axis. You might argue that, to determine the amplitude error, we should use a triangular wave, because a triangular wave has a constant slope. Were we to do so, the result would be a simple ratio. It is true that the results are different. It is false that a triangular wave should be used. The Nyquist theorem indicates the reason. If that clue doesn't help you figure out why you mustn't use a triangular wave, read on.

If you work out the same problem at the Nyquist limit (two samples per cycle), you will find that the worst case error is 90°. The sine of 90° is 1. In other words, you can have an amplitude error of 100%, which corresponds to no signal at all! In Figure 20.5, you can see how this large error occurs: the sample points happen to lie 180° apart – right on the X axis.

Of course, the error doesn't *have* to be 100% – but it might be. However, think about what happens if we sample a 11,010-Hz signal at 44k samples/s. Before, we had an exact integer ratio between the sampling rate and the signal; the samples always landed at exactly the same signal points in every cycle (clearly an ideal case). But, with a noninteger sampling ratio, each successive sine wave is sampled at different points. The error in amplitude will vary back and forth from the minimum possible (0) to the maximum possible (] 30% error). The effect is called (among many other things) amplitude modulation.

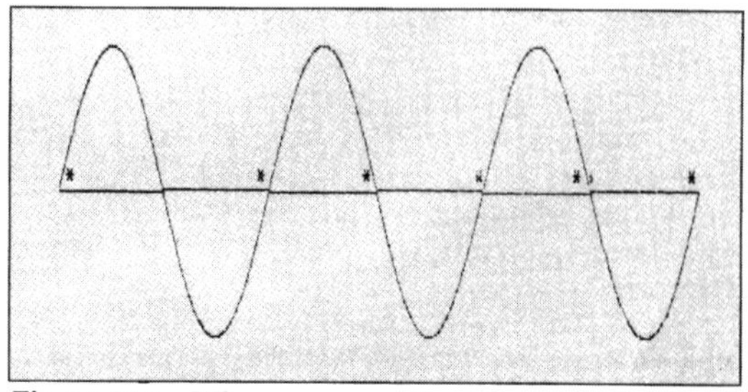

Figure 20.5 When you sample at exactly twice the input frequency, if the samples are positioned just right (perhaps 'just wrong' is a better description), the sampled wave can have an amplitude of zero, a 100% error

Amplitude modulation creeps in

The period of this modulation is the time it takes for the input signal and the sample points on the signal to return to the relative positions they had when they started. Our sampling ratio for a signal at 11,010 Hz is still virtually 4. This means that a 90° change in the signal will bring the sample points back in line with their starting position. So, multiply the input signal frequency by 4 (360°/90°) and heterodyne that with the sampling rate. In this case 11,010 × 4 = 44,040. Only the difference (40 Hz) is within our band of interest. If you sample faster, the modulation frequency goes up.

In summary, the worst amplitude error is equal to the sine of ½ the sampling rate measured in degrees of the input sine wave.

Maximum amplitude error = sin(°/sample × ½).

The amplitude modulation has a frequency equal to the difference between the sampling rate and the input frequency times the number of samples per input cycle.

Modulation frequency = sampling rate minus input frequency × (360/(°/sample)).

As the sampling rate increases, the amplitude error decreases and the modulation frequency increases.

Now look at the possible phase error. Normally the phase error isn't much of a concern. But examination of this special case will point out a frequency error, which is very much of interest.

Look back at Figure 20.1. Earlier we noted that the stepped waveform was offset to the right of the input signal, and we said that the offset was important. Because one corner of the step is always to the right of the input wave, there is a phase shift or delay. That such a delay exists should not come as a surprise; the sampling system takes a finite amount of time between samples.

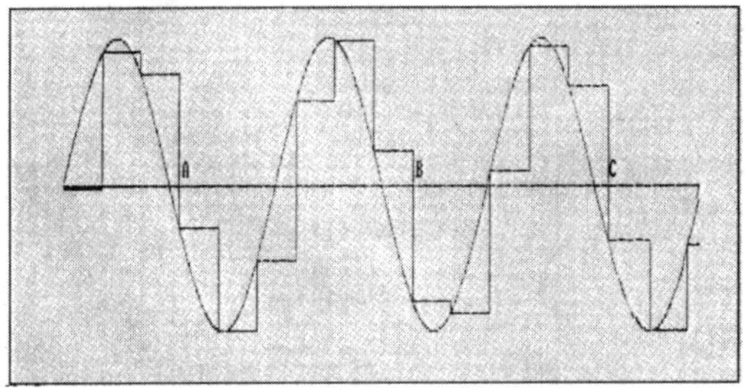

Figure 20.6 A sampled wave can have a period (and hence a frequency) different from that of the input signal. Here, period A–B is clearly longer than period B–C. Neither period is equal to that of the input wave

The phase error could be as large as one sample period. For example, look at the beginning of Figure 20.2a. At the start of the wave, the sampled wave and the input wave have the same value. There is no phase error at this point. However, as time progresses, the sine wave moves upward while the sampled wave doesn't change. There is a maximum difference between the input wave and sampled wave at the instant just before the second sample. This is the worst phase-error point.

Phase error is measured in degrees of the input waveform. So, to figure out the worst phase error, convert the sampling rate to °/sample. We did this before in Figure 20.2. With 8 samples/cycle, the worst phase error was 45°. What is the phase error at the Nyquist limit of 2 samples/cycle?

At times, phase errors should faze you

'But, what's so bad about phase error?' you might ask. After all, a phase error is really nothing more than a slight delay of the wave. It can't be that important, can it? In music reproduction, it's true that phase error doesn't have much impact. But, suppose you were digitally controlling a disk drive that used stepper motors. In that case, to calculate the location of the medium-to-head gap, you'd need to know the relationships among the voltages on several motor windings. Here, the phase error is critical to the operation of the drive, because the stepper motor is controlled by phase differences in the motor windings. Imagine how such a system would behave if it were sampled at the Nyquist limit.

To this point we have looked at phase errors in the special case of an integer number of samples per input cycle. Although there are no frequency errors, the scenario is unrealistic. It is virtually impossible to have an exact integer ratio of input frequency to sampling rate, and the non-integer ratio is what causes frequency errors. Regardless of the sampling rate, sampled systems *always* exhibit frequency errors. By increasing the sampling rate, you can reduce the errors, but you can't eliminate them. Although that may sound like heresy, remember what the Nyquist theorem says: it talks about reconstructing a frequency without aliasing. Because every sampled system involves discrete samples taken at time intervals, there will always be sources of error.

Look at Figure 20.6. Here is a wave that is sampled every 66° of the input waveform or 5.454545 . . . samples per cycle. Compare the time intervals of A to B and B to C. Clearly they are not the same. The time between A and B is substantially longer than the time between B and C. Because the period of a wave is defined as the time between corresponding points on successive iterations of the wave (in this case, the corresponding points are the X-axis crossings), wave A-B has a longer period than wave B-C. Because the periods are different, the frequencies must be different.

It isn't hard to see how this variation in the frequency of a (supposedly) constant-frequency signal arose. The explanation comes back to two concepts noted earlier: the offset in time of the sampled wave and the asynchronous nature of the sampling. As seen above, the offset causes a phase shift. We also noted earlier that the lack of synchronism between the signal and the samples causes variations in the points in the cycle at which the input wave is sampled. This variation corresponds to a variable phase shift. The variation of the phase shift can occur from one cycle of the input wave to the next. The cycle-to-cycle variation in the sample points results in a cycle-to-cycle variation (phase modulation) in the period and the frequency of the sampled signal.

There is another way of looking at the frequency variation. A sampling system cannot directly determine the frequency of a wave; it is not a counting system – it can't count cycles directly. So, to determine a wave's frequency, the system must measure the wave's period. But the precision of measuring periods depends upon how fast the samples are taken. If the period of a wave is 100 µs and the sample period is 22 µs, the best you can say is that the period is between 88 and 110 µs. Moreover, any uncertainty in measuring the period is equivalent to an uncertainty in determining the frequency.

The bottom line is that you don't know exactly when a wave starts. Therefore, the error in measuring the period of a single cycle of a wave can be as much as one sample period. So, if you sample four times per cycle (11-kHz input signal sampled at 44k samples/s), you have a possible error of 90° or ± 45° relative to the input. This corresponds to a frequency error of ± 45/360, or ± ⅛, or ± 12.5% of the input frequency. In short: given an 11-kHz sine wave and a sampling rate of 44k samples/s, your reconstructed wave would vary in frequency by ± 12.5%. The varying frequency is centered at 11 kHz, but ranges from 9.625 to 12.375 kHz.

These variations are brief, and on the average, the result is 11 kHz. But there is no provision for averaging in the simple sampled system. And averaging works only if there are a large number of input cycles to average. In music, there will be many cycles to average, but with many other signals, there won't be.

Most of the errors are inaudible

So, how do CD systems reproduce music so well? We've seen that an 11-kHz audio signal would be amplitude modulated by 30% and frequency shifted by as much as 12.5%. We don't hear the variations for several reasons. First of all, the highest note in music is played by the piccolo at 4608 Hz. Secondly, human hearing drops off very rapidly at frequencies above 4000 Hz. Typically, human ears are 20 dB less sensitive at 10 kHz than at 4 kHz. Just for reference, middle C on the piano is 256 Hz. Additionally, musical characteristics are related to a note's envelope and not just to its frequency components. (The envelope is the way the note develops over time. Compared with what we are interested in, the length of time is so long that we can assume the envelopes are steady-state signals.)

Let's look at the worst-case amplitude error of the piccolo. Remember that a piccolo is not a sine-wave generator. It has a fundamental frequency that allows 9.5 samples per cycle at 44k samples/s. To make the analysis easy, we'll use 10 samples per cycle, which corresponds to 36° per sample. Take the sine of 90 – (36/2). The result is 0.951, so the worst-case amplitude error is 1 – 0.951 or only 4.9% (0.436 dB). That small variation is inaudible.

The second harmonic (2× the fundamental) would be sampled 5 ×/cycle and would have an amplitude error as large as 19.1% or 1.84 dB. By itself, this error might be barely detectable. But the fundamental overwhelms it. Besides, the response of the ear is rapidly falling off at the second-harmonic frequency. (You can work out the other harmonics if you choose. Remember that the harmonics are usually also significantly lower in amplitude than the fundamental. The reduced amplitudes of the harmonics were not considered, but had they been, the variations over time in the amplitudes of the reconstructed harmonics would have appeared still less important. The reason is that you must compare the harmonics' amplitude variations with the amplitude of the fundamental, which is greater than the amplitude of any harmonic.)

But what about the phase and frequency errors? As noted previously, you can ignore the phase errors for music. The frequency error is ± ½ the sample period, which turns out to be ± 18° or ± 5% and seems significant. After all, a full-note change is only 6.25%. It appears that we should notice if the piccolo hit the wrong note. But remember that although any given cycle may be in error by 5%, a large number of cycles will average out to the input frequency. If the piccolo note lasts for about ¼ s (a fast note), more than 1000 cycles will be played. It turns out that the human ear is a pretty good averager at these frequencies. So, although a spectrum analyzer may show a band of frequencies, we hear the piccolo as a clear, single note. Like movies and TV, the CD system works because the technology takes human nature into account.

Now that you understand the Nyquist theorem, you can properly apply it to your designs. You now know that to prevent aliasing, you cannot allow any frequency into your sampled system that is more than half of the sampling rate. That is why you can't use a triangular wave to determine the amplitude error. A triangular wave consists of a fundamental frequency with loads of harmonics; it's nearly as bad as a square wave.

This caveat points to an extremely important rule. Always filter the input of a sampled system. Harmonics will be treated as signals and can be aliased into the sampled result. This is the *real* Nyquist rule, and why it defines the sampling rate.

Sampled systems can be a dream or a nightmare. Success depends upon a fundamental understanding of basic concepts, attention to detail and common sense. If you understand what a sampled system does, its workings and limitations become clear. Basing designs on inappropriate and worn-out phrases is an invitation to disaster.

21 Technique bolsters dc-to-audio converters

A/D and D/A converters based on delta modulation considered

A very useful introduction to oversampling data converters. These, though slower than SAR types and much slower than flash converters, offer greater resolution, integral and differential linearity, combined with low noise. Current types provide resolution to 20 or even 22 bits.

Anne Watson Swager
Regional Editor

Contrary to what you might have heard, oversampling data converters won't displace converters with traditional architectures in all applications. However, the ADCs and DACs currently available are particularly adept at two tasks: converting high-quality voiceband and audio analog and digital signals, and converting low-frequency analog signals with high resolution.

For audio applications, oversampling converters exhibit extremely low noise and distortion when converting small signals. For low-frequency applications, oversampling ADCs offer high resolution at low cost. And, unlike integrating and V/F converters, they don't require any external components. Both sets of converters benefit from the oversampling technique's inherent filtering. Unfortunately, this filtering causes one major drawback: the converters can require from hundreds of microseconds to hundreds of milliseconds to complete a conversion.

Figure 21.1 compares the response of a conventional sampling ADC to two 16-bit audio-range oversampling ADCs, all operating at 48 k samples/s. While the conventional converter's delay to a step input is around 20 μs (Figure 21.1b, trace B), the step and sinusoidal responses (Figure 21.1b and c, traces C and D) of two different oversampling converters exhibit delay times of 400 to 750 μs. The delays of Table 21.1's low-frequency converters can be as high as 400 ms.

This delay limits the converters' ability to multiplex their inputs and also can destabilize feedback control loops. Certain converters minimize this delay, as Crystal Semiconductor's new generation of dc converters has done, but oversampling converters operate more effectively on continuous streams of data than they do converting single events.

The inherent filtering defines the converters as continuous-time devices. Oversampling converters are essentially huge filters with analog (or digital) inputs and digital (or analog) outputs. If filtering is an advantage for your application, these converters are a potentially good choice. If filtering is a disadvantage – either because of the delay it causes or because you want to be able to change certain filter characteristics drastically – these converters are a bad choice: You can't take the filter out of an oversampling data converter.

Applying the name *'oversampling'* to these converters is somewhat imprecise. Oversampling is just one aspect of this conversion technique, which Crystal Semiconductor calls 'delta-sigma', Analog Devices and Motorola call 'sigma-delta', and Philips calls 'bit-stream'. Generally, you can use any data converter to oversample just by raising the sampling rate above the Nyquist rate. However, in this article, 'oversampling' refers to a technique that combines oversampling, noise-shaping, and digital filtering (see 'Oversampling converters in five languages').

An oversampling converter is the epitome of a mixed-signal device, using about 10% of circuit space for analog functions and 90% for digital ones. A number of the converters listed in Tables 21.1 and 21.2 separate the two functions and accomplish conversions using one analog and one digital device. In some cases, especially if you want to build high-dynamic-range DACs, manufacturers advocate using a building-block approach.

According to Craig Aine, an applications engineer with Philips/Signetics, separating the analog and digital portions of an oversampling converter reduces crosstalk, thereby improving the system's overall performance. Both Sony and Philips produce 1-bit DACs, which when combined with digital filters produce DACs with an overall resolution anywhere from 16 to 20 bits. Philips' TDA1547, when teamed with the noise-shaping

Table 21.1 Representative oversampling ADCs for low-frequency and high-dynamic-range applications

Manufacturer	Part number	Resolution (bits)	Differential linearity (max)	Integral linearity (max) (%)	Filter characteristics: 3-dB cutoff (Hz)	Filter characteristics: Settling time (msec)	Modulator sampling rate (Hz)	Output rate (Hz)	Input range (V)	Power supply (V)
Analog Devices	AD7701/03	16/20	±0.5 LSB no missing codes	±0.0015	0.1 to 10	120	16k	4k	0 to 2.5, or ±2.5	±5
	AD7710/11/12/13	8.5 to 21	no missing codes	±0.0015	2.62 to 262	400 to 4	20k (varies with gain)	10 to 1000	0 to 2.5, or ±2.5	5 to 10, or ±5
	AD79024	20 (90 to 115 dB dynamic range)	NS	±0.003	9.375 to 300	160 to 5	4M	1k	±2.5	±5
Crystal Semiconductor	CS5501/03	16/20	±0.5 LSB no missing codes	±0.0015	0.1 to 10	125	16k	4k	0 to 2.5, or ±2.5	±5
	CS5505/07 CS5506/08	16 20	±0.5 LSB no missing codes to 18 bits	±0.003	17	50	16.384k	20	0 to 2.5, or ±2.5	5 or ±5
	CS5322 and 23	24 (120 to 130 dB dynamic range)	NS	NS	25.7 to 412	16 to 0.125	23 to 375	62.5 to 1000	±10	±5
	CS5324	NS (120 dB dynamic range)	NS	NS	0 to 500	NA (see comments)	256k	32k	±10	±5 *
Sipex	SP4620	20 (110 dB dynamic range)	±0.25 LSB	±0.012	950	32	512k	2k	±2.5	±5

Notes:
NS=not specified
NA=not applicable

block inside the SAA7350 and a separate digital filter, features a typical dynamic range of 108 dB and a nonlinearity of 0.2 dB, with inputs from –60 to –120 dB.

At the other end of the circuit-complexity spectrum, manufacturers are integrating these converters into even larger devices, such as codecs, echo-canceling modems, and DSP system chips. Both Motorola's DSP56156 ($105 (100)) and Analog Devices ADSP-21msp50 ($93 (100)) integrate oversampling codecs with 16-bit DSP μPs.

Give pros and cons their due

When manufacturers first began marketing these converters a few years ago, the front pages of the data sheets listed an impressive set of advantages:

- No S/H amplifier required
- Digital instead of analog filtering
- No laser trimming or factory calibration required

Table 21.2 Representative oversampling converters for audio and voiceband applications

Manufacturer	Part number	Description	Dynamic range (dB)	S/N ratio + distortion (THD+noise) (dB)	Passband frequency (–3-dB corner)	Passband ripple (dB)	Maximum input rate/ output rate (kHz)	Power supply
Analog Devices	AD1879	18-bit stereo ADC	103 typ	98 typ	21.7 kHz	0.001 typ	55	±5
	AD28msp01	16-bit modem front end	NS	80 typ	200 Hz to 4.8 kHz	0.1 max	9.6	5
Crystal Semiconductor	CS4328	18-bit stereo DAC	95 typ 93 min	94 typ 92 min	0 to 23.5	0.001 max	48	±5
	CS5317	16-bit voiceband ADC	84 typ 78 min	THD only=80 typ 72 min	5	NS	20	±5
	CS5326/7/8/9	16- and 18-bit stereo ADCs	95 to 97 typ 92 to 94 min	92 to 94 typ 90 to 92 min	21.6 to 23.5	0.001 max	50	±5
	CS5336/8/9	16-bit stereo ADCs	95 typ 92 min	92 typ 90 min	0 to 22/0 to 24	0.01 max	50	±5
	CS5349	16-bit stereo ADC	90 typ 88 min	85 min 87 typ	0 to 24 kHz	0.01 max	50	5
Motorola	DSP56ADC16	16-bit ADC	96 typ	90 typ	0 to 45.5 kHz	0.001 max	100	5
Philips/ Signetics	SAA7323	16-bit stereo DAC	93 min	–90 max –95 typ	14.5 to 21.8 kHz	0.035 max	48	5
	SAA7350	1-bit DAC	98 typ	–93 max –96 typ	NA	NA	16 to 53	5
	TDA1547	1-bit DAC	108 typ	96 min 101 typ	NA	NA	NA	±5
Sony	CXD2552Q	1-bit DAC	NS	96 min (SNR only)	NA	NA	NS	5
Texas Instruments	TMS320AD50	Voiceband codec	80 min 84 typ	62 min	20	0.5 max	40	5
UltraAnalog	ADC20048	20-bit ADC	108 typ	–96 typ	21.5 or 23.5 kHz	0.00087 max	44.1 or 48	±15 and 5

Notes:
NS=not specified
NA=not applicable

Power dissipation (mW max)	Package	Price (100)	Comments
40	20-pin DIP or SOIC	$15/$18	Filter specs based on a nominal clock of 4.096 MHz. Both parts are second sources for the CS5501/03.
40	24-pin double-wide DIP or SOIC	$15/$16/$14/$17	Family of signal-conditioning ADCs with programmable-gain front ends for sampling various types of transducer outputs. Nominal clock rate is 10 MHz. Filter cutoff and gain settings determine overall resolution, output rate, and filter settling time. 7713 draws only 5.5 mW.
50	28-pin DIP, 44-pin quad flat pack	$52	Designed primarily for biomedical and other high-dynamic-range applications.
40	20-pin DIP or SOIC	$18.20/$27.70	Filter specs based on nominal clock of 4.096 MHz.
4.5	20- and 24-pin DIPs and SOICs	$15.70 $18 $11.20 $13.50	CS5507/08 have single-channel differential inputs. CS5505/06 have four pseudo-differential inputs. Converters will either continuously convert, or only sample on demand.
100	2-chip set, 28-pin LCC	$269.70	Five cutoff frequencies selectable over the specified range. Suits seismic and high-dynamic-range applications.
180	28-pin LCC	$217	User must add two stages of external filtering. High-dynamic-range applications. Nominal clock rate is 1.024 MHz.
290	28-pin double-wide DIP	$295	Nominal clock rate is 1.024 MHz. Designed for biomedical-instrumentation, seismic, and other high-dynamic-range applications.

- Inherent linearity
- Nonexistent or lowered antialiasing requirements.

The first two of these claims are easy to substantiate, but the rest require further explanation. No S/H amplifier is required because of the high oversampling and averaging nature of the technique. A particular ADC's output results not from one sample taken at one instant in time, but from a large number of samples taken over a period of

Power dissipation (mW)	Package	Price (100)	Comments
900 typ	28-pin DIP	$42	Includes stable voltage reference and two 1-bit, fifth-order modulators.
500 typ	24-pin DIP or SOIC	$21.45	Specialized for echo-canceling. Also operates at other standard modem rates.
600 typ	28-pin DIP or SOIC	$37	Serial port supports four different interface modes.
300 max	18-pin DIP or 20-pin SOIC	$29.40	Intended for applications such as modems, sonar, and voice-recognition systems. Includes on-chip phase-lock loop.
450 typ	28-pin double-wide DIP	$70.56/$80.90/ $84.70/$97	The CS5326/8 have filters with CD requirements.
400 typ	28-pin double-wide DIP 28-pin SOIC	$56.40	Same modulator in CS5326 family, but different digital filter. CS5336 has 22-kHz cutoff, 8/9 have 24-kHz cutoff.
425 max	28-pin double-wide DIP	$56.40	Includes 2V reference.
400 max	20-pin DIP	$19.80	Specifications listed for 16-bit output. 12-bit output is also obtainable at a 400-kHz rate.
300 typ	44-pin quad flat pack	$16.51	Lower performance version, the SAA7322, is also available.
375 typ	44-pin quad flat pack	$26	Requires either front-end DSP μP or digital filter to achieve specs listed. External clock typically equals 256 or 384X sampling frequency, which is typically 44.1 kHz, but can range from 16 to 53 kHz.
800 typ	32-pin shrink DIP	$17.50	Performance specs achieved by using a separate digital filter and SAA7350 noise shaper's output.
400 max	44-pin quad flat pack	$7.50	S/N ratio listed requires an accompanying digital filter, such as the CXD1144BP ($36).
45 typ	25-pin LCC	$10	Device is also available as a megacell module in the company's DSP cell library.
2800 typ	2×3-in. module plus 48-pin DIP	$164	Multibit architecture in the analog front end.

time. Also, digital filters do have redeeming characteristics when compared with analog filters – linear-phase digital filters are easier to design than comparable analog filters. Digital filters don't rely on precise component matching.

The advantage of no factory trimming or calibration doesn't apply to all oversampling converters. However, these converters contain far fewer precision analog components and don't require strings of matched resistors or capacitors. In some cases, dc converters include autocalibration circuits. However, these circuits are completely independent of the oversampling architecture itself. And ADCs designed to perform in the audio range may require trimming if you want to achieve accurate offset and gain specs at data rates higher than those of the dc converters.

1-bit DAC has ideal linearity

The linearity advantage stems from the fact that most oversampling converters perform a 1-bit D/A conversion in the feedback of the analog-modulator block. This 1-bit converter switches between only two reference points. As Dave Welland, a designer of oversampling converters at Analog Devices, explains it, the modulator picks two points and the filter draws a straight line between them. Thus, 1-bit DACs exhibit ideal integral linearity. If the remainder of the converter's design is perfect, the overall integral linearity of a 1-bit architecture will be close to ideal.

Although the majority of oversampling converters are based on 1-bit architectures, multibit samplers, such as the Sipex's SP4620, also exist. You can't make the same case for inherent integral linearity for multibit systems as you can for monobit systems. A multibit DAC just doesn't have the ideal integral linearity that a 1-bit DAC does. The integral nonlinearity of the SP4620, ± 0.012 at full scale, is higher than that of the other converters in Table 21.1. This higher number is due to the device's performance at high signal levels: The S/N-plus-distortion performance tapers off between 84 and 90 dB for low-frequency signals greater than –20 dB. However, according to Eric Blom, the converter's designer, this device's linearity is very close to that of a 1-bit architecture at lower signal levels.

Oversampling converters' differential nonlinearity and monotonicity are more closely related to the noise shaping and filtering (averaging) functions than with the resolution of the modulator feedback. Thus, all of these converters, whether mono- or multibit, feature differential nonlinearity better than ± 0.5 LSB. The SP4620's maximum differential nonlinearity is ± 0.25 LSB. Manufacturers arrive at these numbers by using statistical measures, such as histograms, rather

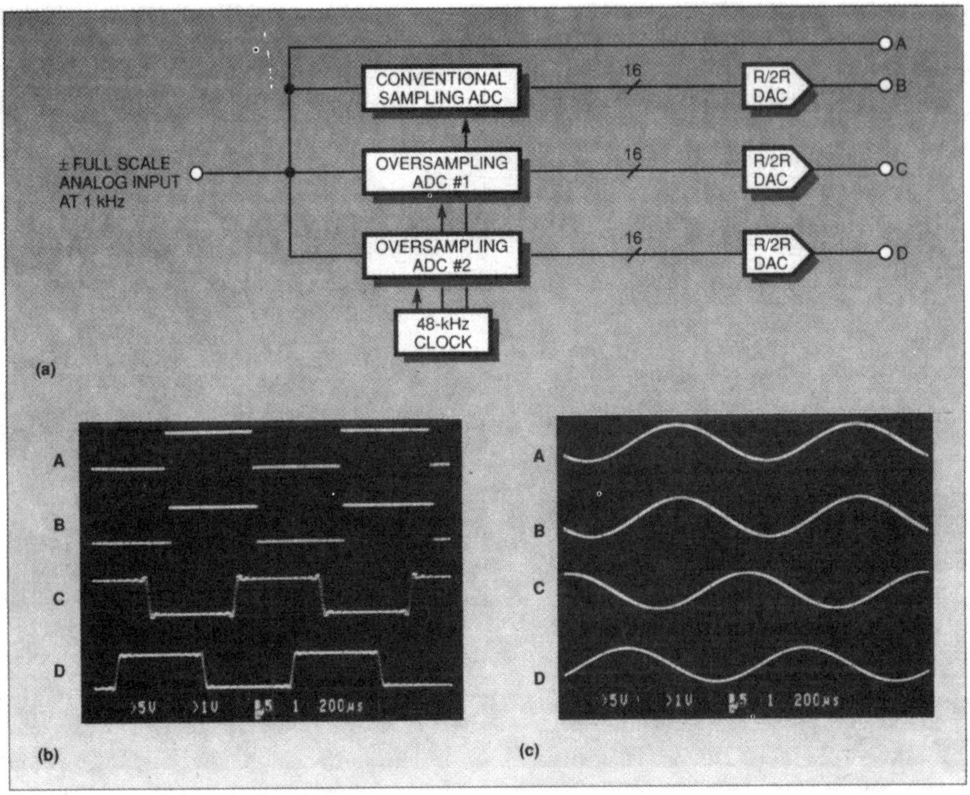

Figure 21.1 Oversampling ADCs can't easily multiplex between two inputs because of the time required for an entire conversion. Compared with a conventional sampling ADC (a), two different oversampling converters exhibit two different delayed responses to step (b) and sine-wave inputs (c). The horizontal scale for these two photos is 200 μs/div and the vertical scale was normalized so that one division equals ± full scale. (Photos courtesy Analog Devices)

than the more traditional measurements normally performed on successive-approximation converters.

Be careful not to confuse multibit architectures with something called the MASH architecture. MASH refers to multistage noise shaping, not to multibit feedback. This architecture is most common in audio converters produced by companies such as Matsushita and Sanyo for use in their own audio products.

The modulator portion of the converter isn't the only element of the converter that can affect linearity. Every digital filter in these converters has an accumulator. If the word length of the device's accumulator isn't long enough to handle overflow bits, the filter will truncate the result. This truncation can cause missing codes.

Don't ignore aliasing

The final advantage, and one of the most misunderstood features of these converters, is the reduced or nonexistent antialiasing-filter requirements of oversampling ADCs. Although it's true that the internal digital filter will remove a certain band of unwanted frequencies – the filters' job is to get rid of out-of-band quantization noise – it's also true that oversampling converters are *not* completely free from aliasing problems.

In general, the digital filter does nothing to remove the input spectrum that repeats at integer multiples of the internal sampling rate. For example, those converters with a 10-Hz passband sampled at a 16-kHz rate, will pass to their outputs and not attenuate any components of the input signal near 16, 32, and 48 kHz. In some cases, the converter will also alias input frequencies greater than one half the output word rate.

Unfortunately, because each of these converters' digital filter is different, it's impossible to generalize what type of external filtering will be required with each converter for every application. In some cases, no filter will be necessary. In others, a simple RC filter will provide the necessary attenuation.

The oversampling nature of these converters can also relax input ADC filtering requirements. As the sampling rate increases, the repeated input spectrums spread out more in frequency, and you can use a filter with an even more gradual rolloff to attenuate any unwanted components.

Front-end filtering not only prevents aliasing but also prevents noise spikes riding on the input signal from saturating the analog modulator and digital filter. An input capacitor to ground also prevents those converters with switched-capacitor inputs from kicking spikes back to the driving circuitry.

Concentrate on the filter

Thus, these converters don't let you completely forget about the effects of aliasing, and they even have a few disadvantages of their own:

- Antialiasing filter considerations are still required
- Filter delay disrupts multiplexing
- Filter relay disrupts loop stability.

Note that the word 'filter' is common to all the entries in this list. Besides being familiar with an oversampling converter's architecture, understanding the filtering inherent to the conversion technique is the key to properly selecting and applying one of these converters. The oversampling ADCs in Figure 21.1 have the same speed, resolution, and bandwidth, but the differences in their filters caused very different delays.

Every filter inside every oversampling converter is unique, and is not necessarily a brickwall lowpass filter. The converters accomplish the filtering in multiple stages, each of which can have different characteristics. For example, some converters combine comb filters, which have a sin(x)/x response, with a more conventional lowpass filter. The notches of the comb filters in the CS5505 and AD771x families of low-frequency converters (Figure 21.2) help to reject 50- and 60-Hz line frequencies. The CS5505 filter's notch at 60 Hz provides a minimum attenuation of 58 dB over line fluctuations of 1%.

Evaluating the filter characteristics of a particular oversampling converter is the most important part of data sheet scrutiny. Most of Table 21.2's audio converters have filter bandwidths approximately equal to one half the input or output rate. However, the same is not necessarily true for the low-frequency ADCs in Table 21.1. It's particularly important to realize that either the quoted output data rate or modulator sampling rate divided by two is *not* necessarily equivalent to the device's input-signal bandwidth, as it would be in a traditional converter. There are separate specifications for the filter's 3-dB bandwidth, the modulator's sampling rate, and the output rate.

For example, the CS5501/3 and AD5501/3 data sheets quote the filter cutoff as the clock frequency divided by 409,600. Using the recommended clock rate of 4.096 MHz, the maximum bandwidth of the device is 10 Hz. If you based your assumption of these devices' bandwidths on a Nyquist treatment of the maximum output rate of the device, which is 4 kHz, you would erroneously think that the bandwidth is 2 kHz, not the actual value of 10 Hz. In one low-frequency case, the SP4620, the 3-dB bandwidth of the filter happens to be close to half of the device's 2-kHz output rate. However, this Nyquist-type relationship between the filter cutoff and output rate is rare.

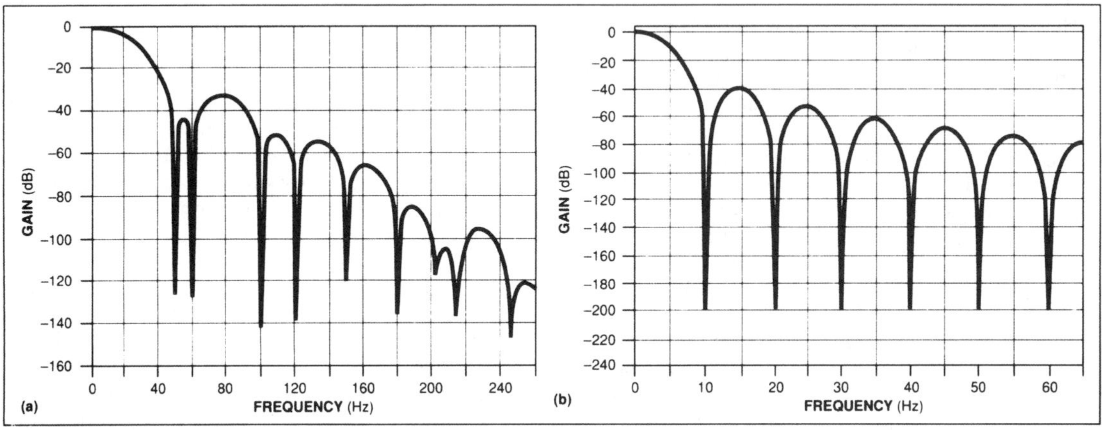

Figure 21.2 No two oversampling converters' filters are alike. Both Crystal Semiconductor's CS5505 (a) and Analog Devices' AD7710 (b) ADCs use comb filters, but each has slightly different notch characteristics

The filters' cutoff points scale with the master clock frequency that you provide, but in many cases the converter's designer has optimized the device's performance for a particular frequency. Although most of these converters were primarily designed to operate over a limited range of clock frequencies – an attribute common to almost all oversampling converters, and particularly true of audio converters – many devices in Table 21.1 let you program the filter characteristics. As the filter cutoff changes, so does the filter's settling time and the converter's resolution.

For example, the AD771x family of signal-conditioning ADCs lets you program the filter cutoff from 2.62 to 262 Hz via a 12-bit control register. Data-sheet tables relate the internal $(\sin(x)/x)^3$ filters' first notch to the 3-dB frequency and the final resolution. This converter family also contains programmable-gain amplifiers, so that another table relates the above characteristics to your chosen gain. At the maximum master-clock frequency of 10 MHz, the minimum cutoff frequency is 2.62 Hz, and the maximum programmable cutoff frequency is 262 Hz. The effective resolution of the device – measured as the magnitude of the output rms noise to a full-scale input – varies between 21 and 8 bits as you change these cutoff points. The resolution is highest at the lowest cutoff frequency, and lowest at the highest cutoff frequency.

The filters resident in the audio converters listed in Table 21.2 are fairly similar in their 3-dB cutoff characteristics because they're all compatible with digital-audio requirements. The filters exhibit a standard lowpass response with 3-dB frequencies around 20 kHz, and are compatible with output rates from 32 to 48 kHz.

All digital filters perform repetitive multiply and accumulate cycles to complete an entire conversion. These cycles cause the delay that Figure 21.1 illustrates. The data sheets specify the delay in terms of filter settling time. For these converters, the filter settling time is more of a last-output-change to present-out-put-change number, or simply the inverse of the output data rate.

Figure 21.1's photos make a strong point, not only about the converters' delay or phase shift, but also their step response. The photos show that the step response of each oversampling ADC exhibits different rise time and ringing characteristics. These varying characteristics stem from each converter's digital filter, which has different passband ripple, transition, and stopband characteristics. Again, these photos underscore the need to evaluate each converter's filter characteristics.

These delays don't cause a particular problem for audio or other systems that typically work from continuous streams of data. However, they do cause substantial phase shifts at relatively low frequencies and can cause major problems in feedback control systems. In Figure 21.1's examples, one converter's sine-wave response shows slightly less than 180 degrees of phase shift at 1 kHz, while the other shows more than 180 degrees. You can almost guarantee instability unless you factor these delays into your loop analysis.

By using a different filter, it is possible to minimize the delay. Crystal Semiconductor addresses the multiplexing limitation in its second generation of low-frequency parts. By 'softening' its predecessor's filter characteristics, the CS5505 family achieves settling times of 50 ms. The IC also includes a convert-control pin and a data-ready flag similar to a successive-approximation converter. The device also includes a 4-channel multiplexer.

Converter vendors also argue that the commercial availability of ADCs in the $10 to $20 range challenges the traditional data-acquisition design approach. If you use one ADC per channel, the multiplexing problem doesn't exist.

Idle tones exist

Although the multiplexing issue is an important system consideration, oversampling converters exhibit other subtle effects that can influence your system's performance. Low-frequency noise, known as spurious or idle tones, is a potential problem unique to the oversampling-converter architecture. Like any other electronic component, the behavior of oversampling converters isn't ideal. The converters produce a large amount of quantization noise that ideally has no relationship to the input signal. In reality, however, the nature of the input can have an affect on the performance of the converter's analog modulator.

For example, an input signal whose dc level lies very close to the transition point of the modulator's sampler creates a very small-error feedback voltage. The modulator has to integrate many of these small errors to change states. A very low-frequency signal results that persists through the digital filter to the output.

This signal manifests itself as a low-frequency tone in the passband. The converter's designer can reduce the production of unwanted tones by designing higher-order modulators, but higher-order modulators extract their own design penalties because of instability. Multi-bit architectures generally have lower tonal effects than do single-bit architectures.

Unfortunately, there isn't an easy way to evaluate the existence or level of these tones from the data sheet. Depending on their level, these idle tones may not be a problem for your application. Also, you may be able to shift your dc operating point away from the transition region.

Once aware of their drawbacks, you can decide whether oversampling converters will benefit your dc-to-audio converter requirements. However, don't expect the designers of these converters to forever be content with just dc and audio frequencies. According to Randy Skinner, a marketing manager at Burr-Brown, which will soon offer audio DACs based on this technique, 'everything that can be invented using oversampling technology, hasn't been invented yet', and oversampling converters will invariably work with wider bandwidths in the future.

Oversampling converters in five languages

After working with traditional converters, it's probably difficult to place faith in a device that contains a 1-bit or low-resolution DAC. However, oversampling converters are close relatives of some familiar devices, such as dual-slope integrating converters and V/F converters, all of which use charge-balancing techniques. Figure 21.3 describes the conversion in five levels, from block diagram to s- and z- domain equations.

Oversampling-conversion techniques rely heavily on one basic fact of all data converters: the conversion process itself introduces inherent error – the quantization noise. The oversampling technique capitalizes on quantization noise. In fact, it intentionally produces a gross amount of the noise, shapes it, and then uses huge filters to dispose of it. ADCs and DACs perform the same basic steps, but in reverse.

For example, the two-step A/D conversion consists of an analog modulator/noise shaper front end and a digital filter (Figure 21.3). Two crucial actions take place in the modulator block: oversampling and noise shaping. The combination of a large data rate compared with the bandwidth of the signal and the integrating architecture of the modulator produce noise that increases with frequency. As the s-domain equations and frequency-domain plots show, the modulator block essentially lowpass filters the signal and highpass filters the noise.

The digital-filter block also has two roles: to remove the high-frequency noise produced in the modulator, and to decimate the oversampled bit stream down to the converter's specified output rate.

This simplistic presentation belies the complexity of both the modulator and digital filter. Most real converters contain 2nd-order or higher-order modulators. Instead of the more common 1-bit feedback DAC in the modulator, some converters use multibit feedback. The modulator order, the number of feedback bits, and the oversampling rate all determine the converter's achievable S/N ratio. Unfortunately, converters with higher-order loops can be unstable and susceptible to overloads, so their designs usually include circuitry to sense overload-induced instability.

Just as no two modulators are alike, no two digital filters are alike. These converters usually

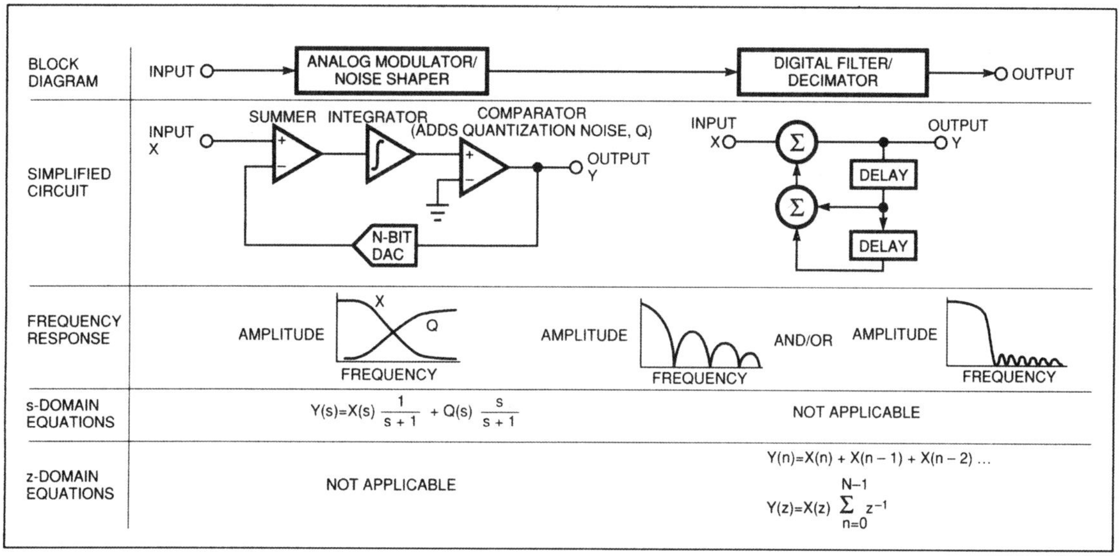

Figure 21.3 Oversampling data conversion requires two basic steps: analog modulation and digital filtering. You can describe these functions using simplified circuit diagrams, frequency-response plots, and s- and z-domain equations

perform the filtering in many stages: those stages can be FIR (finite impulse response), IIR (infinite impulse response) filters, or both. The overall filter characteristics determine the delay, passband, and stopband levels, as well as antialiasing requirements.

Manufacturers of oversampling converters

Analog Devices
Box 9106
Norwood, MA 02062
(617) 329–4700
FAX (617) 326–8703

Burr-Brown Corp
6730 S Tucson Blvd
Tucson, AZ 85734
(602) 746–1111
FAX (602) 746–7401
Randy Skinner

Crystal Semiconductor Corp
4210 S Industrial Dr
Austin, TX 78744
(512) 445–7222
FAX (512) 445–7581
Brad Fluke, audio products
Mike Paquette, low-frequency products

Motorola Inc, DSP Operations
6501 William Cannon Dr W
Austin, TX 78735
(512) 891–2030
FAX (512) 891–2947
DSP Marketing

Philips Components-Signetics
811 E Arques Ave
Sunnyvale, CA 94088
(408) 991–2000
FAX (408) 991–2069
Craig Aine or Ken Wong

Sipex Corp
22 Linnell Circle
Billerica, MA 01821
(508) 671–1944
FAX (508) 670–9001
Bill Lundgren

Sony Corp
10833 Valley View St
Cypress, CA 90630
(714) 220–9100
FAX (714) 229–4333

Texas Instruments Inc
8390 LBJ Freeway
Dallas, TX 75265
(214) 997–3389
FAX (214) 997–5962

UltraAnalog Inc
47747 Warm Springs Blvd
Fremont, CA 94539
(415) 657–2227
FAX (415) 657–4225

22 High-resolution DAC uses coarse/fine control

High resolution comes cheap, but high absolute accuracy still costs

This scheme will provide you with a cheap way of getting high resolution with a couple of inexpensive 8 bit DACs. The fine DAC can fill in between steps of the coarse DAC, but remember that the latter will still only have 8 bit accuracy. So to avoid missing codes, arrange the fine DAC to cover, say, two steps of the coarse DAC, giving 15 bit resolution. Splendid for fine tweaking in a closed loop system, but not for setting an accurate open loop output, where there is no substitute for absolute accuracy.

Terence S Finnegan
Carlisle, UK

The circuit in Figure 22.1 is a high-resolution DAC that provides symmetrical bipolar output current. The resolution can extend to 21 bits with the appropriate components. The design uses two DACs that operate from a common reference in a simple coarse/fine control arrangement. The circuit resistively divides the output current from the fine DAC and then adds this current to the coarse DAC. The circuit's accuracy and resolution are therefore controlled by passive resistors and are independent of the active elements.

The two DACs provide the coarse and fine control through a 4-transistor Wilson current mirror. DAC A provides the coarse control, and its current mirrors connect directly to the output. DAC B provides fine control, and this DAC's output current affects the output only by the resistor ratio $R_1/(R_1 + R_2)$. Because both DACs operate in push-pull between the input and output circuits, the operation is symmetrical about zero when the input code to both DACs is 80_{HEX}. The output current will then vary symmetrically about zero between $+ I_{OUT(MAX)}$ and $-I_{OUT(MAX)}$ as you vary the input code about 80_{HEX}, between 0 and FF_{HEX}.

Both DACs can operate at the same reference current. This symmetry minimizes the errors due to DAC leakage and zero-scale currents. Ultimate accuracy is limited by the resistor ratio, DAC voltage offset V_{OS}, and the differential V_{BE} of Q_1 and Q_2.

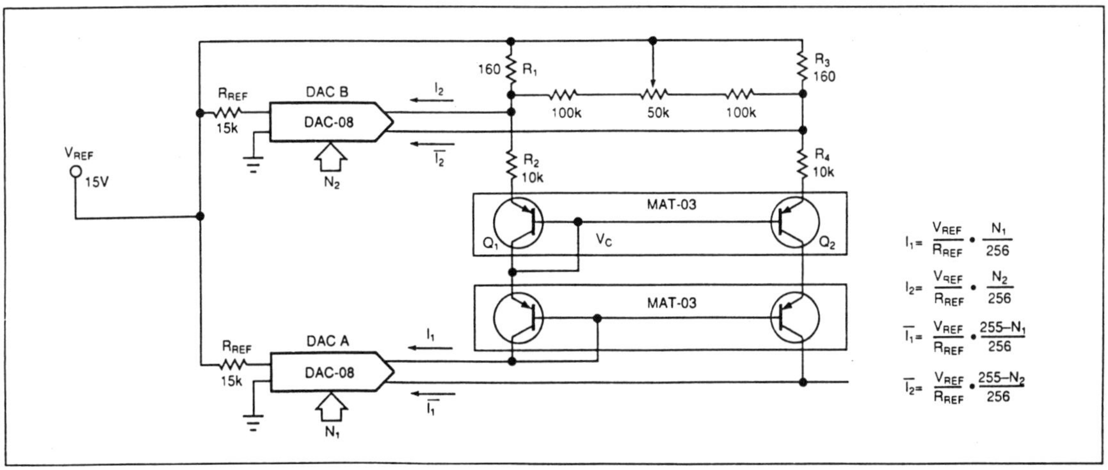

Figure 22.1 DACs A and B provide coarse and fine control through a Wilson current mirror to implement an overall DAC with 14 bits of resolution. The design's maximum possible resolution is 21 bits

You can derive the equation for output current by first equating the voltage drops in the left-hand and right-hand resistor chains up to the common voltage point, V_C. Substituting the right-hand side of the equations shown in Figure 22.1 for I_1, I_1, I_2, and I_2, yields the equation

$$I_{OUT} = V_{OS} + \frac{I_{REF}}{256}\begin{bmatrix} N_1(R_1 + R_2 + R_3 + R_4) \\ + N_2(R_1 + R_3) \\ - 255(2R_3 + R_4) \end{bmatrix},$$

where N_1 and N_2 are the decimal input codes for the coarse and fine DACs. If you let $R_3 = R_1$, $R_4 = R_2$, and set $R_2/R_1 = k$ and $V_{REF}/R_{REF} = I_{REF}$, the expression becomes

$$I_{OUT} = \frac{V_{OS}}{R_1(1 + k)} + \frac{I_{REF}}{256(1 + k)}\begin{bmatrix} 2N_1(1 + k) \\ +2N_2 - 255(2 + k) \end{bmatrix}.$$

The ratio k controls the overall operation by controlling the ratio between the fine and coarse DAC's least significant bits (LSBs); thus, k can set the overall bit weighting to any desired value. For the maximum-length DAC, you must choose resistor tolerances that make the ratio k accurate to 0.19%, limiting the system error to ½LSB. If the resistor ratio is not accurate, there may be a dead band between the end of the fine-DAC control range and the start of the coarse-DAC control range. Choosing a value for k so that the fine DAC overlaps the coarse DAC eliminates this dead band.

For instance, setting k to 126.49 makes the combined system act like a bipolar 15-bit DAC with a total range of 32,765 bits. The fine DAC overlaps the coarse DAC by 1 bit, allowing the use of less accurate resistors and ensuring that the circuit generates all output states without any missing codes. You can set other ranges and overlaps as you like by choosing k appropriately. Figure 22.1 implements a 14-bit DAC with k = 62.25.

If you need more resolution, you can expand this circuit to include combinations for 8- and 12-bit DACs, which will increase the control range.

Part Four

Oscillators, Generators, and Clocks

Introduction to Part 4

For those of you who need to generate signals, rather than condition or process them, here's a pot-pourri of clock, oscillator, and generator material. The first article deals with the thorny problem of clock-edge skew in high-speed, dual-clocked systems. For sine-wave needs, several articles provide useful techniques for configuring and controlling voltage-controlled oscillators (VCOs). One article shows you how to control phase in a phase-locked loop (PLL), a circuit block commonly used in conjunction with a VCO. In one 'Design Idea', a clever circuit uses CMOS logic to generate precision triangle and ramp waveforms.

23 ASIC design techniques synchronize dual clocks in high-speed designs

Avoiding clock skew within an ASIC is as important as elsewhere

The problems which can result from clock skew in boards using several or many digital ICs have long been recognized. Although the clock timing differences within a single ASIC are smaller than between different ICs, they must still be taken into account if problems are to be avoided.

Rick Hansen and Robert Deming
Vitesse Semiconductor Corp

Designers of high-speed synchronous digital ICs commonly clock some sequential logic with a divided-down version of the primary clock. Synchronization problems can easily occur in this type of design because of the skew between the primary and divided-down clock signals. Possible problems in such dual-clock designs include setup and hold-time violations and clocking incorrect data.

Just like discrete logic devices, sequential or registered elements in an ASIC have specific setup and hold-time requirements for the incoming data and clock signals. The setup time defines the required amount of time before the active clock edge when the incoming valid data must be stable – that is, not switching. The hold time specifies a period of time after the active clock edge when the valid data must be stable. If data are changing or are not valid during the period of time defined by setup and hold, and the incoming, active clock edge, then the resulting output is not guaranteed to be valid data. In addition, the register's output could become metastable for an undefined period of time.

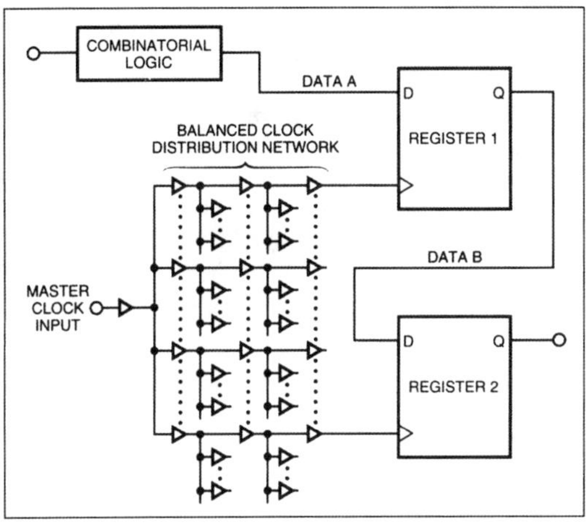

Figure 23.1 Because of on-chip skew, buffers on each branch of a clock tree will exhibit *slightly* different propagation delays, possibly leading to 'double clocking'

Setup and hold violations can occur because, in any ASIC, identical types of logic elements at different locations on the chip can exhibit skew. This condition is true for any semiconductor technology because of the inherent variations in processing across a single die. You must consider this on-chip skew when designing timing-dependent paths, because on-chip skew can render your circuit nonfunctional. In fact, on-chip skew can potentially cause problems in a circuit even if the design has passed best- and worst-case ac simulations.

The cumulative effects of on-chip skew can become a problem in synchronous designs. Consider the example in Figure 23.1, which has a single clock source distributed throughout the entire circuit via a balanced network. In this example, the Q output of Register 1 connects to the D input of Register 2.

Assume that all metal interconnections in the clock paths feeding each register are the same length and that each path contains identical buffering elements. By design, these two registers should pipeline data at each clock cycle because the two branches of the clock network are supposedly balanced. But because of on-chip skew, the buffers on each branch of the clock tree will exhibit *slightly* different propagation delays.

Suppose that on-chip skew becomes great enough so that the active clock edge reaches Register 1 early enough so that data A becomes present at the D input of Register 2 before the clock reaches Register 2. Then Register 2 will erroneously clock data A instead of the intended data B. This problem is called double clocking.

Dual clocks magnify problem

A circuit that uses two clocks at different frequencies magnifies the problems arising from on-chip skew. Of necessity, the two clocks will require their own separate clock-distribution networks, adding to the signal-skew problem. To understand why, begin with the common node where the divided-down clock branches from the main clock. This common node between the clocks will typically have many buffers between it and the point at which the clocks actually arrive at sequential elements. Because of on-chip skew, each layer of logic or buffering in the two paths will add a degree of ambiguity to the arrival of clock edges at the sequential elements (Figure 23.2). This ambiguity will build up over the clock-distribution network, increasing the amount of potential skew between the high-and low-frequency clock edges.

Figures 23.3 and 23.4 show the problems that can occur in a dual-clock design. In Figure 23.3, both a 25- and a 100-MHz clock drive two consecutive registers. The 25-MHz clock triggers flip-flop U_1, and its Q output connects to the D input of flip-flop U_2. The 100-MHz clock triggers U_2. Ideally, the active edge (in this case a rising edge) of the 25-MHz clock aligns with the active edge of the 100-MHz clock (Figure 23.4a), so that data A clocks into register U_2.

In Figure 23.4b, clock skew causes the active edge of the 100-MHz clock to be slightly late relative to the 25-MHz clock. A simulator will show a hold-time violation on U_2 even though

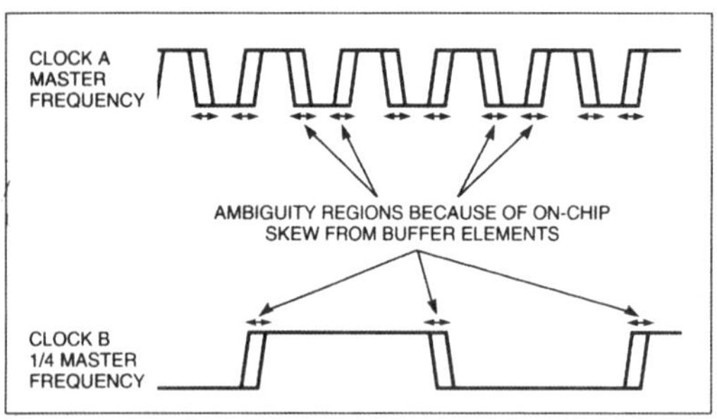

Figure 23.2 A circuit that uses two clocks at different frequencies magnifies the problems arising from on-chip skew. Each layer of logic or buffering will add a degree of ambiguity, increasing the potential skew between high- and low-frequency clock edges

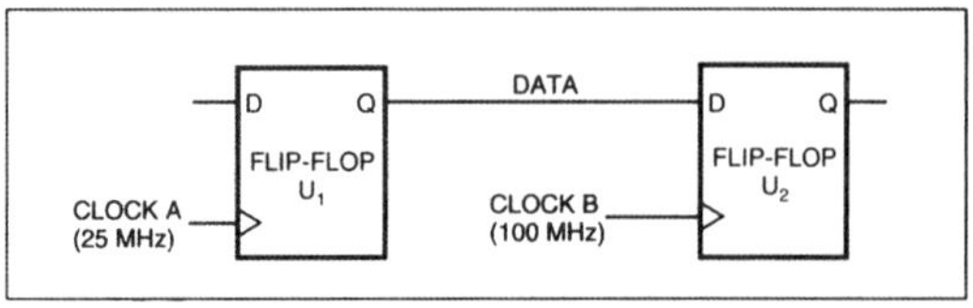

Figure 23.3 Here, both a 25- and a 100-MHz clock drive two consecutive registers. Because of the buffering elements on the clock-distribution networks, the clock signals to the registers accumulate some differing amount of on-chip skew. Serious problems can result from this clock skew

flip-flop U_2 may clock the correct data A. Figure 23.4c shows a more severe situation in which the active edge of the 100-MHz clock occurs much later than the 25-MHx clock's edge. In this case, the desired data A gets missed entirely. Flip-flop U_2 may clock through erroneous data B, and a simulator will flag a setup violation. In the extreme example Figure 23.4d, where clock skew is the greatest, U_2 may clock erroneous data B, but a simulator *will not* flag any timing violations for this error.

Similar problems to all four cases can occur when clocking data in the other direction from a register with a 100-MHz clock to one with a 25-MHz clock.

Solutions

A good way to flag potential problems is to use an accurate min/max timing simulator in addition to best- and worst-case ac simulations. The min/max timing simulator can detect potential setup and hold violations. At Vitesse, all designs must go through a back-annotated min/max simulation to flag any setup and hold violations before releasing the design to manufacturing (Teredyne's (Agoura Hills, CA) Lasar is the timing simulator we use).

Next, to reduce the probability of timing violations, you can insert a delay element between the two sequential elements (Figure 23.5). To ensure functioning over the desired frequency range, your design must meet the following two inequalities. Note that if your design does not satisfy Equation (1), the device will not work at any frequency.

$$T_{QC(min)} + T_{C(min)} + T_{RC(min)} - T_{HOLD(max)} > T_{SKEW} \quad (1)$$

$$T_{QC(max)} + T_{C(max)} + T_{RC(max)} + T_{SETUP(max)} + T_{SKEW} < t, \quad (2)$$

where:

- T_{QC} = signal propagation delay from clock input to data output for register U_1,

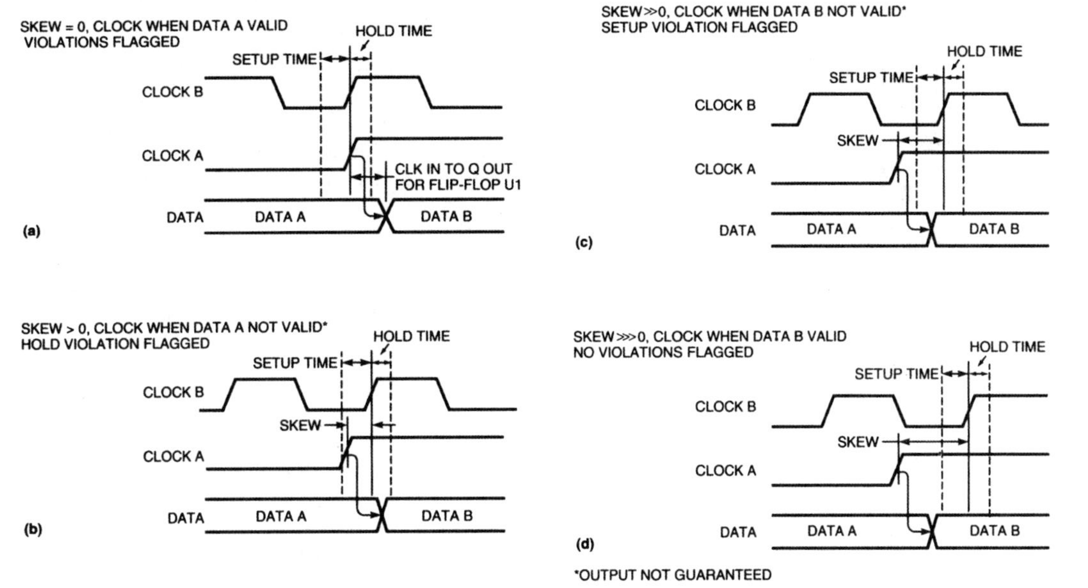

Figure 23.4 Depending on the amount and direction of on-chip skew, your circuit may or may not work properly, and your simulator may or may not flag your design error. In (a), the circuit is working properly. In (b), clock skew causes the active edge of the 100-MHz clock to be slightly late relative to the 25-MHz clock. A simulator will show a hold-time violation for U_2 even though flip-flop U_2 may clock the correct data A. In a more severe situation (c), the active edge of the 100-MHz clock occurs much later than the 25-MHz clock's edge, missing the desired data A entirely. In (d)'s extreme example, U_2 may clock erroneous data B, but a simulator *will not* flag any timing violations

- T_C = signal propagation delay because of added delay element,
- T_{RC} = signal propagation delay because of metal-interconnect RC effects between registers,
- T_{HOLD} = data-valid hold-time requirement for register U_2,
- T_{SETUP} = data-valid setup time requirement for register U,
- T_{SKEW} = maximum amount of skew between clock signals, and

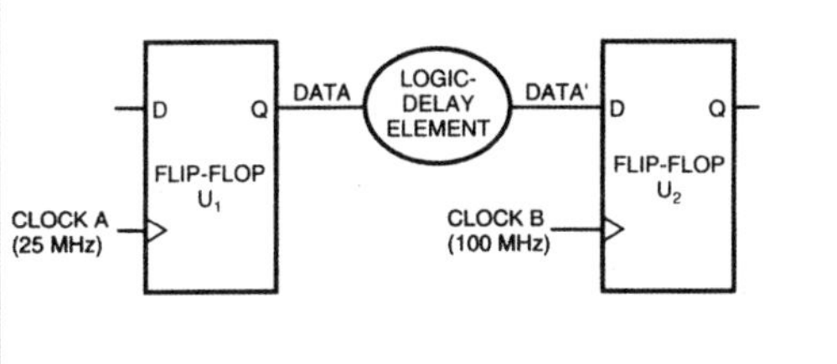

Figure 23.5 One way to reduce the probability of timing violations is to insert a delay element between the two sequential elements

- t = time for one period of the high-speed clock.

When you use the previous approach, you may have to add a delay element for each high-speed-to-low-speed (or vice versa) interface. In addition, for each instance you must estimate the maximum amount of skew that on-chip skew could introduce into the clock signals. (A reputable ASIC vendor should state the maximum value for on-chip skew in its product's design manual, usually as a percentage of the elements' propagation delay). But, these calculations could be quite substantial for large designs. Another approach exists, though, which is easier to implement and is much more tolerant of variations in clock skew.

An alternative to the previous method is a synchronous-control signal for each interface register. Figure 23.6 illustrates this approach. Here, a flip-flop with a 2-into-1 multiplexer at its input replaces the original flip-flop U_2. This enhancement now lets the LOAD/$\overline{\text{HOLD}}$ input hold flip-flop U_2 in either a hold or load mode. As the timing waveforms in Figure 23.6 show, flip-

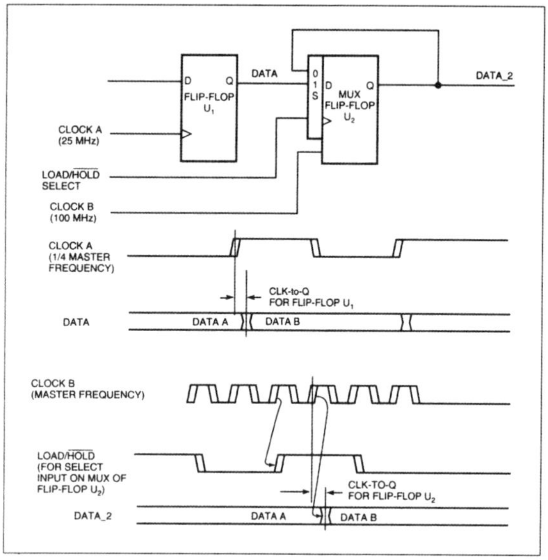

Figure 23.6 An alternative to reduce the probability of timing violations is a synchronous-control signal for each interface register. This enhancement now lets a LOAD/$\overline{\text{HOLD}}$ input hold flip-flop U_2 in either hold or load mode

flop U_2 remains in hold mode during the critical active edge of the low-speed clock. This flip-flop goes into load mode during the nontransitioning period of the data. Note that you must generate the LOAD/$\overline{\text{HOLD}}$ signal synchronously from the high-speed clock.

The advantages of this approach are that the design tolerates considerable clock skew, and it is relatively simple to implement. The main drawback is that you must add a multiplexer to each interface.

Figure 23.7 illustrates one way to generate the dual-clock synchronizing logic that provides both the clock signals and the LOAD/$\overline{\text{HOLD}}$ select signal from Figure 23.6. Releasing the RESET input begins the synchronizing of the LOAD/$\overline{\text{HOLD}}$ select block with the low-frequency clock-generation block. After the reset is released, the LOAD/$\overline{\text{HOLD}}$ select block will track the low-frequency clock. The tracking occurs 90° out of phase to ensure that the receiving flip-flop sees the data at the proper time.

The main disadvantage of this approach is that you must balance the reset-delay path to the LOAD/$\overline{\text{HOLD}}$ select block with the high-speed-clock path. This balancing is required, so that the

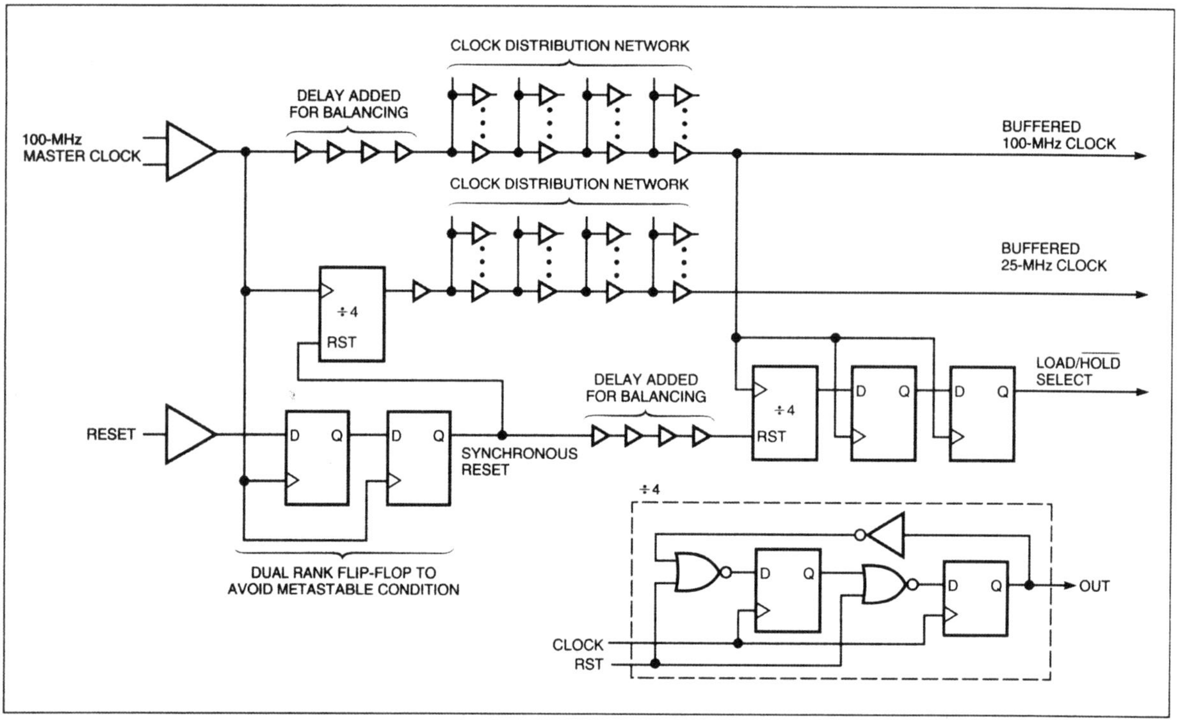

Figure 23.7 This circuit is one way to generate the dual-clock synchronizing logic that provides both the clock signals and the LOAD/$\overline{\text{HOLD}}$ signal from Figure 23.6

reset signal is not transitioning when the active edge of the high-speed clock reaches the divide-by-four block.

This synchronizing-logic-block approach requires minimal overhead and does not impact the maximum performance of a circuit significantly. You can easily implement the technique for very-high-speed devices. After your circuit has been placed and routed, you should use a min/max timing simulator to verify that no potential timing hazards exist. The result will be a robust, high-yielding part.

On-chip skew differs from K factor

Do not confuse on-chip skew with the variation in propagation delay for a given type of logic element across all die. These variations are the so-called K factor for derating performance because of process, temperature, and voltage variations. As Figure 23.8 shows, the range of on-chip skew is smaller than the variation in propagation delay expected for a given logic element across all die. The actual amount of on-chip skew that you may encounter in a given chip design depends upon the die's size, the distance between the macros, and process variation across a single die. In fact,

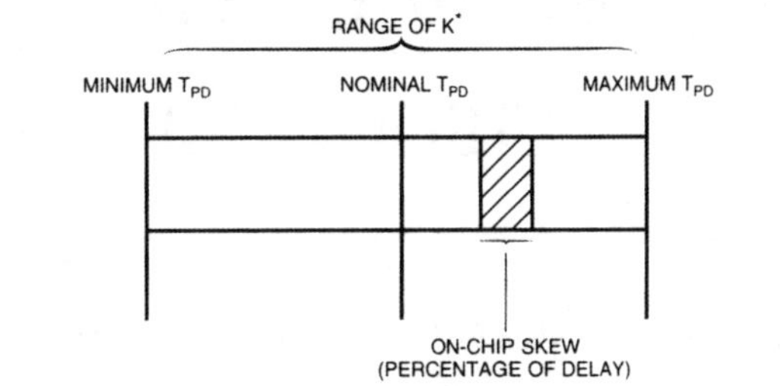

Figure 23.8 The range of on-chip skew is smaller than the variation in propagation delay expected for a given logic element across all die from a given wafer

temperature and voltage variations are not likely to contribute significantly to the on-chip skew of a single die.

Reference

1. Proffit, S, 'Application Note 6: Metastable Behavior of GaAs DCFL Registers', Vitesse 1991 product Data Book.

24 CMOS logic creates precision waveforms

Accurate trianglewave generator is simple, elegant

A number of different circuit designs for ultra linear low frequency ramp/triangular wave generators have appeared in *EDN* over the years. This is one of the simplest and most elegant.

Michael A Wyatt
Honeywell SSO, Clearwater, FL

The circuit in Figure 24.1 generates three different waveforms having frequencies less than 1 Hz: triangle waves, positive ramps, and negative ramps. At very low output frequencies, the stability of the circuit's input frequencies almost completely determines the output waveform's linearity.

XOR gate IC_3 beats input frequency f_{IN} against reference frequency f_{REF}, thus producing a train of pulses whose periods increase gradually until the frequency sources are completely out of phase. Then, the pulses' periods decrease until the sources are again in phase. Flip-flops IC_{1A} and IC_{1B} produce 50% duty-cycle inputs for XOR gate IC_3.

IC_5 and its surrounding components form a third-order, lowpass filter (f_c = 1 kHz). The filter averages the output of pulse buffer IC_4 to pro-

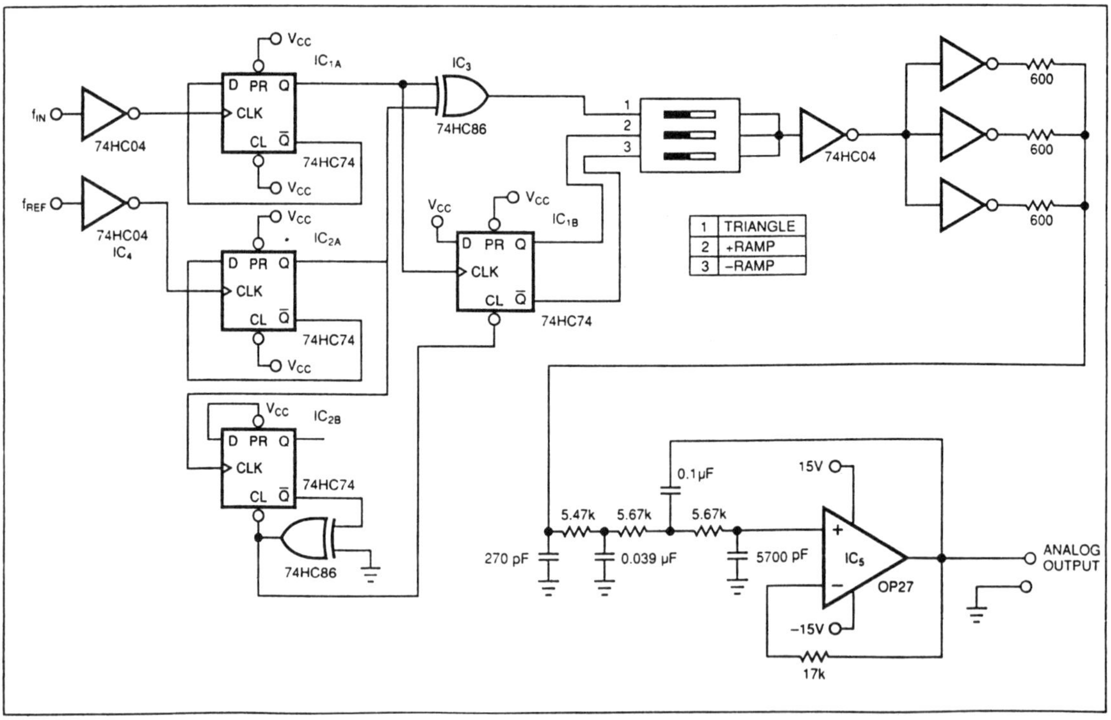

Figure 24.1 By beating two high-frequency input-pulse frequencies against each other and then integrating the resulting beat-frequency pulse train, this circuit produces low-frequency analog waveforms having good linearity

duce a triangle waveform having a peak amplitude of V_{cc} and a frequency of $| f_{IN} - f_{REF} | \div 2$. Be sure to select low-dielectric-absorption capacitors for the filter circuit.

The circuit generates ramp waveforms in a similar manner except that the SR flip-flop phase comparator formed by IC_{2A} and IC_{2B} replaces the XOR gate. The phase comparator sets on every other negative transition of f_{IN} and subsequently resets on every other negative transition of f_{REF}. If f_{IN}'s frequency is greater than f_{REF}'s, then the width of IC_{2B}'s Q output pulse will gradually increase. This increase produces a positive-going ramp at the circuit's output. If f_{IN}'s frequency is less than f_{REF}'s, the output will be a negative-going ramp. Note that the filter's step response controls the ramp's reset time. Selecting a frequency greater than 100 kHz for f_{IN} and f_{REF} attenuates the pulse's ripple, which relaxes the reset-time restrictions.

25 Current sink widens VCO's frequency range

1000:1 frequency range from CD4046

With the usual voltage control via pin 9, the VCO section of that old favourite the 4046 offers a control range of about one decade. This 'Design Idea' explains how that can be extended to about three decades by substituting an adjustable current source at pin 11 (in place of the usual fixed resistor) for the voltage source at pin 9.

Antonio Tagliavini
ADA – Applicazioni Digitali e Analogiche, Bologna, Italy

The familiar HC4046 voltage-controlled oscillator (VCO) circuit (Figure 25.1a) has an output frequency span of about one decade with fairly good linearity over an input voltage range of 1 to 4.75 V. The circuit in Figure 25.1b widens this frequency span to three decades.

Figure 25.1b's circuit replaces the single frequency-determining resistor from pin 11 to ground with a precision voltage-controlled current sink comprising an LM358 op-amp and transistor Q_1. The current sink overcomes the limitations of the integrated current sources normally responsible for charging and discharging the timing capacitor. The circuit applies a fixed

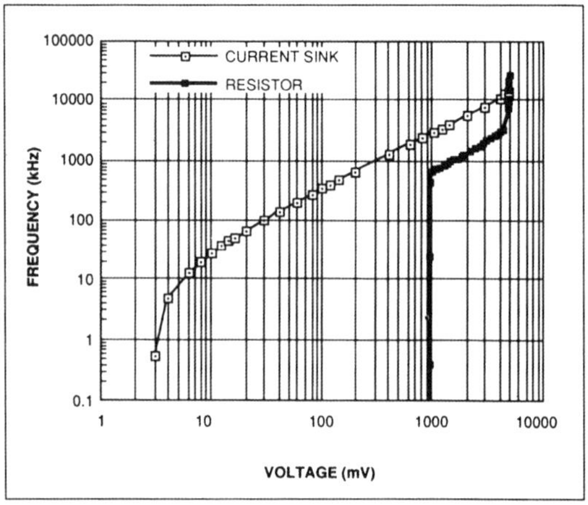

Figure 25.2 The current-sink circuit lengthens the frequency span of the standard circuit from about one decade to three decades

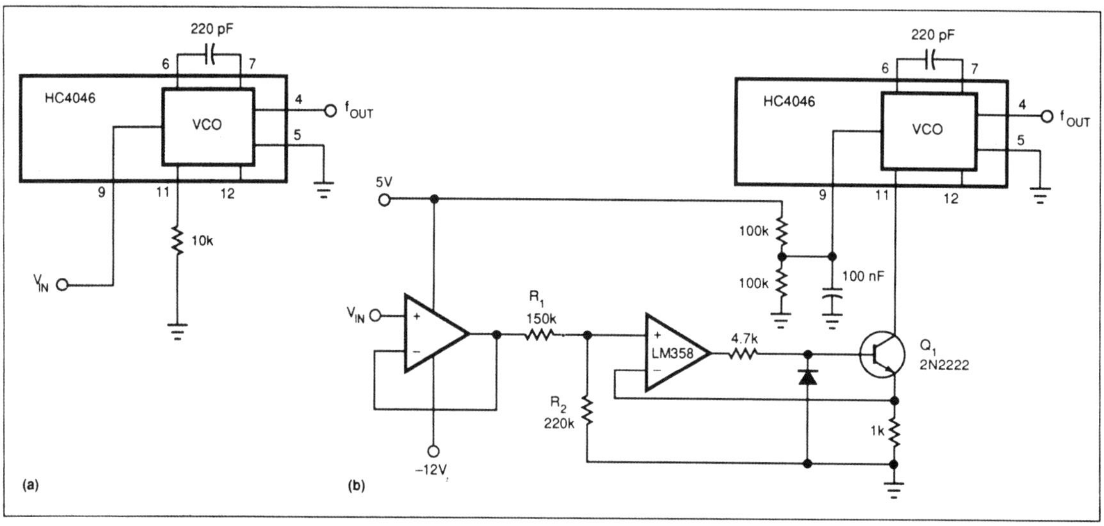

Figure 25.1 By substituting a voltage-controlled current sink for the standard circuit's fixed 10-kΩ resistor (a), the circuit in b extends the HC4046 VCO's frequency range

2.5 V to the VCO's input at pin 9. Because the voltage on pin 11 can't exceed 2 V, the current sink must operate below this level. To meet this requirement, resistors R_1 and R_2 divide the input signal before it reaches the current sink's input.

The graph in Figure 25.2 compares the frequency-vs-voltage characteristics of the standard circuit with those of the new circuit. By using the voltage-controlled current sink, the linear tuning range spans three decades. At the low-frequency end, output phase noise is quite noticeable because the current sink operates at very low current levels. When the loop is locked to a clean reference, the feedback reduces this noise. A better way to get rid of this output phase noise is to operate the VCO near its maximum frequency and then divide the output digitally. This technique reduces the phase noise by the amount of the division.

26 Programmable oscillator runs without μP

Look, no micro

Many chips designed to run under μP control would be very useful in applications where a μP is not included. Here is a scheme for running one of them, the ML2035 programmable sinewave generator, in a stand-alone application.

Jon Klein
Micro Linear, San Jose, CA

The circuit in Figure 26.1, using a clever scheme adaptable to other programmable devices, allows you to operate the ML2035 programmable sinewave generator, IC_3, without a controlling μP. IC_1, a 74HC4060 counter, provides both the sinewave generator's clock as well as a gating pulse to shift register IC_2. When IC_1's pin 5, Q_5, goes high, IC_2 begins shifting eight hard-wired bits into the

Table 26.1 Shift register values and frequency errors for standard crystal values

$f_{CRYSTAL}$ (MHz)	f_{OUT}	D_{10}	D_{HEX}	74HC165 Code ABCD	EFGH	Error
4.00	50	105	69	1001	0110	0.14%
4.00	60	126	7E	1000	0001	0.14%
4.194304	50	100	64	1001	1011	0.00%
4.194304	60	120	78	1000	0111	0.00%
6.00	50	70	46	1011	1001	0.14%
6.00	60	84	54	1010	1011	0.14%
8.00	50	52	34	1100	1011	-0.82%
8.00	60	63	3F	1100	0000	0.14%

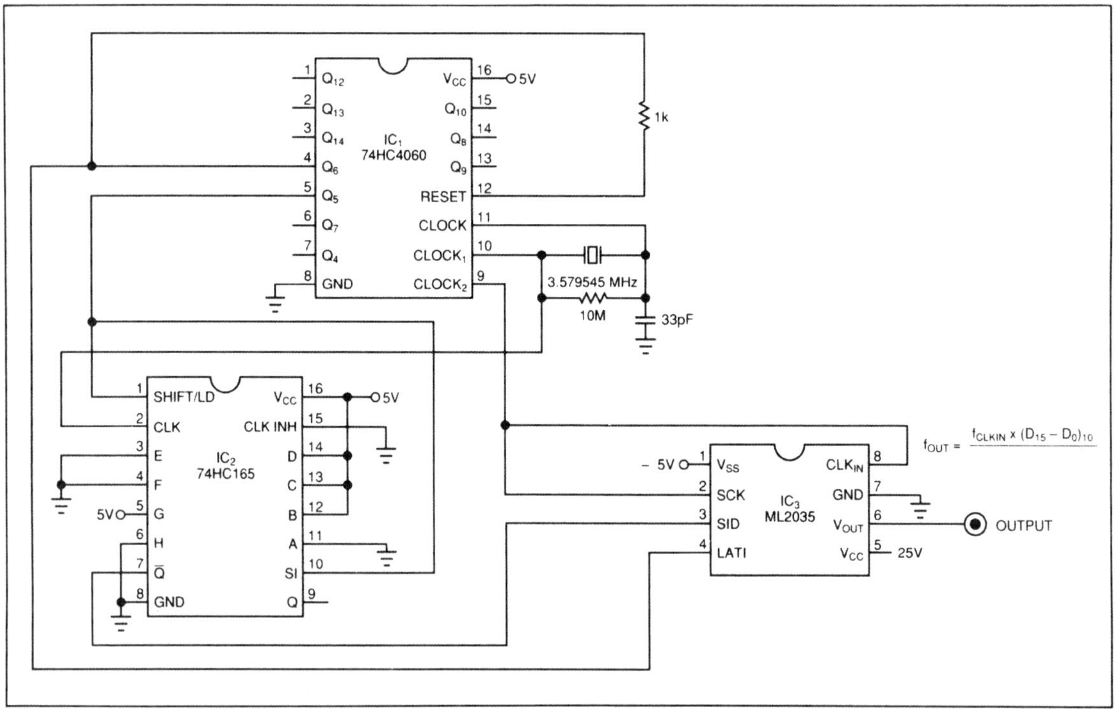

Figure 26.1 Counter IC_1, first clocks in an 8-bit programming code via shift register IC_2, subsequently clocking sine-wave generator IC_3

sine-wave generator to program it. After IC_2 shifts the 8 bits, Q_5 goes low, enabling normal operation. The circuit can produce both 50- and 60-Hz outputs from a NTSC colour-burst crystal (3.579545 MHz). Table 26.1 lists binary codes for other crystal frequencies. The sine-wave generator's output exhibits a maximum of 0.5% THD.

27 High-frequency VCOs top 100 MHz

Saw a filter, made an oscillator

SAW filters are available from American, European and Japanese manufacturers in an increasing range of frequencies and styles. They permit the direct implementation of VHF and UHF high-stability oscillators without the doublers or triplers needed with crystals, and with a wider pulling range to boot. Here are two examples.

Di Paolo Franco
Ericsson Fatme, Rome, Italy

VCOs that use surface-acoustic-wave (SAW) filters have higher operating frequencies and higher pull ranges than do circuits that use crystal oscillators. Figures 27.1 and 27.2 present two practical realizations of SAW-filter-based VCOs. The circuits' operating frequencies are 140 and 181 MHz, respectively. The differences between the two figures stem from that fact that two types of SAW filters are available. Figure 27.1's circuit uses a SAW filter that has 180° of phase shift. The filter in Figure 27.2 has a phase shift of 0°. Both circuits draw about 20 mA and operate from a 5 V supply.

Each circuit's operating frequency is solely dependent on the SAW filter's passband centre frequency, which can be higher than 1000 MHz. The SAW filters have a pull range near 500 ppm when the BBY31 varactor diode's control voltage, $V_{CONTROL}$, varies by 4 V.

These circuits' typical p-to-p voltage when driving a 50 Ω load is 600 mV. The spectrum of the output signals is such that all the harmonics are below 25 dB with respect to the carrier. The variation with temperature when the circuits run in the free-running mode is 100 ppm, which is typical for SAW filters.

Crystal oscillators tend to be more stable over their operational frequency range than are SAW-based oscillators. However, that range is limited compared with the range of SAW-based oscillators. SAW filters are available with center frequencies starting at 120 MHz. The components cost of Figure 27.1 and Figure 27.2, without the SAW filter, is about $1 each. SAW filters cost about $25.

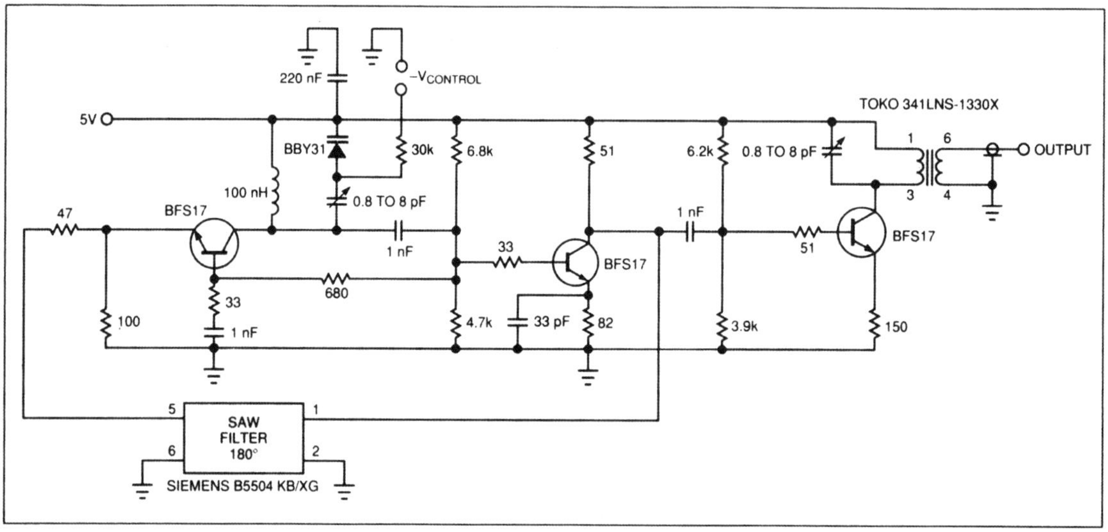

Figure 27.1 This oscillator is based on a 180° phase-shift SAW filter and has a free-running frequency of 140 MHz

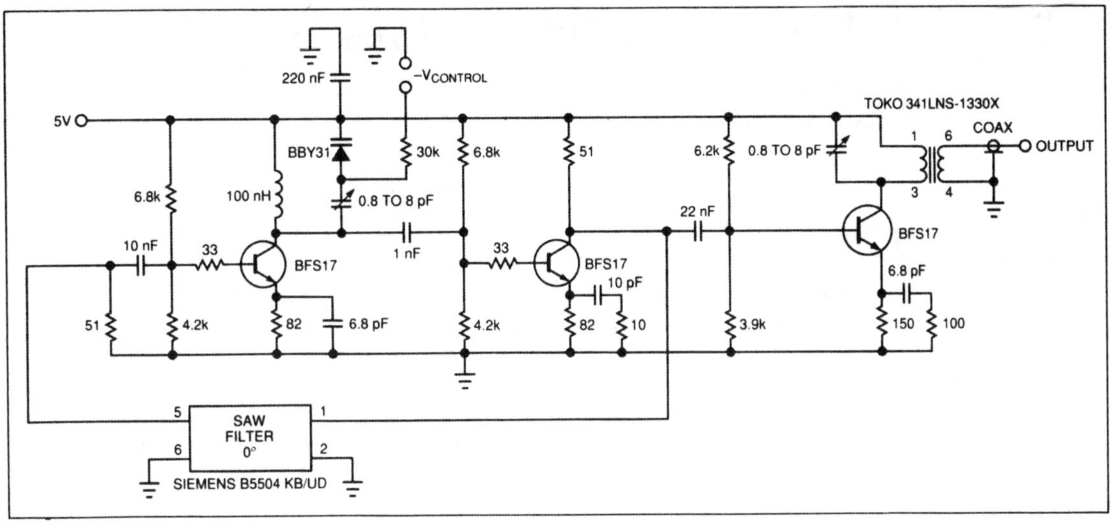

Figure 27.2 This oscillator is based on a 0° phase-shift SAW filter and has a free-running frequency of 181 MHz

28 Offset varies PLL's phase shift

Handy 4046 circuit

This circuit provides an output signal whose phase relationship is variable over ±160° relative to a reference signal of the same frequency. If the signals applied to pins 3 and 14 of the CD4046 are both divided by 2^N first, the available phaseshift range becomes 2^N times 320°.

Donald G Stefani
LeRoy, NY

By deliberately introducing a dc offset into the error-signal path of a phase-locked loop (PLL), you can add phase shift between the input reference and the feedback signal that drives the phase detector. The circuit in Figure 28.1 has a phase-shift range of ± 160° over a reference-signal range of 1 to 10 kHz. Because the loop uses an integrator, once set, the phase shift remains constant over the entire frequency range.

The positive-edge-triggered frequency/phase detector inside the CD4046B compares the frequency and phase of the input reference signal with the feedback signal from the 74HC193 counter. Since this phase detector is edge triggered, the duty cycle of the reference can be arbitrary. IC_{1A} level shifts the three-state output of the phase detector. When in lock, if the positive edge of the reference leads the positive edge of

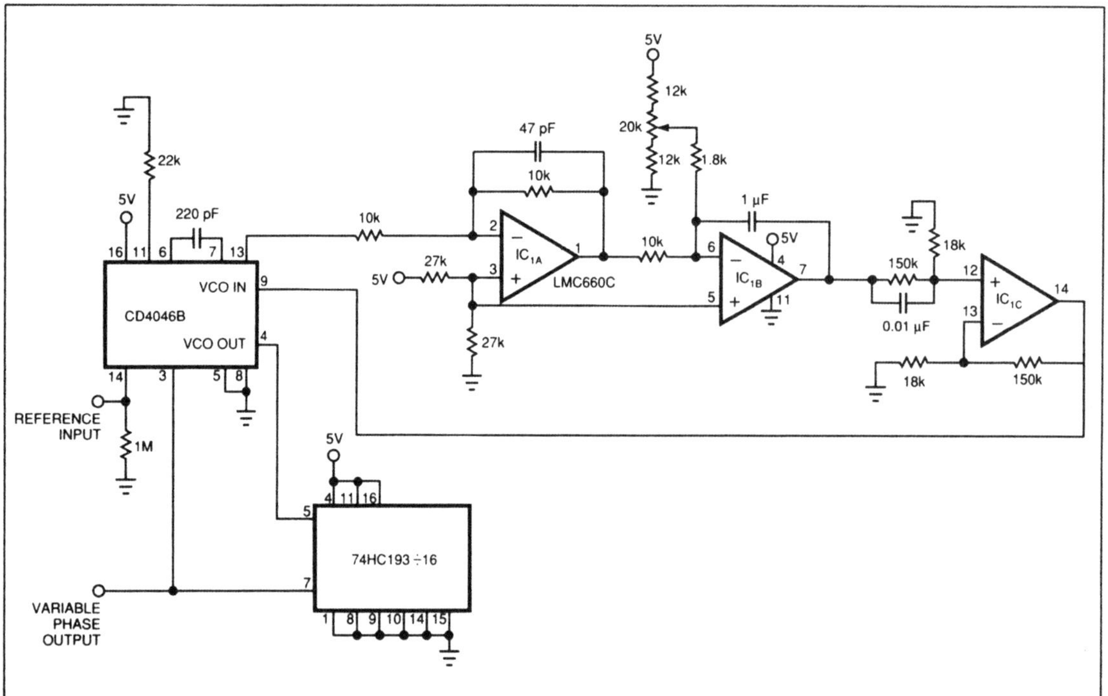

Figure 28.1 By introducing dc offset into this PLL's feedback, you can vary the phase shift between its input and output by ± 160 degrees

the feedback signal, the output of IC_{1A} swings to its lower rail for the time interval between the two edges. IC_{1A}'s output then returns to 2.5 V until the next positive edge of the reference occurs. If the positive edge of the reference lags the positive edge of the feedback, the output of IC_{1A} remains at the upper rail for the time interval between the two edges. It then returns to 2.5 V until the next positive edge of the feedback occurs.

Using the quad LMC660C CMOS op-amp for IC_{1A} permits dynamic-loop, error-voltage variations over almost the entire range of 0 to 5 V. Adjusting the 20-kΩ potentiometer causes the phase shift to change because the average voltage of the pulses at IC_{1A}'s output must change to maintain the constant-feedback-forced value of 2.5 V at IC_{1B}'s inverting input. A unity-dc-gain lead network stabilizes the loop. This design divides the output of the VCO by 16 before applying it to the phase-detector input.

29 Servo loop controls oscillator amplitude

Low distortion high-level output from fundamental crystal oscillator

Whilst not providing the ultimate in low noise crystal oscillator design, the following article describes a circuit offering a sufficient output level (combined with a good waveform having low harmonic content) to drive a double balanced mixer directly. Crystal drive level may be set to optimize either long term or short term stability, as required, and the necessary design steps are spelt out in detail.

Thomas P Hack
Comlinear Corp, Fort Collins, CO

The high-performance, fundamental-mode crystal oscillator in Figure 29.1 uses an AGC amplifier and a crystal to form a very-narrow-band filter at the crystal's series-resonant frequency. The design exhibits reasonably low phase noise and jitter because it places the crystal between two low-impedance points of the CLC520 AGC amplifier (IC_1). The oscillator can drive a 50 Ω load easily and has a well-controlled output impedance. The design exhibits low distortion and is adaptable to a variety of fundamental-mode crystals.

Unlike most oscillators, which use limiting to set the amplitude, this design uses a servo loop to control amplitude. D_1 and C_1 are the key components of a clamping circuit that produces an average voltage proportional to the peak-to-peak oscillator amplitude. The larger the amplitude, the more positive the dc component.

The design configures an LF356 (IC_2) as an integrator that compares the dc signal against the reference voltage of D_2. If the oscillator's amplitude is too high, the integrator's output voltage drops, as does IC_1's gain and the oscillator's loop gain. When the loop gain drops below unity, the oscillator output amplitude begins to drop until it reaches the loop's desired amplitude. If the amplitude is too low, the integrator output voltage increases, thereby increasing the loop gain and increasing the amplitude to the loop's desired value.

When the oscillator amplitude is stable, the average current flowing into the integrator capacitor (C_2) is zero. The average current through R_3 is equal in magnitude and opposite in sign to the current flowing through R_4 (assuming that IC_2's bias currents are negligible). And the oscillator loop gain is exactly equal to one. Because a leveling loop, and not circuit limiting, sets this oscillator design's amplitude, distortion is low. The amount of distortion is mostly set by IC_1. Because IC_1's bandwidth (typically 140 MHz for large signals) is approximately four to five times higher than the highest oscillation frequency of most fundamental-mode AT-cut crystals, the effect of IC_1's bandwidth is negligible.

Designing this oscillator requires six major steps. The first step is to determine the range of the crystal's equivalent series resistance. You should use a range consistent with the distribution of crystals that you use. If you'll be tuning the range, find the equivalent resistance of the crystal combined with the tuning network at the new series-resonant frequency. In the case of a crystal and a tuning capacitor in series, the highest overall series resistance exists at the lowest tuning capacitance and highest crystal series resistance.

The second step is to choose the output amplitude. To determine the output voltage at pin 10 of IC_1, first convert from dBm to watts as follows:

$$P_{OUT} = 10^{(0.1\,\mathrm{dBm}\,-\,3)},$$

where P_{OUT} is the power delivered to the load in watts, and dBm is the power delivered to the load in dBm. The rms voltage delivered to the load is

$$V_{OL} = (R_{LOAD} \times P_{OUT})^{0.5}.$$

To account for a doubly terminated load, use the following equation:

$$V_{OP\ AMP} = 2 \times V_{OL} = (4R_{LOAD} \times P_{OUT})^{0.5},$$

where V_{OP} is in volts rms.

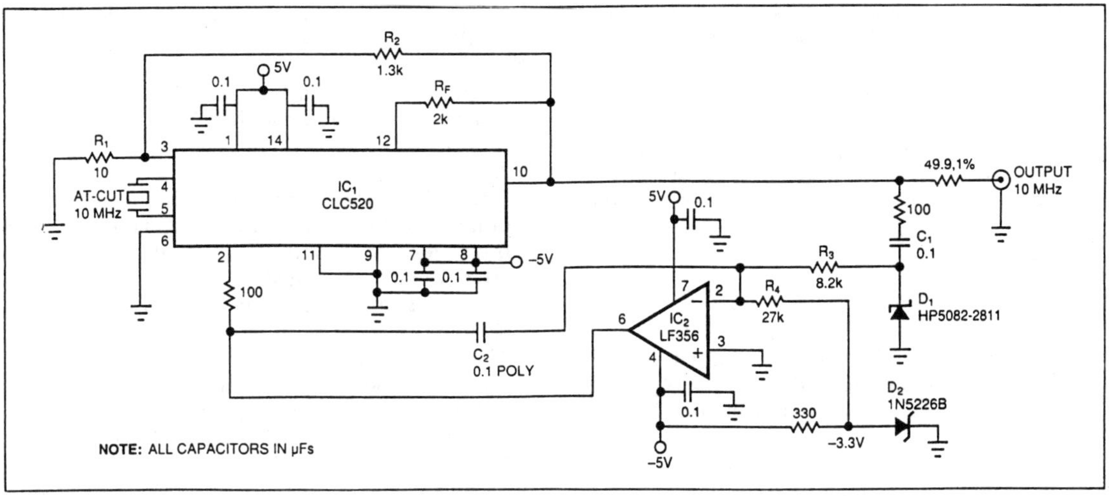

Figure 29.1 Unlike most oscillators, which use limiting to set the amplitude, this 10-MHz oscillator uses a servo loop to control amplitude. Six steps are necessary to tailor the design to your requirements

The third step is to select the crystal drive level. Drive levels should lie between 1 and 20 µW for good long-term stability and between 100 and 500 µW for good short-term stability. Because the equivalent series resistance of the crystal affects the drive level, be sure that the drive level is reasonable for all expected values of this resistance. One way to start is to choose the maximum crystal drive level (D_{MAX}) and see if the minimum drive level is acceptable using the following equation:

$$D_{MIN} = D_{MAX}\left(\frac{R_{S(MAX)}}{R_{S(MIN)}}\right)\left[\frac{R_{S(MIN)} + 3}{R_{S(MAX)} + 3}\right]^2,$$

where $R_{S(MAX)}$ and $R_{S(MIN)}$ are the maximum and minimum series resistances, respectively. If this calculated value of D_{MIN} is acceptable, you need to determine whether or not IC_1 can deliver D_{MAX}. IC_1 will be most limited at the minimum series resistance, as follows:

$$D_{LIMIT} = (0.9113 \times 10^{-6})\, R_{S(MIN)},$$

where D_{LIMIT} is the maximum drive available from IC_1 in watts, and $R_{S(MIN)}$ is the minimum crystal series resistance in ohms. If D_{LIMIT} is greater than D_{MAX}, IC_1 can deliver the targeted maximum drive level. If not, substitute D_{LIMIT} in place of D_{MAX} in Eq 4 to determine the lowest drive that will occur. D_{LIMIT} and the new D_{MIN} set the new drive-level range.

The fourth step is setting the forward gain of the oscillator. First determine the input voltage to IC_1's pin 3 at the maximum series resistance as follows, with D_{MIN} in watts and V_{IN} in volts rms:

$$V_{IN} = \left(R_{S(MAX)} + 3\right)\left(\frac{D_{MIN}}{R_{S(MAX)}}\right)^{0.5}.$$

This equation accounts for the crystal's loading of IC_1's buffers (pins 4 and 5). You can now determine the voltage gain of IC_1 at the highest series resistance and highest gain-control voltage (A_v) as follows:

$$A_V = \frac{V_{OP\ AMP}}{V_{IN}}$$

To achieve this gain, set R_F as follows:

$$R_F = \frac{A_v\left(R_{S(MAX)} + 3\right)}{1.85}.$$

In general, the value of R_F should be between 1 and 2 kΩ. Somewhat higher values are acceptable if the oscillator is running below 10 MHz. If R_F needs to be lower than 1 kΩ, refer to IC_1's data sheet for the output-amplifier loop-gain reduction techniques.

Fifth, you need to calculate the values of the feedback network comprising R_1 and R_2. To keep the noise low at IC_1's input and provide reasonable resistor values, you should make $R_1 \geq 10\,\Omega$

and ≤ 1 kΩ. You should set the loss in the network equal to $B = 1/A_V$, which means that

$$R_2 = R_1 (A_V - 1).$$

The sixth and final step requires setting up the leveling loop. The average voltage from the clamping circuit is as follows, where V_{PK} is the peak output voltage of IC_1, and V_D is the forward voltage drop for D_1:

$$V_{DC} = V_{PK} - V_D = 1.414\,V_{OP\ AMP} - V_D.$$

Once the amplitude of the oscillator is stable, the current flowing through R_3 and R_4 must cancel at C_2. For this condition to be met,

$$R_4 = R_3 \left(\frac{V_{D2}}{V_{DC}}\right),$$

where V_{D2} is the zener voltage of D_2. To ensure the stability of the amplitude-control loop, make C_2 equal to $0.01 \times F$, where C_2 is in μF and F is in MHz.

For the values in Figure 29.1's design, the results of the six steps are as follows:

1. The crystal has a measured equivalent series resistance of approximately 7.3 Ω. The range of R_S is 5 to 25 Ω.
2. The design's output-power requirement is 7 dBm into 50 Ω so that the oscillator can drive a double-balanced mixer directly. This requirement translates into an output voltage at the op-amp of approximately 1 V rms.
3. For a 5 Ω minimum equivalent series resistance, IC_1 limits crystal drive level to 4.56 μW. At R_S of 25 Ω, the drive level falls to 1.86 μW. These numbers produce good long-term stability.
4. The input voltage at IC_1's pin 3 is 7.64 mV rms. The voltage gain is 131. Thus, R_F must equal 1.98 kΩ (use 2 kΩ).
5. R_1 is set to 10 Ω. Thus, R_2 must equal 1.301 kΩ.
6. Assuming a forward drop of 0.4 V for D_1 yields approximately 1 V dc from the clamping circuit. R_3 must then be approximately 8.3 kΩ (use 8.2 kΩ). C_1 is set to 0.1 μF because this design is for a 10-MHz oscillator.

Part Five

Power Sources and Motor Control

Introduction to Part 5

Distributed power is making big inroads in electronic systems. It's useful to be able to distribute relatively loosely-regulated power to various points in the system, then use regulators or dc/dc converters to provide clean and accurate supply voltages at these points. The first article in this part shows you how to select optimum inductor types for your dc/dc-converter designs. Three 'Design Ideas' present some novel twists in designing regulators. One, for example, combines linear regulation with a switching topology to provide both tight regulation and high efficiency. The low-dropout regulator in one 'Design Idea' is useful in 3 V systems, rapidly becoming the norm in portable computers. A very novel circuit simulates a diode, and exhibits only 0.04 V forward drop at 1 A. One article shows how to use recently available motor-control ICs to control dc motors that have no Hall-effect sensors.

30 Careful inductor selection optimizes dc/dc converters

Choose your inductor wisely

The choice of an inductor can make or break a converter design. This article looks at the considerations involved.

Bruce D Moore
Maxim Integrated Products

Battery-operated systems and distributed-power-system concepts have greatly increased the use of low-power dc/dc converters and their associated magnetic components. Designers typically use transformers in dc/dc converter designs. However, unless transformer isolation is required, an inductor makes sense for a low-power switching-converter applications. This is especially true if the circuit must be more efficient than a capacitor-based charge-pump circuit. Compared with transformers, inductors are easier to specify, procure, and mount. Single-transistor-driver schemes are more effective with inductors than with transformers. Best of all, there's a variety of inductor types available, each in a range of inductance values.

Because standard inductors suit most applications, you can usually avoid expensive custom-magnetics development. Selecting the optimum inductor requires some knowledge of the available inductor geometries and core materials, however. (Note that for some unknown reason, inductor data sheets are often cagey about the product's core-material type.)

In the context of designing dc/dc converters that have an output power capability of 10 watts or less, a boost regulator serves as a general example. The boost regulator used as an example, also called a flyback regulator, features pulse-frequency modulation (PFM) to first establish its L value and then calculate its maximum ratings. By examining the inductor further, you can find certain specifications and characteristics that are a function of its shape and materials.

Figure 30.1 illustrates the three basic topologies for single-inductor switching regulators – the buck (or step-down), the boost (or step-up), and

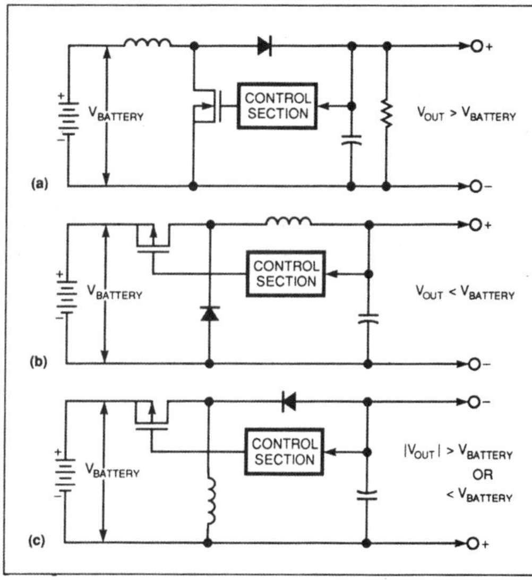

Figure 30.1 By arranging the same basic components in different topologies, you can obtain a boost (a), a buck (b), or a buck-boost (c) dc/dc regulator

buck-boost (or inverting) topologies. Low-power regulator applications generally combine one of these topologies with a PFM or PWM (pulse-width-modulation) controller circuit. The different combinations of topology and controller place similar demands on the inductor. However, you must take extra care in calculating the inductance value in pulse skipping controllers.

When selecting an inductor for a simple buck-regulator circuit, you must take several factors into account – inductance value, saturation current, dielectric strength, dc resistance, EMI, and stray capacitance.

The converter's switching frequency and the voltage drop across the inductor determine the correct inductance value. Both factors affect the

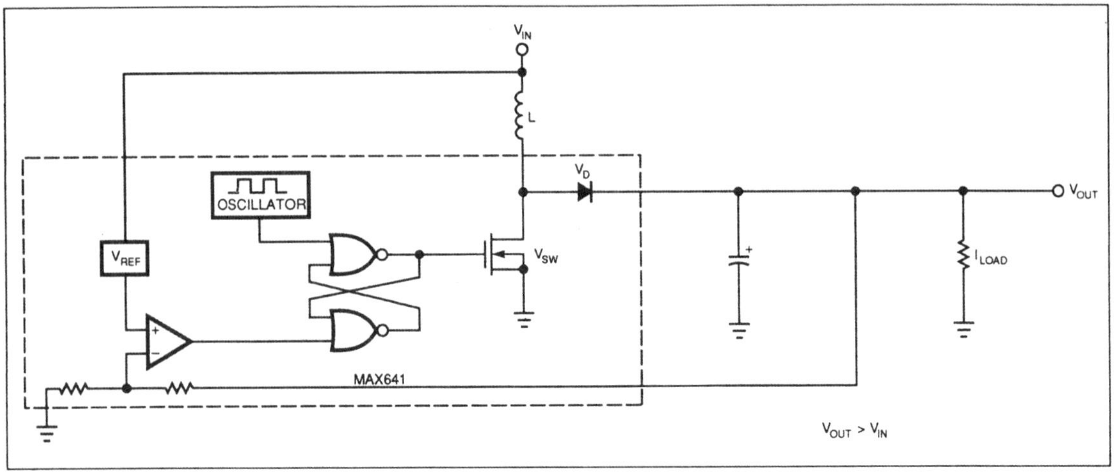

Figure 30.2 To regulate the output voltage, the MAX641 boost regulator employs clocked pulse-frequency modulation or pulse-skipping techniques

inductor current's rate of change in any given application. Erroneous inductor values can generate excessive current in the switching transistor or lower the inductor's energy storage performance. Insufficient energy storage, in turn, slows both converter start-up- and transient-response times. In clocked-PFM regulators, insufficient energy storage will also degrade load regulation.

Economic considerations also provide a strong incentive to reduce inductor values: lower values require less wire and smaller cores. As a rule, you should select an inductor whose value is as low as possible, and yet does not introduce excessive-current, inefficiency, component stress, and high-ripple problems.

Pulse-skipping PFM regulators, such as the boost circuit in Figure 30.2, are a good choice for battery-powered applications because the design draws minimal quiescent current and requires few external components. Here, however, you must be especially careful when selecting the inductor, to avoid load-regulation problems. Studying the regulator's switching waveforms (Figure 30.3) provides some clues about the factors that influence a PFM regulator's output.

Supply voltage and inductance value determine the slope of an inductors' current waveform. Because the regulators in question operate in a discontinuous-current mode (the current returns to zero on every cycle), the regulator's average load current is directly proportional to the peak inductor current. Peak current depends on slope, and the slope depends on the inductor's value. An inductor with an excessive L value, therefore, cannot transfer adequate energy on each oscillator cycle.

Peak inductor current can be expressed in terms of the voltage boost ratio and the load current as

$$I_{PK} = (4I_{LOAD})(V_{OUT} + V_D - V_{IN})/(V_{IN} - V_{SW}),$$

where the factor of 4 is a constant resulting from the 50% duty cycle. Values for I_{PK} and the switch transistors' on time (t_{ON}) let you calculate the optimum inductor value as

$$L = (t_{ON})(V_{IN} - V_{SW})/I_{PK},$$

where V_{OUT} equals output voltage, V_{IN} equals supply voltage, V_{SW} equals saturation voltage of the switch transistor, and V_D equals the forward voltage of the rectifier diode.

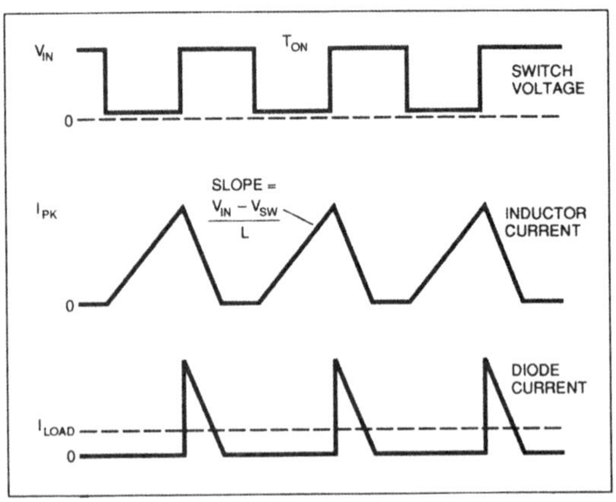

Figure 30.3 To ensure ac stability, the inductor current in Figure 30.1's pulse-frequency-modulation regulator must return to zero during each switching cycle

Coil inductance must not be so low that peak currents saturate the core or overstress the switching transistor. This rule applies to both PFM and PWM regulators. Excessive inductor current causes many odd symptoms, including low efficiency, rattling heat sinks, whining coils, and increased output ripple. Very low inductance can lead to burned windings and shattered, smoking transistors and ICs. The worst-case I_{PK} in the previous PFM-regulator calculations occurs when load current, supply voltage, and the diode's forward voltage are all maximum values while inductance, switch on-resistance, winding resistance, and switching frequency are all minimum values.

As you strive to extract higher output current from a PFM regulator, the minimum- and maximum-allowable inductor values tend to converge. An actual convergence indicates the need for a power transistor capable of higher currents. But excessive peak currents can result from the discontinuous-current mode in which the inductor operates. This I_{PK} limitation makes clocked-PFM regulators impractical for power levels exceeding 10 W.

The following example uses earlier equations for I_{PK} and L to convert an input of 5 V (± 10%) to an output of 15 V using a 1N5817 diode and a MAX641 regulator IC. The IC has a ± 10% tolerance on its 50-kHz oscillator frequency. The output-current capability must be 15 mA. First, calculate the maximum-allowable inductor value using the equation

$$I_{PK} = 4(15\,\text{mA})(15\,\text{V} + 0.4\,\text{V} - 4.5\,\text{V})/(4.5 - 0.75) = 174\,\text{mA}$$
$$L = (9\,\mu\text{s})(4.5\,\text{V} - 0.75\,\text{V})/174\,\text{mA} = 194\,\mu\text{H}.$$

Next, you must calculate the minimum-allowable inductor value. Here, let I_{PK} equal 450 mA – the maximum current rating for the switch transistor in the IC. With those criterion,

$$L = (11\,\mu\text{s})(5.5\,\text{V} - 0.25\,\text{V})/450\,\text{mA} = 128\,\mu\text{H}.$$

From off-the-shelf inductors, you can choose a standard value like 150 µH – a value which lies between the calculated minimum and maximum values.

Selecting the inductor for PWM regulators

Unlike pulse-skipping PFM regulators, most PWM regulators exhibit worst-case peak currents at minimum supply voltages. Because PWM regulators generally operate in a continuous-current mode at high duty cycles, their inductance values are limited only by winding constraints and the need for reasonable start-up and transient-response times.

In continuous-current mode, the inductor current fluctuates, but never returns to zero. Because the current may increase in staircase fashion over a period of several cycles, its rate of increase (determined by the inductance value) does not constrain the maximum level attained by the inductor current or the average load current. Thus, PWM regulator designs do not impose a hard limit on the maximum inductance value. The minimum value depends on the inductor's I^2R loss and the switching transistors' current capability.

The exact inductance value is seldom critical for regulation in a PWM circuit, but some control schemes require a particular value for other reasons. The MAX743 dual-output regulator IC, for example, which features current-mode feedback control, achieves ac stability by compensating the inductor current's rate of change, or slope. To ensure ac-stable operation, which is free from subharmonic noise, you must tailor the slope compensation to a given supply voltage and inductance value.

Don't overlook saturation effects

In some specialized circuits the design deliberately saturates the magnetic core (for example, Royer-type self-oscillating transformer-drive schemes and saturable magnetic reactors that provide regulation in multiple-output, transformer-based power supplies). In most applications, however, you select the inductance and the core material (using air gaps in the core if necessary) to avoid saturation. For PWM regulators in particular, you must take care to avoid magnetic saturation.

Saturation-current ratings measure an inductor's ability to handle high concentrations of magnetic flux. Strong magnetizing forces put an inductors' core at risk of saturation when the peak current rises to a high level. When a core saturates, the apparent inductance value falls off and current begins to rise exponentially. I^2R losses cause a drop in circuit efficiency, and the inductor cannot store additional energy.

High-current spikes resulting from saturation can endanger power transistors and cause noise and efficiency problems. To avoid saturation, the inductors' worst-case peak current must not exceed its peak current or incremental-current rating. Note that inductors lacking a dc-current rating are usually prone to saturation, as are those with an ac amps rating.

Energy stored in an inductor determines the output power available at a given operating frequency. You can calculate energy storage as $E = LI^2/2$, where E equals energy in joules, L equals nominal inductance in henries and I equals peak or incremental current rating in amps. The energy storage requirement for the previous clocked-PFM design example is therefore

$$E = (194\,\mu H)(174\,mA)/2 = 2.9\,\mu J.$$

Air gaps greatly amplify an inductors' ability to store energy by extending the effective length of the cores' magnetic path. Air can store a tremendous amount of energy. A small air gap, whether built into the material or created by grinding or machining operations, can more than double the power output from a given core by preventing saturation at high current levels.

Ferrite and other highly permeable core materials are very susceptible to saturation, and you must treat them accordingly. The closed magnetic path of a ferrite toroid, for instance, is extremely good for containing EMI. However, the short path combined with the ferrite's high permeability make toroids prone to saturation. Ferrite toroids make good transformer cores, but they are less suited to simple boost regulators featuring large dc-offset currents.

For applications of 5 W or lower, powder-type cores with distributed air gaps are often a better choice than ferrite toroids or pot cores. After cutting an air-gap slot in the core of a ferrite toroid, the machining costs and EMI level will be so high that you may have been better off choosing a ferrite bobbin, with its large inherent air gap, in the first place. Pot cores, though self-shielding, are more expensive because they require more manufacturing steps than the other types.

Iron-powder and molypermalloy-powder toroid cores offer the best combination of cost, size, and EMI performance for many low-power applications. These materials have built-in air gaps that allow the core to saturate gradually as the magnetizing force increases. The large number of tiny air gaps are created by the binder material. Each air gap saturates at a slightly different level of magnetizing force.

The cylindrical or bobbin-core geometry, although noisy, is best for low-power ferrite applications because such shapes are easy to wind, and are therefore inexpensive to manufacture. Ferrite pot cores are preferable for applications higher than 5 W. The pot core's low EMI emissions are a benefit in applications that involve high levels of current and magnetic field strength.

A look at core-material tradeoffs

There are no clear cut best-choice winners when you look at today's available core materials: each has a distinct advantage in certain areas (Table 30.1). Ferrites are attractive because they combine low cost with high volumetric resistivity, which

Table 30.1 Common inductors that suit dc/dc converters

Type	EMI	Comments
Ferrite bobbin	high	Makes compact, low cost, axial-lead (cylindrical) inductors. Low core losses support high efficiency.
Ferrite bobbin with ferrite shield	low	Efficient but prone to saturation.
Ferrite pot core	low	Efficient. Easily gapped to the correct value. Best for high-current or high-frequency applications.
Molded (low cost)	high	OK for light loads. Prone to saturation and often inefficient. Observe current ratings carefully.
Silicon-steel toroid	low	Tape wound; similar to iron powder. Use thinner tape for higher frequencies.
Ferrite toroid	low	Prone to saturation.
Molypermalloy-powder toroid	low	Best available for frequencies less than 400 kHz. Low EMI, low losses, compact, and expensive. Use high-flux type.
Iron-powder toroid	low	Specify core material carefully to achieve low losses.

minimizes eddy-current losses. Ferrites are the only choice for switching frequencies of 500 kHz and higher. On the other hand, ferrites' high permeability usually calls for an air gap and the associated complications: high EMI for bobbins and extra assembly steps for pot cores.

Molypermalloy powder (MPP), the Porsche of powder core materials, combines good saturation characteristics with low hysteresis losses. MPP is expensive, however; it contains scarce ingredients (nickel) and requires many processing steps. Iron powder and silicon-steel tape, despite their tendencies to sustain eddy-current and hysteresis losses, are inexpensive materials also suited to general-purpose applications.

For small size combined with good EMI performance, it's hard to beat high-flux MPP toroid cores. Standard MPP cores, formulated for RF applications, contain 80% nickel, plus iron and molybdenum. The high-flux variety contains 50% nickel and doesn't work as well for RF applications. However, high-flux MPP cores provide tremendous flux-handling that is useful in switching-regulator circuits.

For comparison, ferrite and standard MPP materials handle flux densities of 4500 and 7500 gauss, and the high-flux MPP material is good for 15,000 gauss. High-flux cores can handle switching frequencies ranging to 300 or 400 kHz before eddy-current losses become excessive. Like all MPP cores, the high-flux types are expensive. For low-power miniaturized applications, however, the high-flux MPP cores are often more cost-effective than ferrite types because they eliminate the need for precision gapping.

Size reduction is possible with high-flux core materials, as can be seen by comparing two prototype boards for the MAX743 switching regulator. Both boards generate 3 W at ± 15 V from a 5 V source. However, the surface-mount version uses a high-flux core material to achieve a much higher power density – 18 W/in^3 vs 2 W/in^3.

The core material affects the power level for a given inductor size, but dc resistance in the windings will waste some of that power. In the step-up circuit of Figure 30.2, the approximate average inductor current is given as

$$I_{AVE} = I_{LOAD}\,((V_{OUT} + V_D - V_{IN})/(V_{IN} - V_{SW}) + 1).$$

High winding resistance produces an L/R effect in the inductor-current waveform. The resulting I^2R losses degrade circuit efficiency and cause the core temperature to rise.

For pulse-skipping PFM-type regulators, the dc-winding resistance can have the same effect as an overly high inductance value. By limiting I_{PK}, dc resistance limits the available output current. Note that the calculation for maximum inductance value does not account for dc resistance. This dc resistance is significant for battery-powered and low-voltage applications of 3 V or less. In these cases, the inductance values must be low enough to achieve an acceptable slope for the inductor-charging current. Ideally, this slope is determined by the input voltage and the inductance value of $I(t) = V_{IN}(t)/L$. The peak current is therefore a function of t_{ON}, or $I_{PK} = V_{IN}t_{ON}/L$. Finally, dc-winding resistance (R_{DC}) limits inductor current as the inductance value approaches zero, as given by the expression

$$I_{PK} = (V_{IN}/R_{DC})\,(1 - e^{(-R_{DC}t_{ON})})/L.$$

Another current-related topic to consider is temperature rise. Inductor specifications generally include two current ratings – continuous (or rms) and dc saturation, which is sometimes referred to as peak or incremental current. A continuous rating accounts for the temperature rise caused by winding resistance, the inductor's operating temperature range, and its insulation or potting-material properties. The continuous rating is usually higher than the dc-saturation rating. Often, however, just the opposite is true for higher-valued inductors. As a rule of thumb, make sure that the inductors' average current is less than its continuous current rating. In high-frequency applications, be sure to include a safety margin for additional temperature rise due to core losses.

High-frequency losses consist of three major components – losses due to hysteresis, losses due to eddy currents in the core, and losses due to eddy currents in the wire. Legg's equation defines losses within the core as

$$R_{AC}/L\mu = Xf + YB_mf + Zf^2,$$

where R_{AC} equals the equivalent loss resistance in ohms, L equals inductance in henries, μ equals magnetic permeability, X equals residual-loss coefficient, f equals frequency, Y equals hysteresis-loss coefficient, B_m equals maximum flux level in gauss, and Z equals eddy-current-loss coefficient.

Magnetic hysteresis, which occurs as flux density nears its saturation point, becomes a problem in iron-powder cores at switching frequencies of 100 kHz or lower. A simple cure for this problem is to enlarge the core volume, which will reduce the peak flux density at high currents. Larger cores, however, exacerbate the eddy-current problem by providing more low-resistance paths for current. Eddy current in the core is a function of f^2, and rapidly becomes a problem as the frequency approaches 300 to 400 kHz. To combat the eddy-current problem, consider switching to another core material rather than changing the core size.

Eddy current in the windings (circulating currents within the wire) can also be a problem at frequencies of 500 kHz and higher. The solution here is to select a wire thickness equal to the skin depth of copper – a value determined by the switching frequency. Litz wire (ultra-thin, multi-stranded wire) or windings made from pc-board traces can help reduce eddy current. Positioning these windings within the core as far as possible from the air gap also helps.

Eddy currents appear in cores that have low volumetric resistivity. Low-resistance paths in the core cause the core to behave like a length of metal in a changing magnetic field. Circulating currents in the low-resistance path dissipate power. High switching frequencies (above 100 kHz) can develop significant eddy currents in iron-powder and steel-tape cores and these currents generate a core-temperature rise (Table 30.2.). Because the regulator may appear to be operating correctly, the problem can be difficult to detect.

You should settle EMI issues early in the design process, before they can affect the pc layout and component placement. Shielded inductors, for example, tend to be larger, more expensive, and more difficult to mount than unshielded types.

Table 30.2 Frequency limits for standard core materials

Frequency	Core Material
To 100 kHz	Standard iron powders and steel tape
To 200 kHz	Low-permeability, high-frequency iron powders
To 400 kHz	High-flux molypermalloy powders
To 500 kHz	Standard molypermalloy powders
To 1 MHz	Manganese-zinc ferrite
To 10 MHz	Nickel-zinc ferrite

Electronic circuitry can often tolerate moderate levels of EMI. This is particularly true for digital circuits. Even analog circuits, if they involve general-purpose ICs such as the LM324 op-amp, can often weather the noise associated with unshielded inductors. In these cases, it's worth the effort to try an unshielded bobbin-type inductor in place of a pot-core or toroid inductor. The bobbin type unit will cost half as much and be twice as small as an electrically equivalent pot-core or toroid inductor. Bobbin inductors generate their highest magnetic fields near the ends along the axis, so point them away from sensitive nodes and mount them at 90° angles to other magnetic components.

Another point to consider when selecting inductors is the problem of EMI. The air-gap spacings in powdered-material cores are so small that EMI is seldom a problem as long as the core is a toroid or has a geometry that features a closed magnetic path. Fringe fields are much greater for core geometries that include a cut, or have a large inherent gap such as that between the ends of a bobbin inductor wound on a cylindrical core. Pot cores and other clever mechanical designs make it possible to have air gaps in the ferrite material while keeping the EMI problem minimized.

The final problem area present is the stray inter-winding capacitance in inductors with values of several millihenries and larger. This capacitance combines with the coils' inductance to form a tank circuit, which rings at the inductors' self-resonant frequency (SRF). In low-SRF coils, a sudden jump of coil current, which occurs when the power switching transistor turns on, precedes the normal linear ramp of current. Large values of stray capacitance, therefore, cause switching losses that lower the regulator circuits' efficiency.

To avoid this problem, set the SRF at least five to ten times the oscillator frequency, thereby minimizing the interwinding capacitance. You can minimize this capacitance during the toroidal-winding phase by overlapping the ends of the winding somewhat, or by leaving a gap between the winding ends rather than ending the winding with one full layer.

31 ICs provide control for sensorless dc motors

Special ICs control dc motors without Hall sensors

Size, cost and efficiency constraints are driving Hall effect sensors out of small dc motors. So special techniques are required for speed control, and particularly for starting. Specialist ICs for the purpose are now appearing and this article describes one such.

Dave Peters and Jeff Harth
Silicon Systems Inc

As dc motors get smaller, there is less space for the Hall-effect sensors you normally use for motor control. As a result, miniature sensorless motors are becoming common. Eliminating the Hall-effect sensors removes the need for sensor-control signals, reduces the size of the connector, and eliminates connections. Hall-effect sensors are also sensitive to temperature variation, RFI, and other types of noise commonly experienced in motor applications. All of these factors reduce motor reliability. Finally, Hall-effect sensors increase motor cost and make the motors larger. As a result, the motors consume power – an undesirable trait in battery-driven applications.

While eliminating Hall-effect sensors obviously has its advantages, it also has some disadvantages. Without Hall-effect sensors, brushless dc motors are somewhat more difficult to start and control. Fortunately, a variety of semiconductor manufacturers now offer ICs that are specifically designed to control sensorless dc motors. These chips monitor the back-EMF the motors generate across specific sets of windings and appropriately adjust the amplitude and polarity of the voltages across the motor windings. In this manner, the ICs switch the motor from one commutation state to the next in a programmed manner, automatically setting and regulating the speed.

In addition to speed control, you should also consider the question of motor starting. Starting is a unique problem for the sensorless motor, because back-EMF exists only when the motor is already turning. For start up to occur, the motor must rotate in the proper direction and attain a speed that lets back-EMF sensing begin. Today's sensorless motor-controller chips switch the amplitude and polarity of the voltages across the motor windings so that the motor starts turning slowly and then accelerates, in a programmed manner, to a speed where back-EMF speed control can take over.

If start up always involves identical conditions, you can preprogram a single self-start acceleration profile for the motor. If the load or other conditions will vary, start up may require adaptive microprocessor control. In this case, the μP will adjust the start profile to begin with the unknown load.

You can control motor speed by varying the current in the motor windings. You can vary the current in a linear fashion by varying the voltage applied to the windings, or you can use pulse-width modulation (PWM) techniques and keep the voltage constant.

PWM-control techniques typically vary the duty cycle of the voltage applied to the motor windings to control the current in them. PWM control is most appropriate in applications where

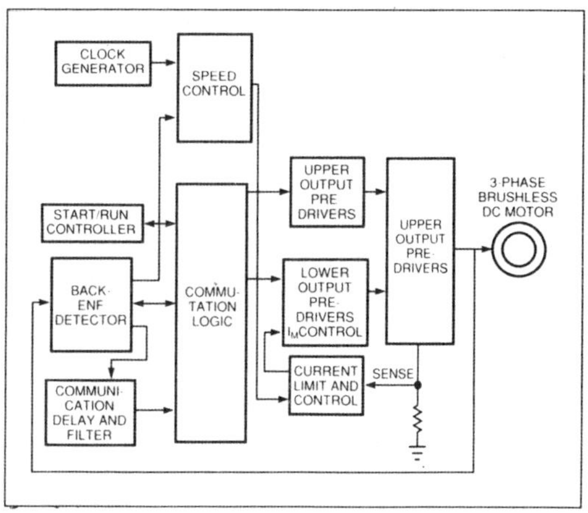

Figure 31.1 The speed-controller chips available today provide all the timing and control functions necessary to control speed, drive, and brake 3-phase, multipole brushless dc motors

low power dissipation is very important. You vary the current to each winding by switching a power transistor on and off at a variable rate, or at a fixed rate with a variable duty cycle. Unfortunately, PWM-control schemes are more difficult to implement than linear schemes.

For either speed-control method to work, the controller must receive an index pulse, which indicates the start of a motor revolution. There are two ways to generate this index pulse. You can use an external optical shaft encoder to generate the pulse, or you can use the on-board speed controller or an external μP to flag the start of a motor revolution by counting changes in commutation states. A 3-phase motor has six commutation states for each pair of poles. Once the

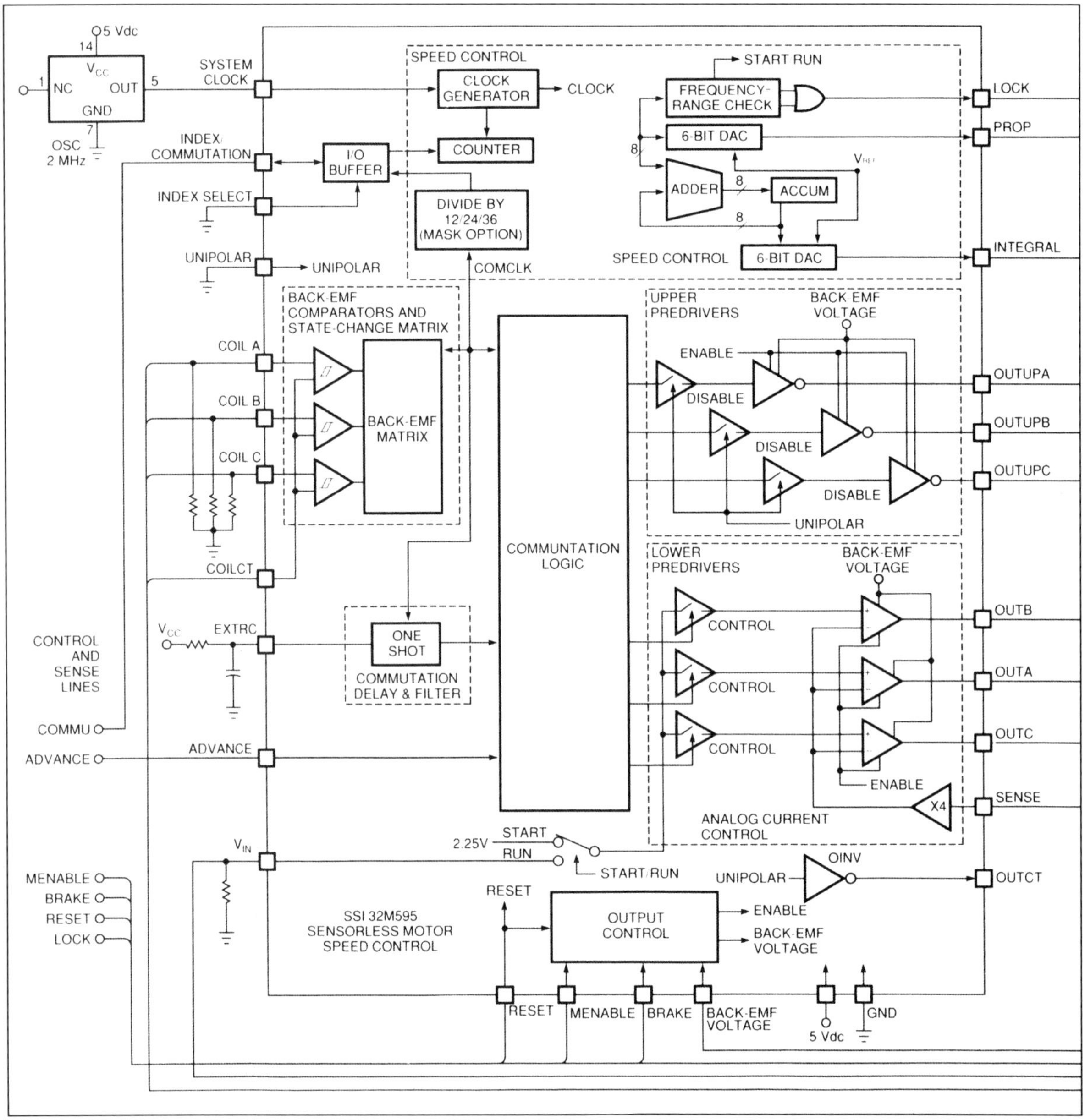

Figure 31.2 The current-limiting resistor R_{SENSE} is a key component in the sensorless motor-controller circuit. The resistor limits the start current to the motor windings and should be chosen to allow a start-up current about 10% higher than what is required

μP knows how many pairs of poles the motor has, it can calculate the number of changes of commutation state in one revolution of the motor. It then simply counts changes of commutation state for each revolution and knows precisely when a new revolution is beginning.

Figure 31.1 illustrates the basic system for brushless dc-motor control. This generalized system appears in two specific examples – one involves the control of a 12 V bipolar motor and the second deals with a 24 V unipolar motor. Both examples use the SSI32M595 sensorless motor-speed control chip – a device that contains on-board speed control. However, other products on the market can do a similar job.

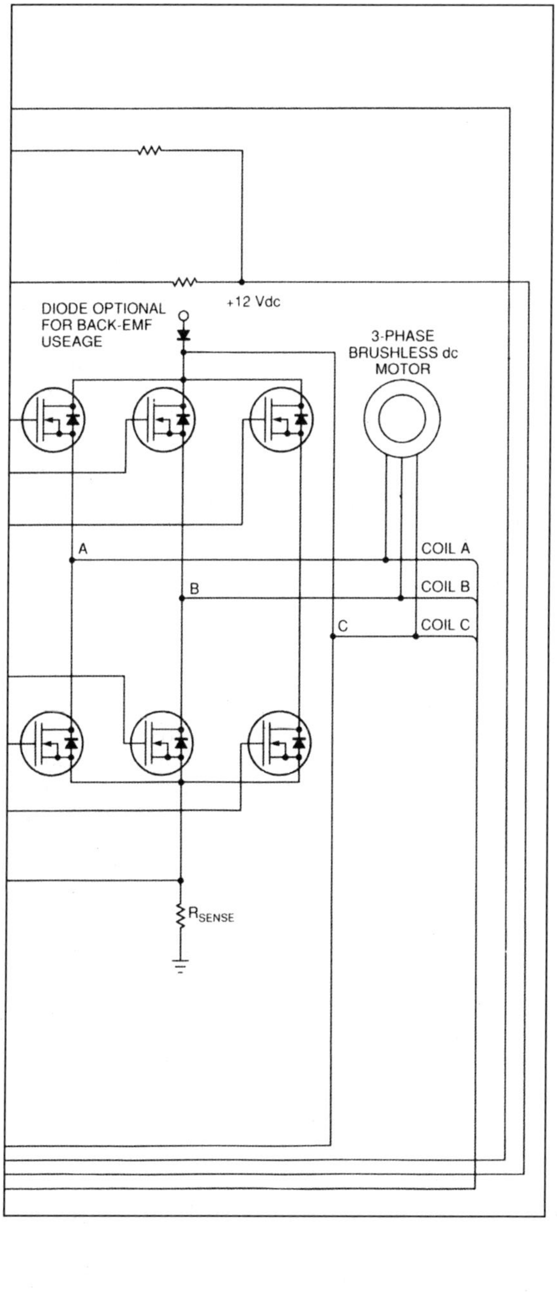

The bipolar approach has the advantage of providing greater torque with the same motor design, while the unipolar scheme allows higher rpm, because all the voltage develops across a single winding. In addition, the unipolar design requires only one power transistor per terminal. However, it also requires a voltage divider circuit at each terminal to bring the voltage to the 12 V range.

Solving disk-drive problems

A disk-drive application using a sensorless motor as the spindle motor best typifies the task of 12 V bipolar motor control. Assume that the drive has four 3½-in disks and eight read/write heads. The disks rotate at 3600 rpm (377 rad/s), and the spindle motor must get the disks up to the speed where the heads fly within 2 s. The heads start in contact with the disk surfaces, yielding 15 g of head load and a friction coefficient of about 1. At a speed of 1000 rpm, the heads start to fly.

Estimating the system inertia is the first step in the design. If you assume that motor inertia is small compared to disk inertia (a valid assumption in most cases), then system inertia will be equal to disk inertia. Assume that disk inertia equals 0.0075 oz-in-s^2 in this case. Calculating torque requirements is the next step. The acceleration rate (α) necessary to get up to 1000 rpm (104.7 rad/s) in 2 s equals the speed divided by the time to reach that speed, or 104.7/2 = 52.35 rad/s^2. The torque (T) required to accelerate the inertia equals the system inertia (estimated earlier to be 0.0075 oz-in-s^2) multiplied by the acceleration rate α, or

$$T = 0.0075 \times 52.35 = 0.393 \text{ oz-in.}$$

To overcome the head friction at start-up, you must add an additional torque element, T_F. Assuming a head-park radius (distance from center of the disk to the track location where the heads are parked) of 0.55 in, then T_F will equal

$$(15\,\text{g}/28.35)(8)(1)(0.55\,\text{in}) = 2.3 \text{ oz-in,}$$

where 15 g equals the head load, 28.35 equals the metric-to-English conversion factor, 8 equals the number of heads, 1 equals the coefficient of friction, and 0.55 equals the head-park radius. The total starting torque required is the sum of T and T_F, or 2.69 oz-in.

Take a worst-case approach

To calculate the starting current for the motor windings, you must first calculate the motor-torque constant. To reflect worst-case conditions in a 12 V system, assume a value of 10.8 for the low-line condition. Allowing 3 V for head room to control speed means that the motor's back-EMF should be no more than 7 V. Thus, the torque constant, K_T, will be

$$(7/377)141.6 = 2.63 \text{ oz-in/A},$$

where 377 equals the disk rotation in rad/s and 141.6 is a factor that converts (volt-s)/rad into oz-in/A. This is the probable upper limit to K_T and is meant to serve as a first approximation. Dividing the total starting torque ($T + T_T$) by the torque constant gives the start-up current for the motor windings as 2.69/2.63 = 1.02 A.

Figure 31.2 details the control scheme for the 12 V bipolar sensorless motor application. Note that the schematic includes a current-limiting resistor (R_{SENSE}) that limits the start current to the motor windings. Select a value for R_{SENSE} that allows a start-up current about 10% higher than necessary. In this example, the current limit is equal to 2.25 (the chip's internal acceleration voltage) divided by $4R_{SENSE}$. If R_{SENSE} is equal to 0.5 Ω, the current limit works out to 1.125 A.

At this point, you can calculate the elements of a start-up table. This table is a list of delay times between advances to the commutation state counter that allow time for the motor to rotate to the location where the next advance should occur. For a typical 3-phase, 8-pole motor, the angle of rotation per state is

$$\theta = 2\Pi/(3)(8) = 0.262 \text{ rad/state}.$$

The amount of time required to complete a rotation of angle θ is either

$$(\omega_{(N-1)} + \omega_N + \alpha T_N/2)T_N,$$

or

$$(\alpha TN^2)/2 + \omega_{(N-1)}T_N - \theta + 0,$$

where T_N is the time to move through an angle of θ in a time period of N; $\omega_{(N-1)}$ is the angular velocity at the end of the last time period and at the start of the N time period; ω_N is the angular velocity at the end of the N time period; and α is the angular acceleration.

You must now calculate α at the current limit I_M. The torque produced equals $I_M K_T$, which equals 1.25 A × 2.63 oz-in/A = 2.96 oz-in. The amount of torque available for acceleration is the torque produced at the current limit less T_F, or 2.96 – 2.3 + 0.66 oz-in. The acceleration rate α equals the torque available for acceleration divided by inertia, or 0.66/0.00785 = 88 rad/s².

You can enter this value of α into the quadratic equation for the amount of time to complete a rotation of angle θ and then solve the equation to determine time per step. The first step, where the previous velocity is zero, is starting the motor from a dead stop. When solving the equation for N = 1 and θ = 0.262 rad/state,

$$(88/2)T^2 + 0T - 0.262 = 0,$$

the time (T) to complete the first step of θT equals 0.077 s, and the instantaneous rotational velocity equals 6.79 rad/s.

Get a computer

You can use a simple computer program to run this calculation and generate a table. The number of steps in the start table must bring the motor above the speed where reliable commutation will occur. Then, the start cycle will end and self-commutation will take control. Assuming that the target speed is 500 rpm, a table for the design in Figure 31.2 will require only about 60 steps (Table 31.1).

The next task is to calculate the maximum motor resistance, R_M. The total available low-line power-supply voltage is impressed across three components – the driver transistors, the current-sample resistor, and the motor. The motor-terminal voltage is simply the back-EMF plus the IR drop across the motor resistance. There are two power transistors driving each motor winding in a bipolar application. Assume that the voltage drop will be the same across each transistor.

For a current of 1.125 A (the current at start-up), the voltage drop across both transistors will be about 2 V. The voltage drop due to motor resistance, V_{RM}, equals the low power-supply-voltage level minus the back-EMF (which is zero at start)

Table 31.1 Motor start up

Interval	T (msec)	T total (msec)	rpm
1	77.2	77.2	64
2	31.98	109.19	91
10	12.52	244.15	204
20	8.74	345.29	289
30	7.1	422.89	354
40	6.14	488.31	409
50	5.48	545.95	457
59	5.04	593.05	497
60	5.0	598.06	501
61	4.96	603.02	505
70	4.63	645.97	541

minus the voltage drop across the driver transistors less the voltage drop across the sense resistor. Thus, V_{RM} works out to be $10.8 - 0 - 2 - (1.125 \times 0.5) = 8.24\,V$. Since $V_{RM} = I \times R_M$, R_M will equal $8.24/1.125 = 732\,\Omega$.

The values for K_T and R_M are a good approximation for a design start. Having some values for K_T and R_M (two key motor specifications) lets you move on to the next step – setting the speed-control compensation. A 2-MHz system clock sets the internal speed control for 3600 rpm. External resistors (between the proportional-error DAC, the integral-error DAC and V_{IN} to ground) provide the compensation for the speed-control loop.

You can determine the required values for the gain-control resistors needed to complete the speed-control compensation design by solving the loop-gain equations. The loop-gain control equation is

$$LG = (11.6K_IK_AK_T/Js^2) + (0.775K_PK_AK_T/Js),$$

where LG is the speed-control loop gain; K_I is the integral gain set by the external resistor (R_I) in the network; K_P is the proportional gain set by the external resistor (R_P) in the network; K_A is the current-control transconductance gain = $1/(4R_{SENSE}) = 0.5\,A/V$; K_T is the motor torque constant = 2.63 oz-in/A; J is the inertia = 0.0075 oz-in-s^2; s is the LaPlace Transform frequency variable, which $j\omega$ can replace.

For 45° phase margin when the loop gain equals 1.0, the loop-gain in rectangular form is

$$-0.707 - j0.707 = (11.6K_IK_AK_T'J\omega^2) - j(0.775K_AK_TK_P/J\omega).$$

Assume you want a 1-Hz bandwidth, then you must solve the above equation when $\omega = 2\pi(1\,Hz)$. Solving the real and imaginary parts separately for K_I and K_P yields values of $K_I = 0.0132$ and $K_P = 0.0327$. Figure 31.3 shows the

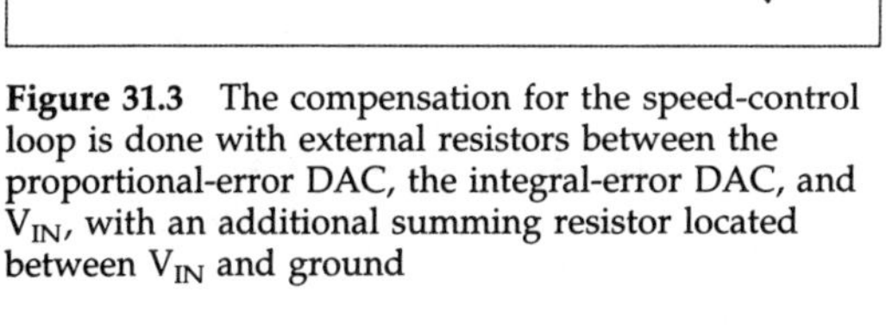

Figure 31.3 The compensation for the speed-control loop is done with external resistors between the proportional-error DAC, the integral-error DAC, and V_{IN}, with an additional summing resistor located between V_{IN} and ground

network that will produce these gains, where $R_O = 10\,k\Omega$, and R_P and R_I are the gain control resistors:

$$R_P = R_O(1 - k_I/K_P) - 1 = 291.6\,k\Omega$$

and

$$R_I = R_O(1 - K_P/K_I) - 1 = 695\,k\Omega.$$

This ends the design process for the circuit in Figure 31.2.

Handling unipolar applications

To illustrate the design process for sensorless motor control in unipolar motor applications (Figure 31.4), consider the following design problem. The application here requires a variable speed control with an external speed sensor and controller, using a sensorless dc motor. The input for the speed control is the transconductance input V_{IN}. Load inertia plus motor inertia will be 0.01 oz-in-s^2, and speed control is required over a 1000- to 10,000-rpm range. Speed must be capable of changing from the low speed to the high speed in 1 s. In this case, assume that you have negligible friction torque or viscous torque to deal with. Also, the motor is a 3-phase, 4-pole wye-connected motor, with centre tap, as in the 12 V bipolar example.

To operate in a unipolar fashion at greater than 12 V, you must develop voltage divider networks for each of the motor's terminals, since the control chip should never have more than 15 V on any input. In addition, you must bypass the motor-controller chip's internal low-rpm switch so that it will not switch motor current control away from V_{IN} to the internal acceleration voltage (2.25 V). This circuitry provides full acceleration if the speed is very low compared to 3600 rpm (the nominal fixed motor speed used in disk-drive applications).

With a 2-MHz clock and an 8-pole motor, the low-rpm switch will activate if the rpm drops below 3492. To bypass the low-rpm switch, select a new clock frequency, which is a function of the number of poles in the motor and the motor's lowest operational control speed. The formula for the new clock frequency is

$$F_C = (\text{number of poles}/8)(\text{low rpm}/3492)(2\,MHz).$$

Because the number of poles is four and the low rpm is 1000, the new clock frequency will equal 286 kHz. The resistor-divider network, connected to the motor windings, is shown in Figure 31.5. The maximum voltage at A, B, and C equals the supply voltage plus the back-EMF.

The acceleration rate necessary to reach full speed in 1 s places a torque requirement on the motor. You can calculate this torque by multiplying the acceleration (from 1000 rpm to 10,000 rpm in 1 s) by inertia. Because the acceleration required is 942.5 rad/s^2, and the inertia is 0.01 oz-in-s^2, the torque required is 9.425 oz-in.

Torque constant is a function of the amount of back-EMF the system can handle at maximum speed and the motor-winding resistance. The total voltage available is 24 V. This voltage drops across the back-EMF, the driver transistor, and the motor winding. Maximum back-EMF occurs when the motor is spinning at 10,000 rpm (1047 rad/s). Therefore, the back-EMF is related to the torque constant K_T by $(K_T/141.6)(1047 \text{ rad/s})$ and equals $7.394K_T$.

You can determine the IR drop in the motor windings (R_M) and R_{SENSE} by dividing the motor torque by the torque constant and multiplying the

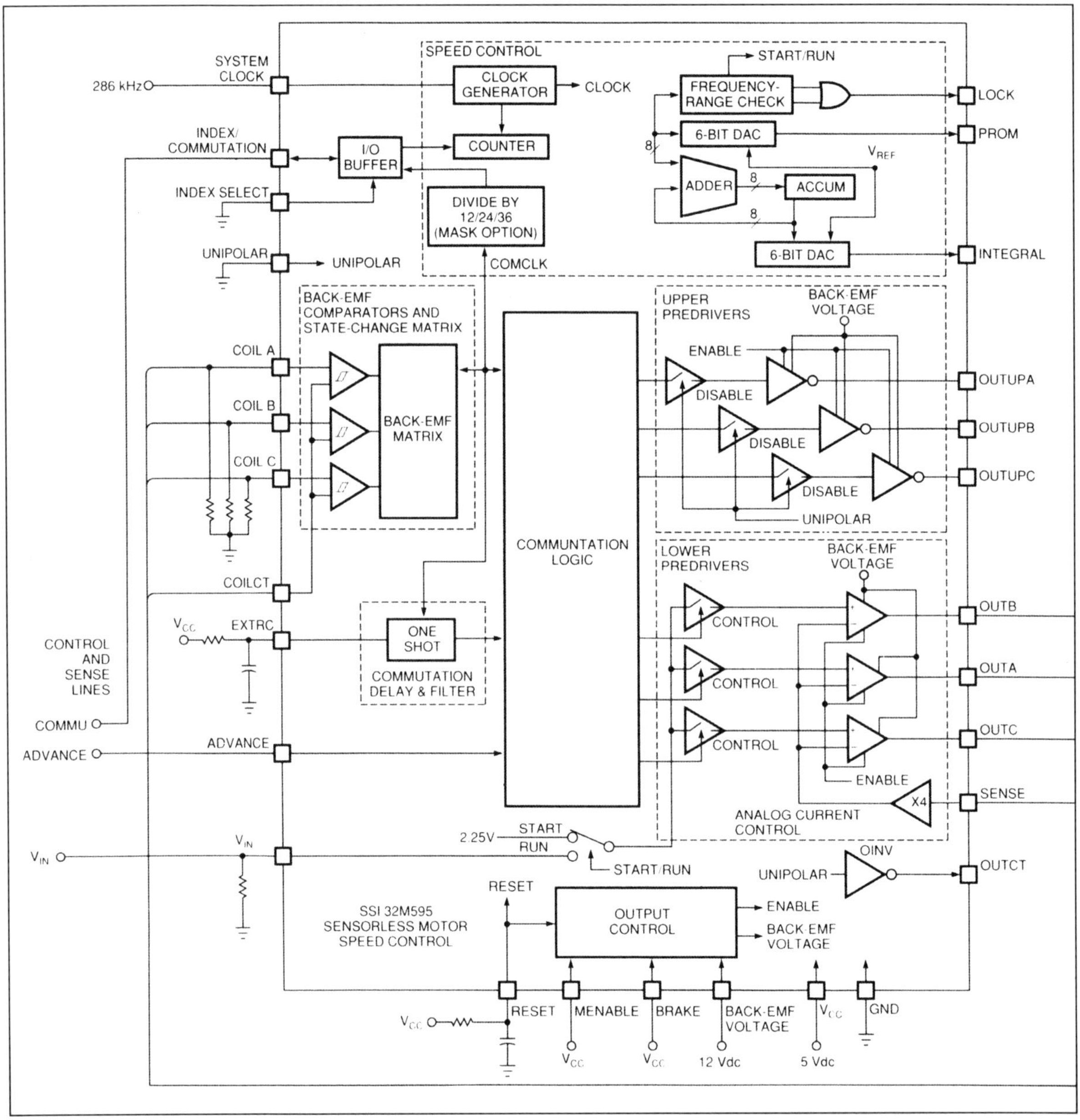

Figure 31.4 Speed control for this unipolar motor-control circuit is provided at the transconductance input, V_{IN}. In this case, the motor is again a 3-phase, 4-pole wye-connected unit with a center tap

result by the sum of the motor-winding resistance (R_M) and R_{SENSE}. To calculate V_{SAT}, the voltage dropped across the driver transistor, divide the motor torque by the torque constant and multiply the result by the effective resistance (R_{DSON}) of the driver transistor – 0.3 Ω, in this case.

The voltage applied across the motor terminals in Figure 31.4 is equal to the back-EMF ($7.394K_T$) plus the IR drop in the motor windings $(9.425/K_T)(R_M + R_{SENSE})$ plus V_{SAT} $(9.425/K_T)$ R_{DSON}. Therefore, 24 V – back EMF – IR – V_{SAT} = 0 is a limit for K_T and R_M. Let $R_T = R_M + R_{SENSE} + R_{DSON}$. Therefore,

$$24 - 7.39K_T - (9.425/K_T)R_T = 0.$$

Now you rewrite the equation to solve for K_T as

$$7.394\ K_T2 - 24K_T + 9.425R_T = 0$$

You can assume a value for R_T and solve the quadratic equation K_T. If the roots are real, K_T can take the higher value and R_T can assume the trial value. To find the highest possible value for R_T, iterate to find the point where the roots for K_T are real and nearly equal to each other. The iteration yields an R_T of 2.065 Ω and a K_t of 1.66:

$$R_T - R_{DSON} = R_M + R_{SENSE} = 2.065 - 0.3 = 1.765\,\Omega$$

The current requirement through the motor windings will be equal to

$$9.425/K_T = 9.425/1.66 = 5.68\text{ A}$$

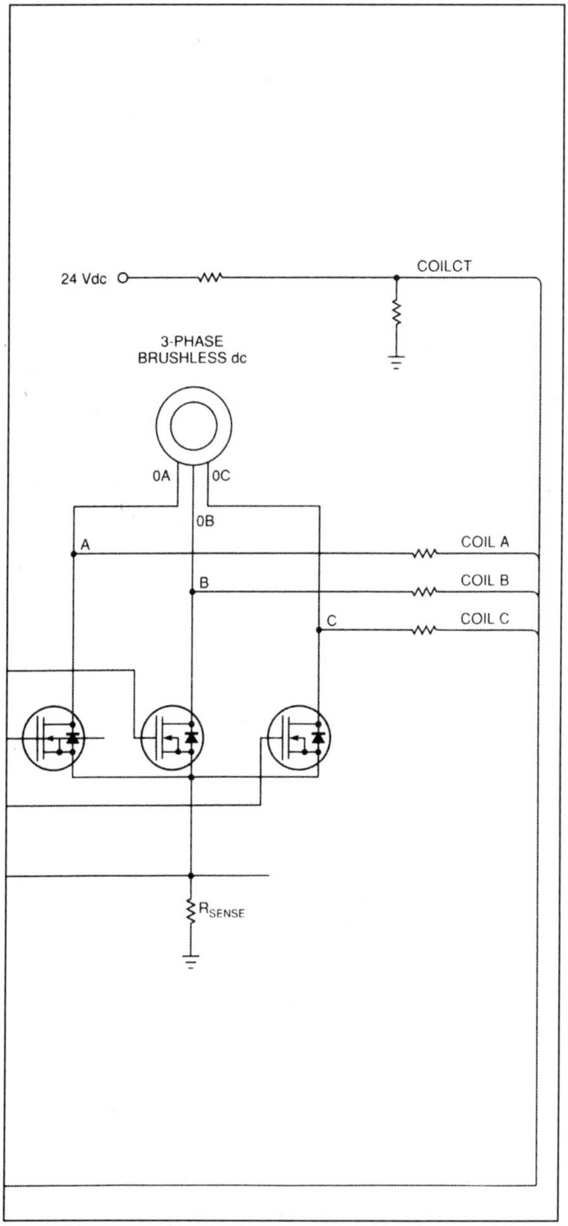

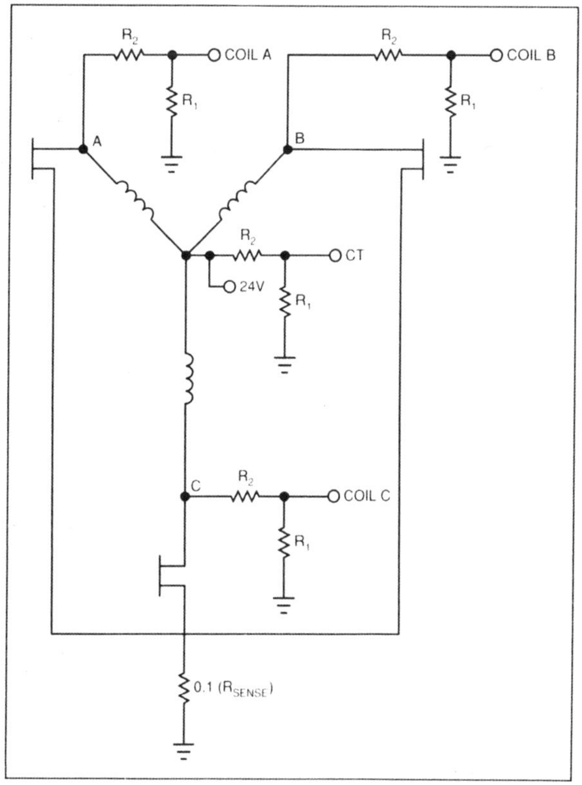

Figure 31.5 To operate in a unipolar fashion at voltage levels greater than 12 V, you must develop voltage-divider networks for each of the motor's terminals because the control chip should never have more than 15 V on any input

The value of R_{SENSE} will be equal to the voltage level at the current limit divided by the internal gain of the sense amplifier times the current requirement through the motor windings, or

$$2.25/(4 \times 5.68) = 0.099\,\Omega$$
$$R_M = 1.765 - R_{SENSE} = 1.765 - 0.099 = 1.666\,\Omega$$

Therefore, $K_T = 1.66$ oz-in. and $R_M = 1.666\,\Omega$ max.

For these values, the peak voltage at the motor terminals is the voltage drop across the motor windings plus the voltage drop across the driver resistor, or

$$24 + 7.394K_T = 36.274\,\text{V}.$$

To divide this peak voltage down to 12 V or less at the motor winding, you will have to use a divider network such as that shown in Figure 31.5. You can use the expression

$$36.274(R_1/R_1 + R_2) = 12$$

or

$$R_2 = 2.023R_1$$

to calculate R_1 and R_2. If R_1 equals 1.8 KΩ, then R_2 would equal 3.64 kΩ.

V_{IN} is the input to control speed and 2.25 V is the maximum input voltage. V_{IN} controls the motor current, which can control speed. You need an external speed monitor in this case. The transconductance from V_{IN} to I_M is $1/(4R_{SENSE})$ A/V. Having calculated values for the voltage divider and motor parameters K and R, you have completed the design of the 24 V unipolar sensorless dc motor controller.

32 Switcher boosts regulator's efficiency

Linear regulator always operates at minimum headroom

Circuit operation is controlled by the ripple present across C_2, but this is of course regulated out by the three-leg linear regulator. Remember, though, that at very light loads, conduction through the choke will become discontinuous, and the switching frequency will wander down into the audio band, which could be troublesome.

Brian Huffman
Linear Technology Corp, Milpitas, CA

Placing a linear regulator in the servo loop of a switching regulator yields a circuit that has both the precise regulation of a linear regulator and the efficiency of a switcher. The circuit in Figure 32.1 maintains only the bare minimum voltage necessary for proper operation across the linear regulator, even if you short the outputs. (Note that Figure 32.1 shows a pair of identical supplies configured as a bipolar supply.)

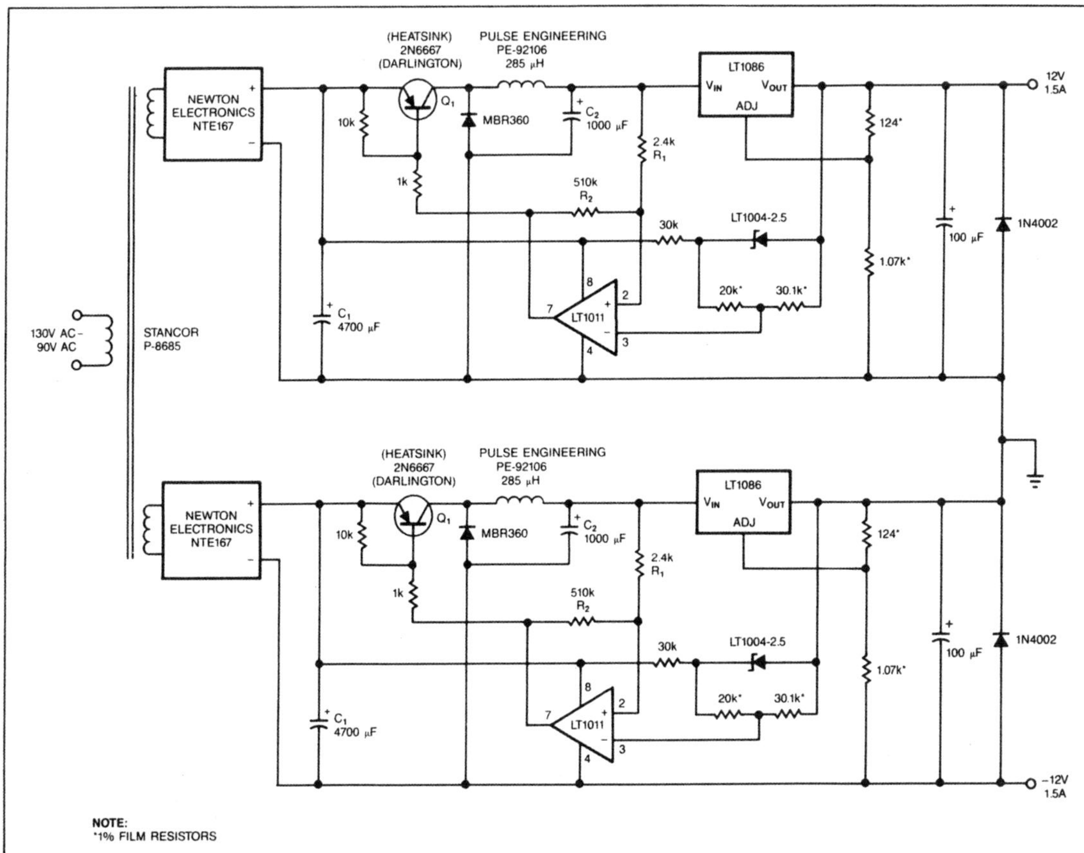

Figure 32.1 Placing a 3-terminal linear regulator within the servo loop of a switching regulator yields a power supply with tight regulation and good efficiency

The voltage drop across the resistive divider, which bridges the 2.5 V, LT1004 voltage reference, determines the differential voltage across the linear regulator. When the comparator senses that the regulator's input voltage is below the differential-voltage setting, the comparator switches low, thus turning on Q_1. With Q_1 on, charge stored on C_1 flows through Q_1, charges up the inductor and C_2, and supplies the load via the linear regulator.

When the linear regulator's input voltage rises above the differential-voltage setting, the comparator's output goes high, thus turning off Q_1. With Q_1 off, the stored energy in C_2 and the inductor supply the load. R_1 and R_2 set the switching loop's hystersis. Protection diodes D_1 and D_2 eliminate common-load problems.

33 Low dropout regulator suits 3 V systems

Regulator needs less than 150 mV headroom

This regulated power supply provides a stabilised 3.0 V supply rail for 3 V systems from three 1.2 V NICAD cells, right down to an end point of 3.15 V.

Chester Simpson
National Semiconductor, Santa Clara, CA

Linear regulators for 3 V systems must have especially low dropout voltages. The cell voltage at the end of a NiCd battery's life drops to about 1.05 V, which only provides about 3.15 V of total supply voltage to generate the 3 V output. This means the dropout voltage of the regulator must be less than 150 mV to maintain a 3 V output.

The circuit in Figure 33.1 specifically suits this very low dropout requirement with a typical dropout of 130 mV while supplying 100 mA of load current. You can achieve higher load currents or lower dropout voltages by using an FET with a lower on resistance or by using more FETs in parallel. The circuit uses an LM358 for the error amplifier and an LM4041–1.2 micropower reference to provide a stable 1.23 V to set the output voltage. Q_1 acts as a current source to supply load current, forcing the voltage at the output to be the value set by the ratio of R_1 and R_2. In this case, $R_2 = 100\,k\Omega$, so

$$R_1 = (100k \times V_{OUT}/1.23) - 100k.$$

C_1, C_2, and R_3 reduce bandwidth and ensure loop stability. The LM358 has sufficient common-mode range to operate with its input 1.8 V away from the upper rail. Also, its output will pull to within 60 mV of ground, which puts maximum available gate drive on the FET.

When choosing an FET, remember it must be a logic-level type that turns on with only about 3 V of gate-drive voltage. Many FETs specify their

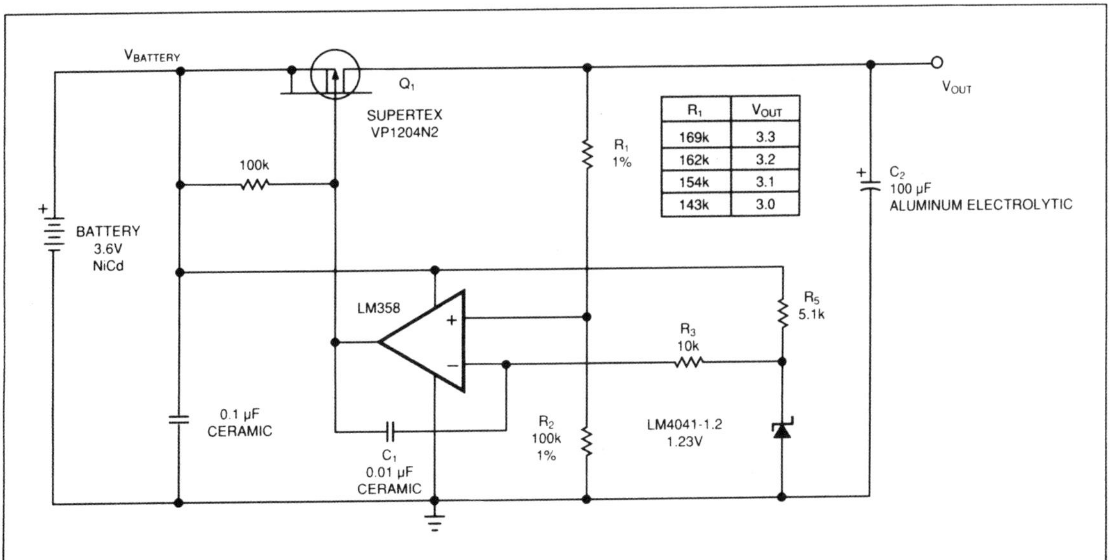

R_1	V_{OUT}
169k	3.3
162k	3.2
154k	3.1
143k	3.0

Figure 33.1 This linear regulator features a very low dropout voltage of 130 mV when powered from a 3.6 V battery and supplying 100 mA of load current

Table 33.1 Low-dropout regulator performance

V_{OUT} (V)	**V_{IN}** (MIN V)	**DROPOUT** (mV)
3.400	3.523	123
3.300	3.424	124
3.200	3.328	128
3.100	3.228	128
3.000	3.133	133
2.900	3.046	146

R_{DS}(ON) with 5 V of gate drive, and the on-resistance will be higher when you drive them with 3 V. Table 33.1 lists the dropout voltage under various conditions, with a load resistor of 31 Ω and an ambient temperature of 22°C. Note: this circuit uses no current limiting to protect the FET.

34 Regulator splits 8 to 40 V input into ± 15 V

Single regulator chip provides dual stabilized supplies

This clever circuit can do so since the IC's switch output at pin 4 can – unusually – swing below ground, to provide the negative output. Meanwhile, the positive output is provided via the 'flying capacitor' C_2.

Brian Huffman
Linear Technology Corp, Milpitas, CA

The power-supply circuit in Figure 34.1 uses one switching regulator IC, two inductors, and a flying capacitor, C_2, to generate a dual-output supply from a single wide-ranging input voltage. For a ± 15 V output, the negative output voltage tracks the positive supply within 200 mV for load variations from 50 to 500 mA. Negative output load current should not exceed the positive output load by more than a factor of 4 because the imbalance causes loop instabilities. For common-load conditions, the two output voltages track each other perfectly. Though the circuit was primarily designed for ± 15 V outputs, you can adapt it to ± 5 V outputs using the component values specified in the table. The output voltage is

$$V_{OUT} = -V_{OUT} = 2.21 \times (1 + R_2/R_3).$$

Efficiency of the ± 15 V output version is greater than 62% with a 0.5 A common load. Efficiency peaks at around 72% with an input voltage between 15 and 20 V. For ± 5 V outputs, efficiency is greater than 60% and peaks at 73% for an input between 20 and 25 V. The two main loss elements are D_1 and D_2, and IC_1's power switch. At low input voltages, the efficiency drops because the switch's saturation voltage becomes a higher percentage of the available input supply.

During the switching cycle, IC_1's V_{SW} pin swings between the input voltage, V_{IN}, and $-V_{OUT}$. The ability of IC_1's V_{SW} pin to swing below ground is unusual – most other 5-pin switching regulator ICs can't do this. While IC_1's power switch is on, current flows from the input voltage source through the switch, through C_2 and L_1 and into the load. A portion of the switch current also flows into L_2. This current recharges C_2 and C_4 to the potential of V_{OUT} during the switch-off time. The current waveforms for both inductors are

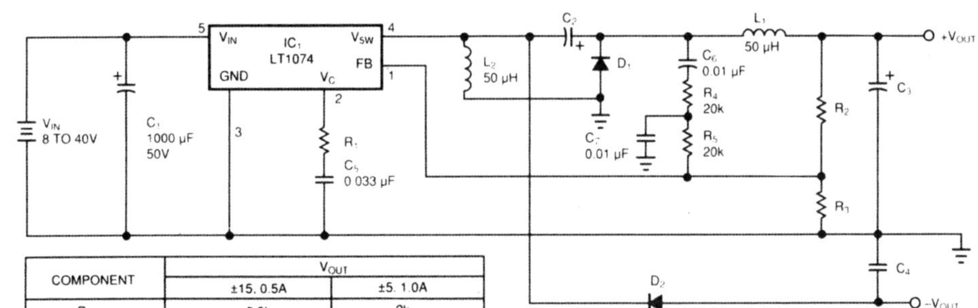

COMPONENT	V_{OUT} ±15, 0.5A	±5, 1.0A
R_1	3.3k	2k
R_2	7.50k, 1%	2.80k, 1%
R_3	1.30k, 1%	2.21k, 1%
C_2, C_3, C_4	470 µF, 25V (NICHICON UPL1E471MPH)	680 µF, 16 (NICHICON UPL1C681MPH)
C_5	0.01 µF	0.033 µF
D_1, D_2	MUR410 (MOTOROLA)	MBR360 (MOTOROLA)

NOTES:
C_1=NICHICON UPL1H102MRH
L_1, L_2=COILTRONICS CTX50-2-52

Figure 34.1 An alternative approach to a switching power supply with a multiple secondary transformer, this circuit generates a dual output using a switching regulator IC, two inductors, and a 'flying' capacitor virtually identical because the inductors have identical values, and the circuit applies the same voltage potentials across them during the switching cycles.

When the switch turns off, the current in L_1 and L_2 begins to ramp downward, which causes the voltages across them to reverse polarity. The voltage at the V_{SW} pin is forced below ground and falls until D_1 and D_2 are forward biased. During this interval, the voltage on the V_{SW} pin is equal to a diode drop below $-V_{OUT}$. L_2's current then circulates between both D_1 and D_2, charging C_2 and C_4. The energy stored in L_1 replaces the energy lost by C_2 and C_4 during the switch-on time. Assuming that the forward voltage drops of D_1 and D_2 are equal, $-V_{OUT}$ will equal V_{OUT}. After the switch turns on again, the whole cycle repeats itself.

IC_1's internal error amplifier controls the output voltage. The amplifier compares a fraction of the output voltage via the R_2 and R_3 divider network with an internal 2.21 V reference and varies the duty cycle until the two values are equal. R_1, C_5, and the network comprising R_4, R_5, C_6 and C_7 provide sufficient compensation to stabilize the control loop.

One advantage of this circuit is that L_1 acts as both an energy storage element and a smoothing filter for V_{OUT}. The output ripple voltage has a triangular wave-shape whose amplitude is determined by the inductor ripple current and the equivalent series resistance of C_3. This type of ripple is usually small, so a post filter isn't necessary.

35 Diode simulator reduces forward drop to 0.04 V

MOSFET pseudo-diode

This circuit acts as a near-ideal diode, provided you keep both electrodes within the comparator's common-mode range. Note that the MOSFET is used in the inverted mode, so that the device's internal source/drain diode is reverse biased when the 'diode' is off.

Isaac Eng, ESTCO Battery Engineering
Ottawa, Ontario

To gain higher power efficiency, a lower than 0.7 V diode drop is sometimes desirable. Figure 35.1's diode simulator exhibits a voltage drop of 0.04 V with 1 A of forward current, a reduction of more than tenfold the usual diode forward-voltage drop. Current is normally switched to flow from drain to source in an N-channel enhancement MOSFET. In Figure 35.1, however, current flows in the opposite direction, which is the same direction as an intrinsic drain-source diode. When the source is more positive than the drain, the comparator turns on the IRFZ40 FET to conduct current. A forward voltage drop of 0.04 V is observed when one amp passes through the simulated diode in the forward direction. When the drain is more positive than the source, comparator LM393 outputs a low value, which turns off the FET's current flow. Hence, current conducts in only one direction. The drain and source voltage must be within 0 and $V_{CC} - 1.5$ V, which is the input range of the comparator. The 210-kΩ resistor is a pullup resistor for the open-collector comparator.

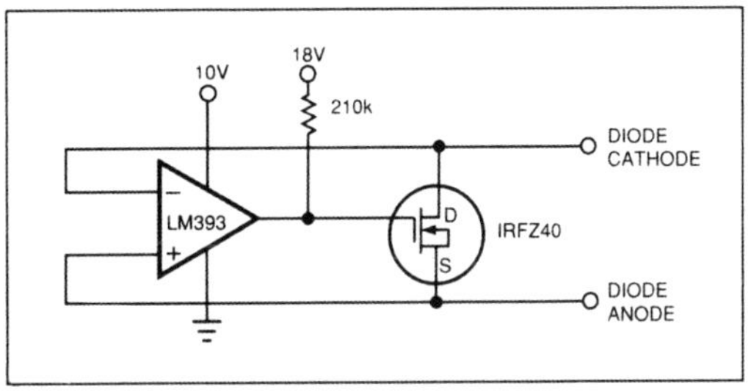

Figure 35.1 To create a circuit element that emulates a diode with a 0.04 V drop, this simple circuit causes current to flow from the source to the drain of the N-channel enhancement MOSFET

Part Six

Test, Measurement, and Standards

Introduction to Part 6

A good portion of *EDN*'s articles, both staff-written and contributed, deal with test instruments and testing techniques. And rightly so – engineering doesn't stop at the design level; you must be able to test systems and devices, to ensure that the designs work, and that the devices do what they're supposed to do. The first three articles in this part deal with A/D converters (ADCs). Two articles present useful techniques for testing and characterizing these complicated devices; the third shows how to use them to configure a fast-transient recorder.

Two articles center on the use of wideband oscilloscopes. One deals with testing digital circuits, and shows the shockingly deceiving results of using a scope with inadequate bandwidth. The other presents some tricky techniques for measuring the signals that can develop in disparate points of a ground system. Not to neglect the analog world, Jerry Graeme gives you techniques for testing distortion in modern op-amps. A staff-written report discusses available instruments for infrared-radiation (IR) temperature testing.

ISO 9000 is rapidly gaining momentum as the governing norm for quality-control systems. A staff-written article tells you the steps your company must take to earn this coveted certification. Another norm of European origin is the European EMC Law, a set of specs that govern EMI/RFI emissions. The staff-written article in this part discusses the arduous and expensive steps you'll have to take to satisfy the law.

36 Time-domain techniques enhance the testing of high-speed ADCs

Frequency domain testing can be incomplete

This article discusses some of the finer points of ADC operation, and how to test the converters.

Michael J Demler
Micro Networks

The increasing availability and applications of high-speed A/D converters has led most manufacturers to include specifications for dynamic performance in their data sheets. Limiting dynamic testing to DSP techniques such as FFTs or statistical modeling such as sine-wave histograms does not portray the complete picture of a device's performance. Designers can employ a sine-wave curve-fitting algorithm or beat-frequency test techniques to add measurements in the time domain for a more comprehensive analysis. Time-domain analysis allows an engineer to evaluate test results without the limitations inherent in transforming sampled signals into other forms.

The most common DSP test uses a fast Fourier transform (FFT) to do a spectral analysis of an ADC's digital output. Figure 36.1 shows the FFT obtained by testing the Micro Networks MN5902, an 8-bit CMOS flash ADC. The ADC was sampling a full-scale 828.75-kHz sine wave at 15 MHz. The test was a 4k-point FFT. The ratio of signal amplitude to noise plus distortion was 44.1 dB.

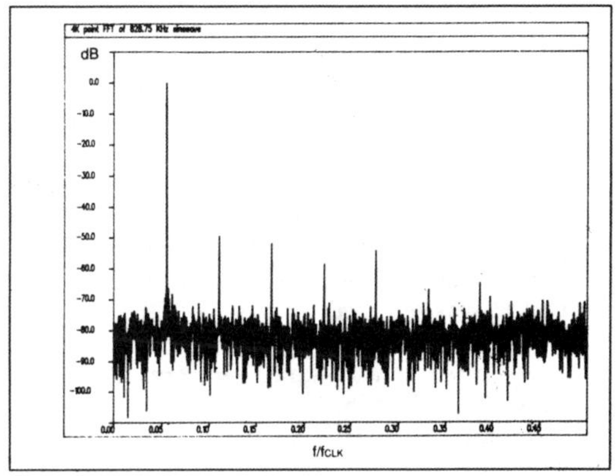

Figure 36.1 This plot shows the result of a 4k-point fast Fourier transform (FFT) performed on the MN5902 8-bit flash A/D converter. The ADC was sampling a full-scale 828.75-kHz sine wave at 15 MHz. FFT plots graphically illustrate parameters such as S/N ratio and spurious-free dynamic range (SFDR)

Spurious-free dynamic range (SFDR) is a measure of the usable dynamic range of an ADC. SFDR is defined as the difference, in decibels, between the amplitude of the fundamental and that of the largest spurious component in the FFT spectrum. In the example, the SFDR measured 49.6 dB between the fundamental and the second harmonic at 1.66 MHz.

An equation that expresses the signal-to-noise ratio (S/N ratio) of an ideal ADC with a resolution of n bits is S/N ratio = 6n + 1.8 dB ([1]). Because the S/N ratio of an ideal 8-bit ADC is 49.8 dB, the spurious-free range of 49.6 dB illustrated in Figure 36.1 depicts an ADC that exhibits a usable dynamic range of 8 bits. A parameter called the effective number of bits (ENOB), which is commonly derived from an FFT, provides a measure of dynamic resolution. To find the ENOB of an equivalent real ADC, the equation for S/N ratio is rearranged so that ENOB = (S/N ratio − 1.8)/6.

Using the data from the example FFT, with an S/N ratio of 44.1 dB, the effective number of bits would be approximately 7. One problem with this calculation is that all manufacturers do not use the same technique for measuring S/N ratio. Many measure the S/N ratio by first eliminating all harmonics and intermodulation products from the FFT. In such cases, the manufacturer should also report these distortion products separately.

In the Figure 36.1 example, the S/N ratio without harmonics was 47 dB, which results in an effective bit calculation of 7.5. The problem exemplified by this erroneous 0.5-bit improvement is that the S/N ratio measurement doesn't reflect the actual resolution and accuracy of the individual samples taken by the ADC. If your interest is only in frequency-domain processing, the error may not matter; but if the accuracy of dynamic measurements is important, you'll want to include the harmonics and other spurious components in the ENOB calculation.

To show how severe the ENOB error can be when relying solely on the frequency domain, the data used to calculate the FFT in Figure 36.1 was edited to create a spurious code at one of the zero crossings in the data record. Figure 36.2 depicts a portion of the edited sine wave. Instead of producing the code 128, the data was changed to produce code 0. Such spurious codes often occur and are called sparkle codes. The high slew rate of the input signal near the mid-scale point of the waveform can cause problems in the encoding circuitry of some flash ADCs. These problems result in glitches in the digital output. Figure 36.3 shows the result of reproducing the 4k-point FFT with the corrupted data. The spectra differ very little. The signal-to-noise measurement of the Figure 36.3 spectrum indicates less than 2 dB of loss, which shows up as a slightly elevated noise floor. The SFDR did not change significantly, and the ENOB was reduced by only 0.3 bits.

Using a long data record, such as the 4k points in the example, is a common way to accurately measure an ADC's S/N ratio. An undesirable result of this method is that small numbers of large errors, such as the mid-scale glitch, are averaged out with no indication of the severity of the error in the frequency domain. Shorter FFTs can have other problems because the higher noise

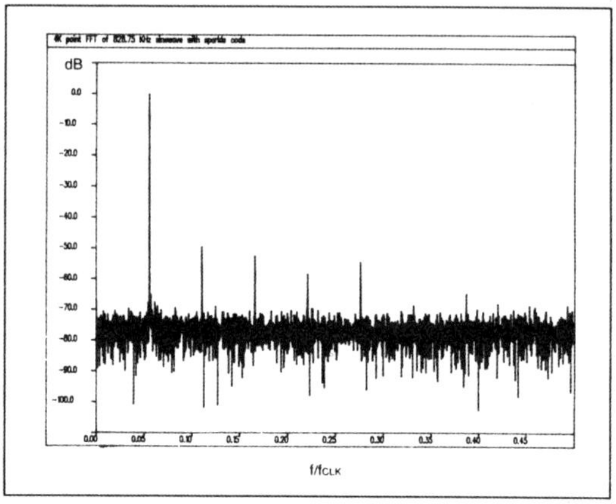

Figure 36.3 The result of reproducing the 4k-point FFT of Figure 36.1 using the corrupted data of Figure 36.2 shows very little difference when compared with the original. The main difference is the corrupted-data plot's slightly elevated noise floor

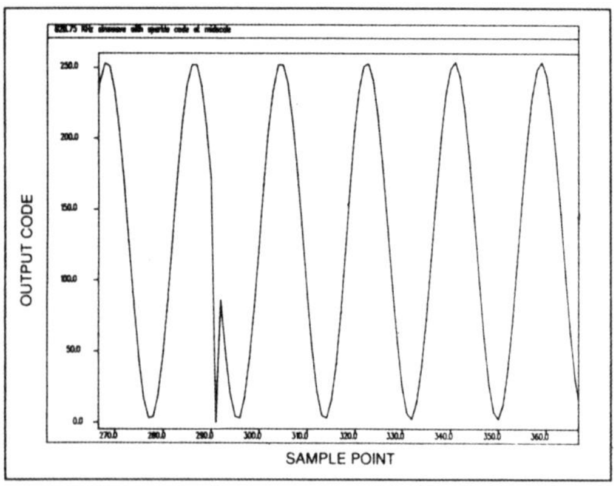

Figure 36.2 The data used to calculate the FFT in Figure 36.1 was edited to create a spurious code (code 0 instead of code 128) at one of the zero crossings in the data record. This plot depicts a portion of the edited sine wave, which exhibits corrupted data in the form of a sparkle code

floor that results from taking fewer samples can mask many small coding errors. For this reason, multiple FFTs are often performed so that the spectra can be averaged together to remove random noise and clearly discriminate harmonics. Of course, this averaging further conceals information about sampling errors unless you examine the actual acquired data.

An alternative method of calculating effective bits is to use an algorithm that will find the sine wave that achieves the 'best fit' to the data from the ADC[1]. A sine-wave-fitting algorithm recursively calculates the amplitude, offset, and phase of an ideal sine wave that results in the minimum rms error when compared with the actual quantized waveform. Using the same flash ADC tested in Figure 36.1, 512 data points were collected to produce an FFT and a best-fit sine wave. Subtracting the best-fit waveform from the original quantized data produced the dynamic-quantization-error plot of Figure 36.4. The plot provides a graphical measurement of time-domain errors under dynamic conditions.

As you can see, the samples were generally within ± 1 bit of the ideal, with worst-case errors of ± 1.5 bits. By including all of the nonlinearity and noise effects of the ADC, this method gives a true indication of the device's dynamic resolution. You should not confuse the dynamic-error plot with 'dynamic linearity', which some manufacturers specify in their data sheets. Dynamic linearity is usually based on histogram techniques, which can hide time-domain errors. In contrast, looking directly at the quantization error plot, both large and small errors, which could be lost in an FFT or sine-wave histogram, are immediately evident.

The rms quantization error (V_e (rms)) in Figure 36.4 is 0.55 bits. In an ideal ADC with a uniform distribution of all codes, the rms quantization error would be $q/\sqrt{12}$, where q is one quantization step, or LSB. This relationship leads to an alternative equation for effective bits:

$$V_e = \frac{V_{REF}}{2^{ENOB} \cdot \sqrt{12}}$$

$$ENOB = \log_2\left(\frac{V_{REF}}{\sqrt{12} \cdot V_e(\text{rms})}\right),$$

where V_{REF} is the full-scale range of the ADC.

Because this method includes all harmonics and spurious components, it is better for calculating the effective number of bits than the FFT-based technique. When you analyze data digitally, as in the quantization error plot of Figure 36.4, you can convert the equation for ENOB to units of bits rather than volts:

$$ENOB = \log_2\left(\frac{\text{MAX COD–MIN CODE}}{\sqrt{12} \cdot \text{rms BIT ERROR}}\right).$$

Converting to the more convenient base-10 logarithm yields

$$ENOB = 3.32 \cdot \log_{10}\left(\frac{\text{MAX COD–MIN CODE}}{\sqrt{12} \cdot \text{rms BIT ERROR}}\right).$$

A disadvantage of sine-wave curve-fitting algorithms is that they may have long execution times and sometimes have difficulty converging to a

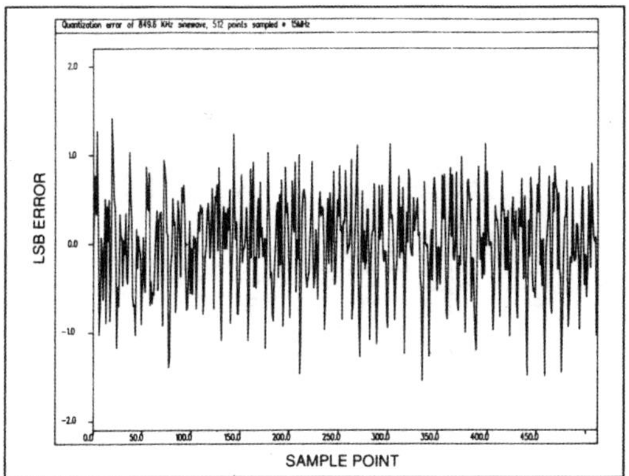

Figure 36.4 This quantization-error plot of the flash ADC first tested in Figure 36.1 provides a graphical measurement of the time-domain errors under dynamic conditions. As is evident, the samples are generally within ± 1 bit of the ideal

solution. An alternative that is fast and can provide a quick graphical interpretation of performance, as well as supply data for more quantitative measurements, is the beat-frequency test. This test is the ultimate stress test of a flash A/D converter in the time domain. The beat-frequency test directly measures input signal bandwidth, reveals sampling errors that result in spurious or out-of-place codes, and provides a measure of aperture errors.

Designers familiar with sampling theory know that the Nyquist criterion requires that you sample a signal at a rate greater than twice the signal's highest frequency component. Signals that do not adhere to this limitation result in 'aliases'. Aliases are the result of high-frequency signals at an ADC's input that are transferred to lower-frequency signals within the output bandwidth of one half the sampling rate. Beat-frequency testing exploits this phenomenon by deliberately producing low-frequency representations of high-frequency signals.

You should note that many flash A/D converters are specified for a 'full-power bandwidth' that is less than the Nyquist rate. Often this specification describes the point where gross errors occur on the ADC's output.

During a beat-frequency test, the ADC digitizes input signals that are multiples of the sampling rate. If you use an input sine wave just slightly higher in frequency than the sampling clock, the sampling point will change in relative phase at a rate equal to the difference in frequency between the two signals. By connecting the sampled data points in a continuous time record the single-tone beat frequency is traced out as Figure 36.5 shows. To guarantee that you exercise all the codes, calculate the input frequency to produce a change of less than one bit in the beat tone at the maximum slewing point. This point is set to the mid-scale point of unipolar ADC's ADCs. The calculation is

$$\frac{dV_{IN}}{dt}(\text{MAX}) = 2\pi \cdot A \cdot f_{BEAT}.$$

Set A to half the full-scale range of the ADC, which is $V_{REF}/2$ for a unipolar converter. For a change of one bit per sample at the maximum slewing point:

$$\frac{dV_{IN}}{dt} = \frac{2\pi \cdot V_{REF} \cdot f_{BEAT}}{2},$$

$$\text{and} \quad f_{IN} = \frac{f_{CLK}}{2^N \cdot \pi} + x \cdot f_{CLK},$$

where x is an integer.

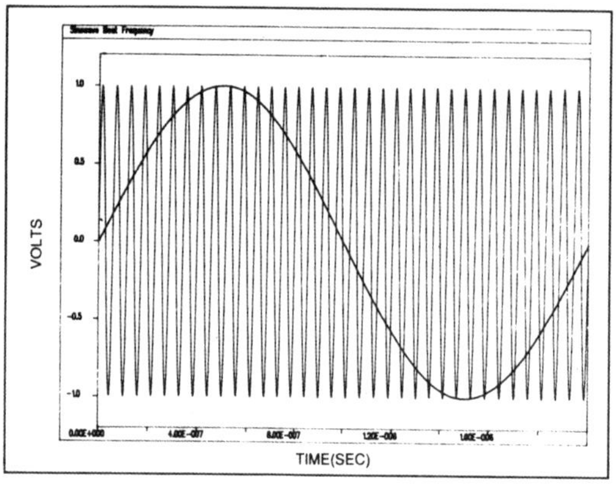

Figure 36.5 This single-tone beat frequency is the result of using an input sine wave slightly higher in frequency than the sampling clock, and connecting the sampled data points in a continuous time record. To guarantee that you exercise all the codes, calculate the input frequency to produce a change of less than 1 bit in the beat tone at the maximum slewing point. This point is set at the mid-scale point of unipolar ADCs

To demonstrate the technique, a series of beat-frequency tests were performed using the MN5902 CMOS flash ADC in the test circuit of Figure 36.6. The beat tests started by digitizing a full-scale input (f_{IN}) of 17.01 MHz with the ADC sampling at a 17.00-MHz clock rate (f_{CLK}). This test produced the 10-kHz beat frequency evident in the Figure 36.7 plot of digital output codes. In this plot, the horizontal axis represents the number of data points, and the vertical axis represents the 8-bit code. The peak-to-peak output-code amplitude was – 0.275 dB relative to full scale. An examination of the beat waveform shows no evidence of spurious codes.

You can also derive the beat waveform from signals that are offset in frequency from harmonics of the sampling rate. In the second test, which Figure 36.8 shows, the input frequency was doubled to 34.01 MHz while the sampling rate was kept the same. Once again the 10-kHz beat was obtained, with a peak-to-peak attenuation of 1.24 dB. In the third test, which Figure 36.9 shows, the input frequency is increased to three times the clock rate for an input of 51.01 MHz. Under these conditions, no spurious codes were present, and the peak-to-peak output was attenuated – 3.4 dB. Although these results include effects caused by the large-signal-bandwidth roll-off from the input-signal amplifier, attenuation in the beat frequency gives an accurate indication of the overall – 3-dB bandwidth of the composite sampling system.

Finally, in Figure 36.10, the input frequency was increased to 68.01 MHz, or four times the sampling rate. This increase reduced the output amplitude by 9.20 dB but did not result in any sparkle codes. Although you may observe increased distortion as you approach an ADC's performance limits, it is more important to users that a device degrades gracefully and has no abrupt onset of gross sampling errors. If more accurate measurements are required under extreme undersampling conditions, you can lock the two signal sources to a common reference to reduce phase jitter, which can corrupt the low-frequency beat. No attempt to do so was made for the tests described here.

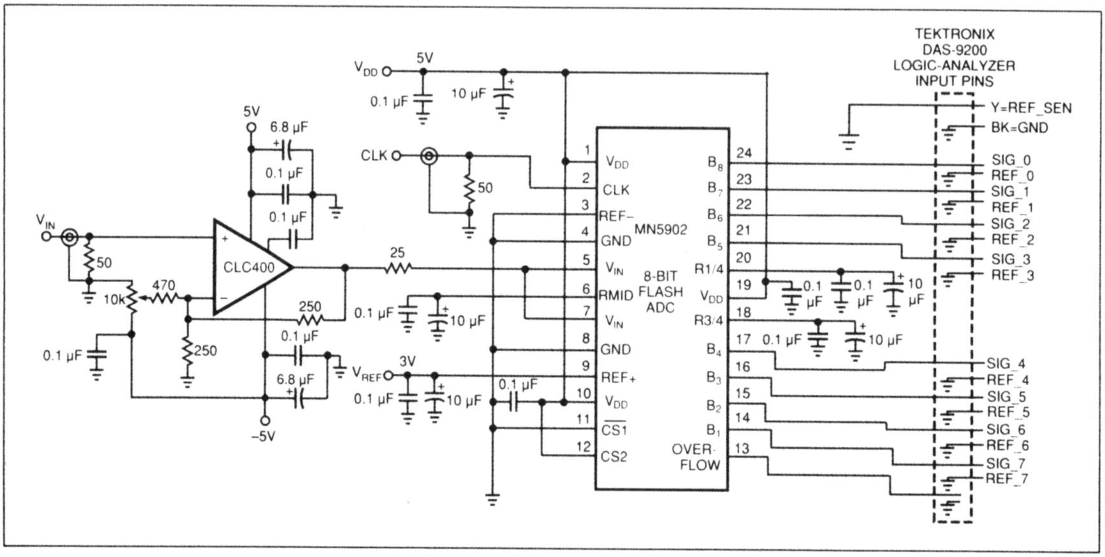

Figure 36.6 You can use this test circuit to demonstrate the ADC beat-frequency measurements Figures 36.7–36.10

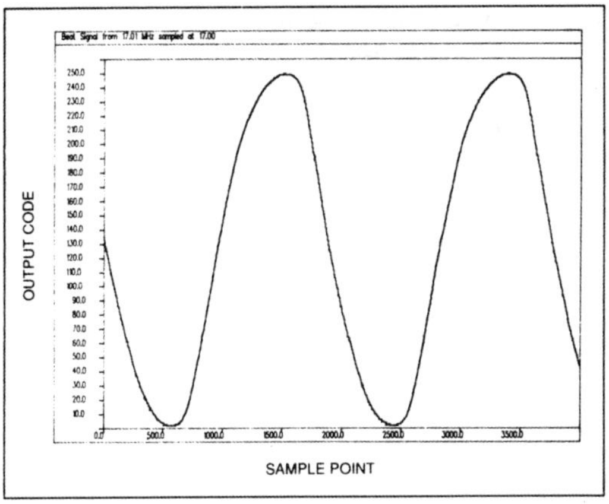

Figure 36.7 This plot shows the 10-kHz beat-frequency results of digitizing a full-scale input at 17.01 MHz with the ADC sampling at 17.00 MHz. The peak-to-peak amplitude of the output code is – 0.275 dB relative to full scale, and the beat waveform shows no evidence of spurious codes

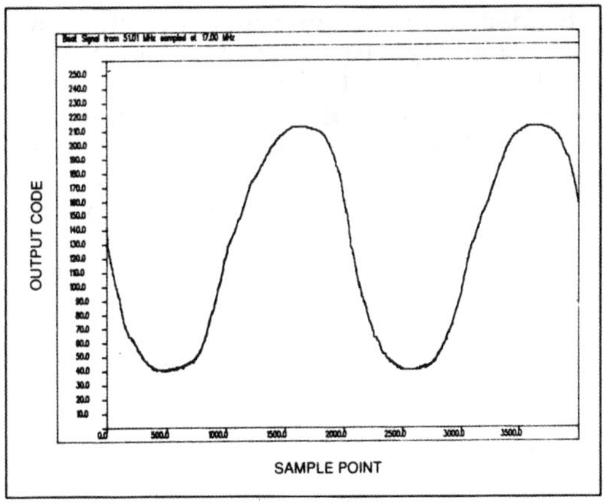

Figure 36.9 In this third test, the input frequency of 51.01 MHz is three times the clock rate. Under these conditions, there were still no spurious codes present, and the peak-to-peak output was attenuated by 3.4 dB. The attenuation of the beat frequency gives an accurate indication of the overall – 3-db bandwidth of the composite sampling system

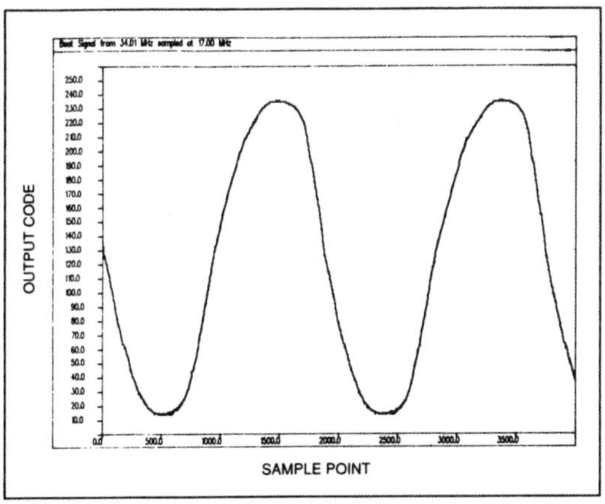

Figure 36.8 In this test, the input frequency was doubled to 34.01 MHz while the sampling rate was kept at 17.00 MHz. Again, the plot shows a 10-kHz beat, but with a peak-to-peak attenuation of 1.24 dB

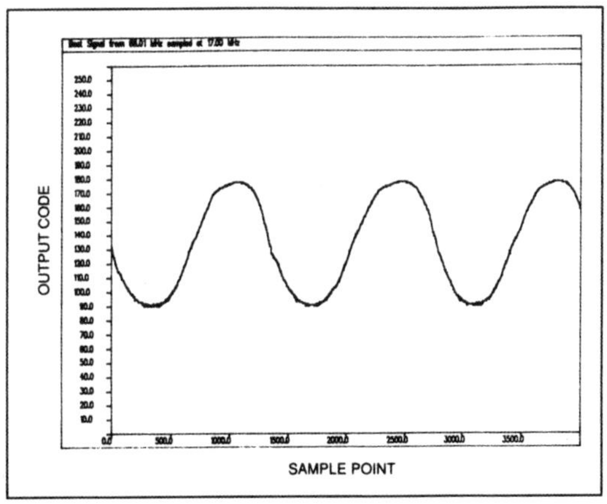

Figure 36.10 In this final test, the input frequency of 68.01 MHz is four times the sampling rate. The frequency increase attenuated the output amplitude by 9.28 dB but generated no sparkle codes

You can use the extremes of input frequency to measure the margin of aperture uncertainty a flash ADC is capable of providing. If there is a small variation in the sampling point of the ADC's adjacent comparators, high-frequency signals will slew through multiple quantization levels while the circuit makes a decision. The resulting ambiguity can cause large errors in the encoding of the digital output. To determine the amount of sampling aperture uncertainty that would produce coding errors greater than 1 bit, use the following equation:

$$\frac{dV}{dt} = 2\pi \cdot A \cdot f_{IN}.$$

To scale in fractions of an LSB, use

$$dt = \frac{dV}{\pi \cdot 2^N \cdot f_{IN}}.$$

For a full-scale signal and an error in the sampling voltage of 1 bit:

$$dt(MAX) = \frac{1}{\pi \cdot 2^N \cdot f_{IN}}.$$

As an example, for the Figure 36.9 input signal of 51.01 MHz with an effective attenuation of – 3 dB, the aperture uncertainty is less than 35 ps.

In addition to providing an immediate indication of large-signal bandwidth and sparkle immunity, sine-wave beat signals can be used to perform the other dynamic tests previously described. For example, you can use sine-wave-fitting algorithms to measure the quantization error of the beat waveforms. You can perform FFTs if you take aliasing into account and provide any required bandpass filtering at the input.

Reference

1. Demler, Michael J, *High-Speed Analog-to-Digital Conversion*, Academic Press, San Diego, CA, 1991.

37 Evaluate your ADC by using the crossplot technique

Simple technique makes ADC errors very visible

This technique, using an oscilloscope display, clearly indicates ADC faults such as integral and differential nonlinearity, missing codes, hysteresis, and code-transition noise.

Ron Knapp
Maxim Integrated Products

Before you commit a particular high-speed A/D converter to your design, you must verify its performance and accuracy at the desired conversion speed. Unfortunately, relying on the spec sheets alone to make your decision may not be enough. Some ADCs are specified very conservatively; others are made to operate right at their limits. Because data sheets give no indication of how rapidly an ADC's performance degrades beyond the maximum conversion speed, you must be sure you never exceed that speed unless you know what to expect. Further, it may be risky to depend on an ADC that's operating close to its limits, particularly when you take the clock-speed tolerance and temperature coefficient (which is often unspecified) into account.

To find out how well an ADC meets its performance specifications, you'll need to perform additional testing. ADCs, however, are difficult to test. A range of ADC inputs may give only one output, a phenomenon that's sometimes referred to as quantization error. To measure the actual transfer function of the converter under test, you must sweep the input through a range of voltages and map output codes to input ranges. You can use a technique known as a 'crossplot' to perform this mapping operation (see 'An overview of the crossplot test').

Crossplot indicates fastest conversion speed

You can construct a crossplot test fixture from readily available components and standard laboratory equipment. The crossplot test allows you to check the ADC's accuracy dynamically at its rated conversion speed. Beyond that, the test will help you determine the fastest conversion speed possible for your ADC. By 'dialing in' any code, you can determine the ADC's integral linearity, differential linearity, and noise, and you can also find other unexpected problems. ('Distinguish ADC resolution from accuracy' defines and clarifies some ADC specifications.)

You can use the crossplot technique to test an A/D converter such as the AD578, a 12-bit, high-speed successive-approximation converter (Figure 37.1). You can easily extend the test method and interpretation to other converters.

Examining some typical crossplot codes will help you diagnose specific converter problems. A

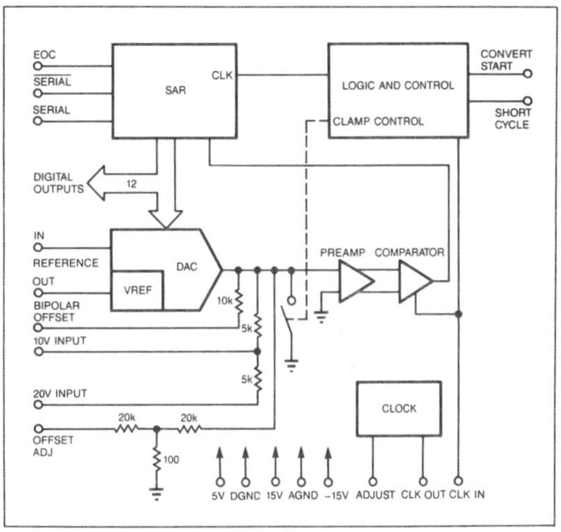

Figure 37.1 You can use the crossplot technique to test an A/D converter such as the AD 578, a 12-bit, high-speed successive-approximation converter from Analog Devices (Maxim is an alternate source for the part)

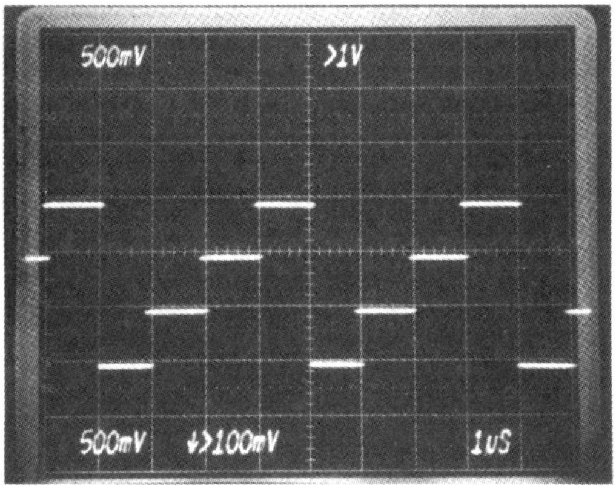

Figure 37.2 In a perfect code, all the code widths are equal, and all the code transitions occur on the scope's vertical lines

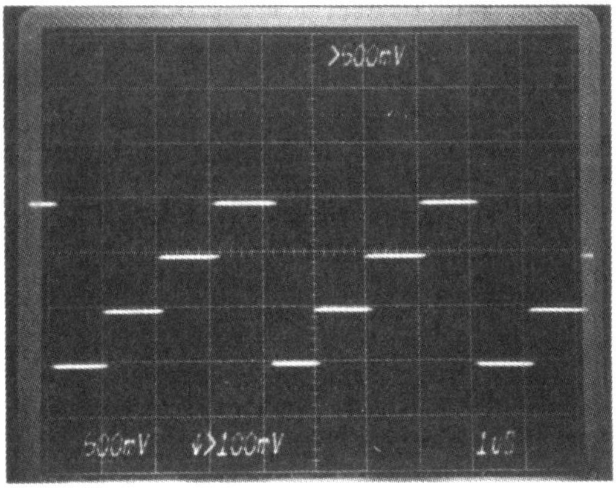

Figure 37.3 An integral-linearity error in the ADC causes all the codes to shift (in this case, they shift left by 1 LSB)

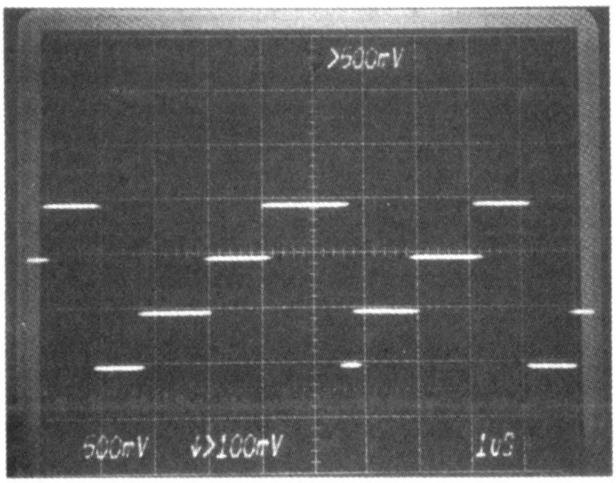

Figure 37.4 A differential-linearity error in the ADC causes one code to narrow while the adjacent code widens

perfect code, shown in Figure 37.2, is exactly one division wide, is the same width as all the other codes, and is centered on the CRT in such a way that the transitions line up on the vertical lines. If your ADC has integral-linearity error, all the codes on the CRT will shift left or right (Figure 37.3). If the data-sheet limit is ± ½ LSB, the maximum allowable shift is ± ½ division. In Figure 37.3, the differential linearity is perfect when all code widths are equal to 1 LSB.

If the integral linearity is perfect, but there are differential-linearity errors, the codes will not be shifted, but one or more codes will be narrow, at the expense of an adjacent one that's wider by the same amount (Figure 37.4). Differential-linearity errors of this type are normally due to dynamic errors, either within the ADC, such as insufficient DAC settling at major transitions, or externally, from insufficient input amplifier bandwidth (and/or settling time) or from improper grounds.

To find whether the errors are dynamic-linearity errors (and not, for example, static or dc errors caused by a mistrimmed bit in the internal DAC), you should decrease the ADC's conversion speed by a factor of two or more. If the codes become good, dynamic errors are degrading the performance. Evaluating your grounding scheme at this point may uncover a layout problem that you can solve easily. Grounding is so critical for high-speed ADCs that it's usually the root of the problem. You should also take another look at the amplifier's bandwidth specification. If the linearity is just as bad at slow speeds as at fast ones, the errors are probably due to the ADC alone, and the only solution is to replace it or correct the errors in software.

Differential-linearity errors greater than 1 LSB are called missing codes, because the ADC output jumps two or more codes for every LSB voltage increment. The code is 'missing' because there is no input voltage that will produce that code. Figure 37.5 shows an ADC with one missing code. A convenient way for you to determine the fastest possible conversion speed for your ADC is to increase the conversion frequency (assuming the ADC has either an external or an adjustable clock), until you first detect missing codes. The clock rate at this point is the device's maximum conversion speed. Data sheets rarely give you this kind of accuracy-vs-conversion-speed information.

You can tell whether the DAC's internal settling time is the primary speed limitation in the ADC by testing various codes and noting which code-test bit is the first to drop out as you increase the speed. If it's bit 2 (second from the MSB at ¼ of full scale) and/or bits 1 and 2 (at ¾ of full scale) (Table 37.1), in most cases it's the DAC that's preventing the ADC from going faster. In some

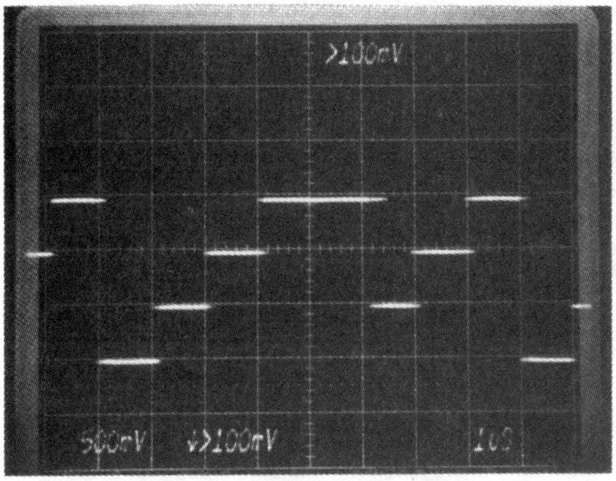

Figure 37.5 Missing codes occur when an ADC's differential non-linearity is > 1 LSB

successive-approximation ADCs, the algorithm allows extra time for the MSB to settle after the DAC is reset to ½ full scale. On the next clock, the DAC will go to either ¼ or ¾ of full scale, depending on whether bit 1 was accepted or rejected. This signal swing is the largest one in the conversion process that must be performed to rated accuracy at full clock speed.

Most likely, the crossplot codes on the oscilloscope's CRT will have both integral- and differential-linearity errors associated with them. The differential-linearity errors are straightforward enough, because the code width is relative to the oscilloscope's horizontal divisions and is measured independently of the integral-linearity shift. However, it's difficult to separate integral-linearity errors in the presence of differential-linearity errors, which displace the code transitions.

Table 37.1 Crossplot test codes and switch settings

FRACTION OF FULL SCALE	BITS	DECIMAL	SWITCH SETTINGS (TEST CODES)
INDIVIDUAL BITS			
1/2	1	2048	1000 0000 0000
1/4	2	1024	0100 0000 0000
1/8	3	512	0010 0000 0000
1/16	4	256	0001 0000 0000
1/32	5	128	0000 1000 0000
1/64	6	64	0000 0100 0000
1/128	7	32	0000 0010 0000
1/256	8	16	0000 0001 0000
1/512	9	8	0000 0000 1000
1/1024	10	4	0000 0000 0100
1/2048	11	2	0000 0000 0010
1/4096	12	1	0000 0000 0001
SUMMATIONS			
1/2	1	2048	1000 0000 0000
3/4	1+2	3072	1100 0000 0000
7/8	1+2+3	3584	1110 0000 0000
15/16	1+2+3+4	3840	1111 0000 0000
31/32	1+2+. . +5	3968	1111 1000 0000
63/64	1+2+. . +6	4032	1111 1100 0000
127/128	1+2+. . +7	4064	1111 1110 0000
255/256	1+2+. . +8	4080	1111 1111 0000
511/512	1+2+. . +9	4088	1111 1111 1000
1023/1024	1+2+. . +10	4092	1111 1111 1100
2047/2048	1+2+. . +11	4094	1111 1111 1110
4095/4096	1+2+. . +12	4095	1111 1111 1111

Figure 37.6 gives an example of how both integral and differential nonlinearity can appear: The codes on one half of the display are perfect, but an integral-linearity error in the codes on the other half produces a large differential-linearity error where the two steps of the staircase waveform meet. This error is often caused by an interquad divider in the DAC inside the ADC. Many 12-bit DACs are designed with interquad dividers, which are three similar sections of four binary-weighted current sources or R-2R ladder sections. These quads are summed together with the output of a current divider. The current divider is laser trimmed so that the total sum of one quad equals the first bit of the next quad minus one LSB. If it's the ADC's internal circuit that's causing the error, you have no choice but to try another device.

Linearity errors represent only part of the picture. You must also take noise into account. Noise makes the voltage ranges overlap at the transitions between adjacent codes. You'll have to make a tradeoff between noise and accuracy; how much noise you can tolerate in your ADC depends on your accuracy requirement.

Manufacturers rarely specify ADCs' inherent noise in the parts' data sheets, and even when they do, they specify it only in vague terms. To measure the noise in an ADC, it's useful to quantify the width of these transition-noise bands. A convenient method of measuring the peak noise is to use a digital storage oscilloscope in the peak-hold or envelope mode (Figure 37.7). Analog scopes don't always give you the true picture of noise, because of the CRT's write-speed and persistence limitations. The rms value of noise is approximately ⅙ of the peak-to-peak transition width.

You can find the noise source by observing where the noise appears in the codes. If the ADC is very noisy at + full scale but acceptable at – full scale when the DAC is off (unipolar mode), the noise is probably coming from the ADC's voltage reference. At – full scale, the reference-voltage noise is attenuated well below the noise of the other components. To find whether the noise is caused by the reference, you place a large capacitor (10 μF) on the reference output or connect a low-noise external reference to the reference input. If the transition noise is reduced, the reference is the problem. If the transition noise doesn't change, there could be noise problems in the internal DAC.

If noise is equally distributed throughout the ADC's range, its cause is something other than the signal level or code. That eliminates the DAC

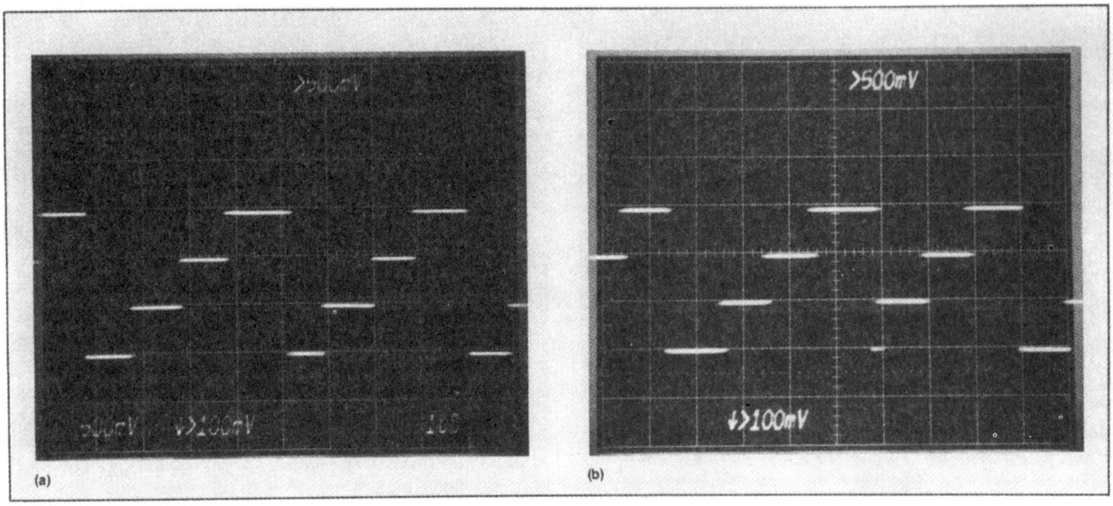

Figure 37.6 Differential-linearity errors can be produced by integral-linearity errors. In both photos, the codes in the right half of the display are correct, but in the left half, codes shifted left cause wide code at the top (a), and codes shifted right cause narrow code at the bottom (b)

and its reference. Most likely the comparator is at fault. Unfortunately, there is nothing you can do to improve it except to get a better ADC.

The crossplot technique can help you verify that your device meets its accuracy and noise specifications, but it doesn't stop there. ADCs sometimes have problems that result in undesirable effects, even though the devices meet all their data-sheet specifications. These problems may be caused by other inherent design inadequacies within the parts, or they may be induced by external factors such as grounding, shielding, coupling, and other components in your system.

The crossplot can reveal such problems as hysteresis, oscillation, and uneven codes, for instance. The crossplot result in Figure 37.8 shows a form of hysteresis, or overlapping codes. What's occurring is that the staircase signal is tracing two different paths when the scope sweeps right and left. In other words, the code width depends on whether the ADC previously converted to a higher input voltage or a lower one. Hysteresis is most often caused by DAC problems, such as excessive dielectric absorption in the output capacitance, or by disturbances introduced at the internal-summing-junction connection between the DAC output and the comparator input.

Normally, you'd expect the transition noise to be purely random – that is, if it's actually caused by component noise in the ADC. But if you can discern any pattern that looks like pulses (Figure 37.9), some comparator oscillation is probably occurring when the comparator inputs are nearly equal. To help avoid oscillation problems, most high-speed comparators in fast ADC applications employ a strobed output. Such an output is

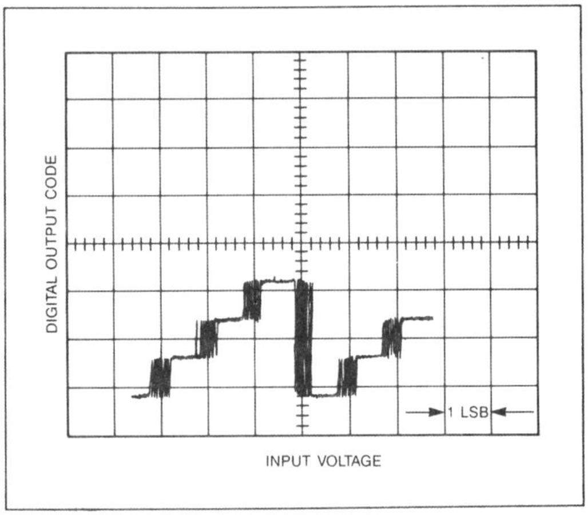

Figure 37.7 You can use the peak-hold function of a digital oscilloscope to measure the code-transition noise of the device under test

strobed on for a minimal time after the DAC's output has settled and the bit decision is ready to be clocked into the successive-approximation register.

Test-fixture layout is critical

You would expect to see differential nonlinearity or code-width variations primarily on the major vertical transitions. But when differential-nonlinearity problems are scattered throughout the codes (Figure 37.10) – and the errors are just as bad at – full scale (when all internal-DAC bits are equal to zero) as they are at + full scale (when all internal-DAC bits are equal to one) – chances are that, barring a serious design problem in the

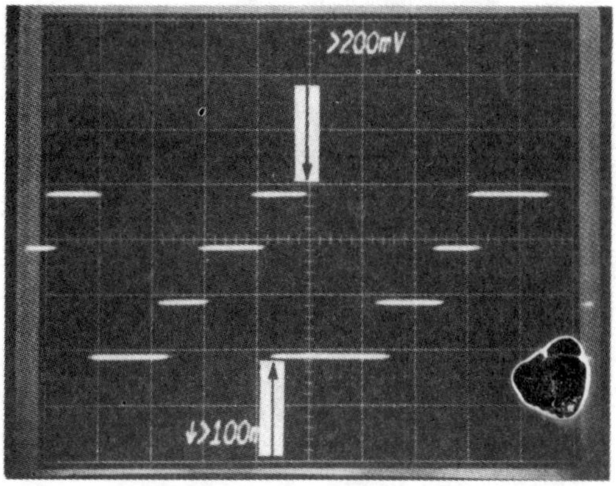

Figure 37.8 In the crossplot, hysteresis appears as an overlap of two adjacent codes

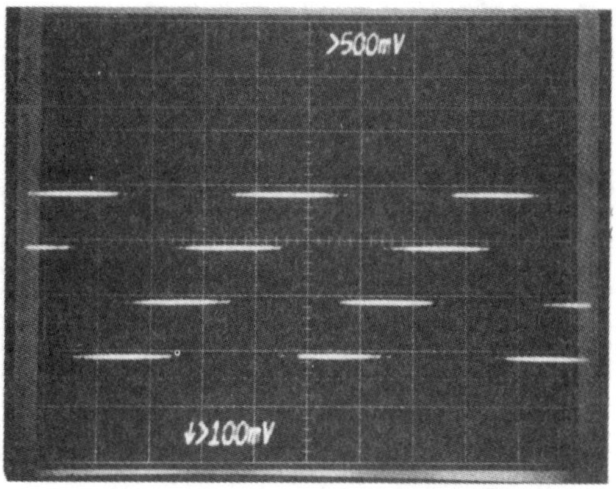

Figure 37.9 Transition pulses appearing in the crossplot results may be caused by comparator oscillation

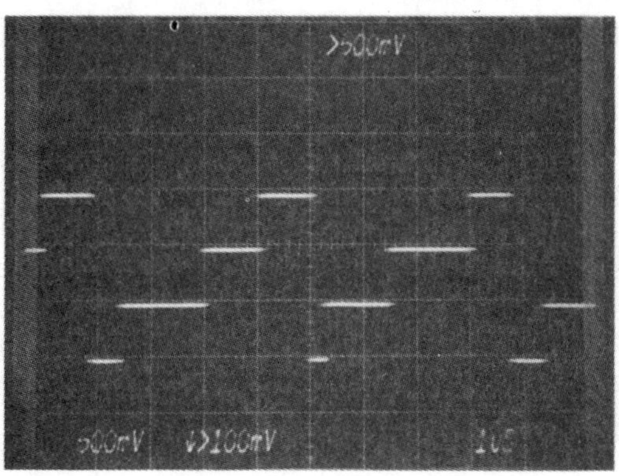

Figure 37.10 A repeating pattern of uneven codes looks like differential nonlinearity, but it's wise to check the grounding and bypassing before blaming the ADC

ADC, the problems are caused by ground loops and/or crosstalk between the analog and digital sections of the system. To confirm that ground loops and/or crosstalk – and not just gross mismatch of the LSBs – are the source of the trouble, you decrease the ADC's conversion speed. If the codes become uniform, the ADC is fine, and you must clean up the grounding and the circuit layout.

The design and layout of your fixture is a critical factor in obtaining accurate test results. A ground-plane pc board is desirable, but it's not necessary if you maintain a star-ground system in which the ADC analog ground is the center. The crossplot fixture shown in Figure 37.11 consists of an input buffer amplifier, two octal-output data registers, a dual one-shot configured as a convert-start-pulse generator, and a reference DAC that's normalized to the ADC under test. Instead of adjusting the offset and gain of the ADC, you can simply force the offset and gain of the reference DAC to conform to the ADC. This procedure gives you a more general test fixture, because different ADCs have different calibration schemes and adjustment ranges, and some don't have any provision at all for calibration.

Further, remember that unless you connect remote offset and gain potentiometers as closely as possible to the ADC's inputs, you might degrade the ADC's performance. Although you'd normally use the reference DAC with its mode switch in the Variable position, it also has a Calibrate (Cal) position. When the mode switch is in the Calibrate position, the gain and offset are preset for zero error.

The D/A converter you use as the reference DAC must have much better linearity than the ADC under test. The test-fixture design in Figure 37.11 uses the Datel DAC-HP16B to test the 12-bit AD578 (½ LSB = 0.012%). The DAC is a 16-bit part with 15 bits of linearity (½ LSB = 0.0015%). To test 12-bit ADCs, you use only the top 12 bits of the 16-bit reference DAC. The initial gain and offset of the DAC are not important as long as they're adjustable to zero and they're stable.

You must buffer the reference DAC's output voltage with a high-speed op-amp as close to the ADC's input as possible. You may wonder why a high-speed op-amp is necessary when only manually switched dc test voltages are applied to the ADC's input. The problem is that the input load current in a successive-approximation ADC changes by as much as several milliamps at each clock or bit decision during the conversion. The output of the buffer or S/H amplifier must recover from this abrupt change and settle back to within a fraction of an LSB (0.012% for ½ LSB of a 12-bit ADC) within the clock period. The internal comparator also requires sufficient setup time so it can make a correct decision.

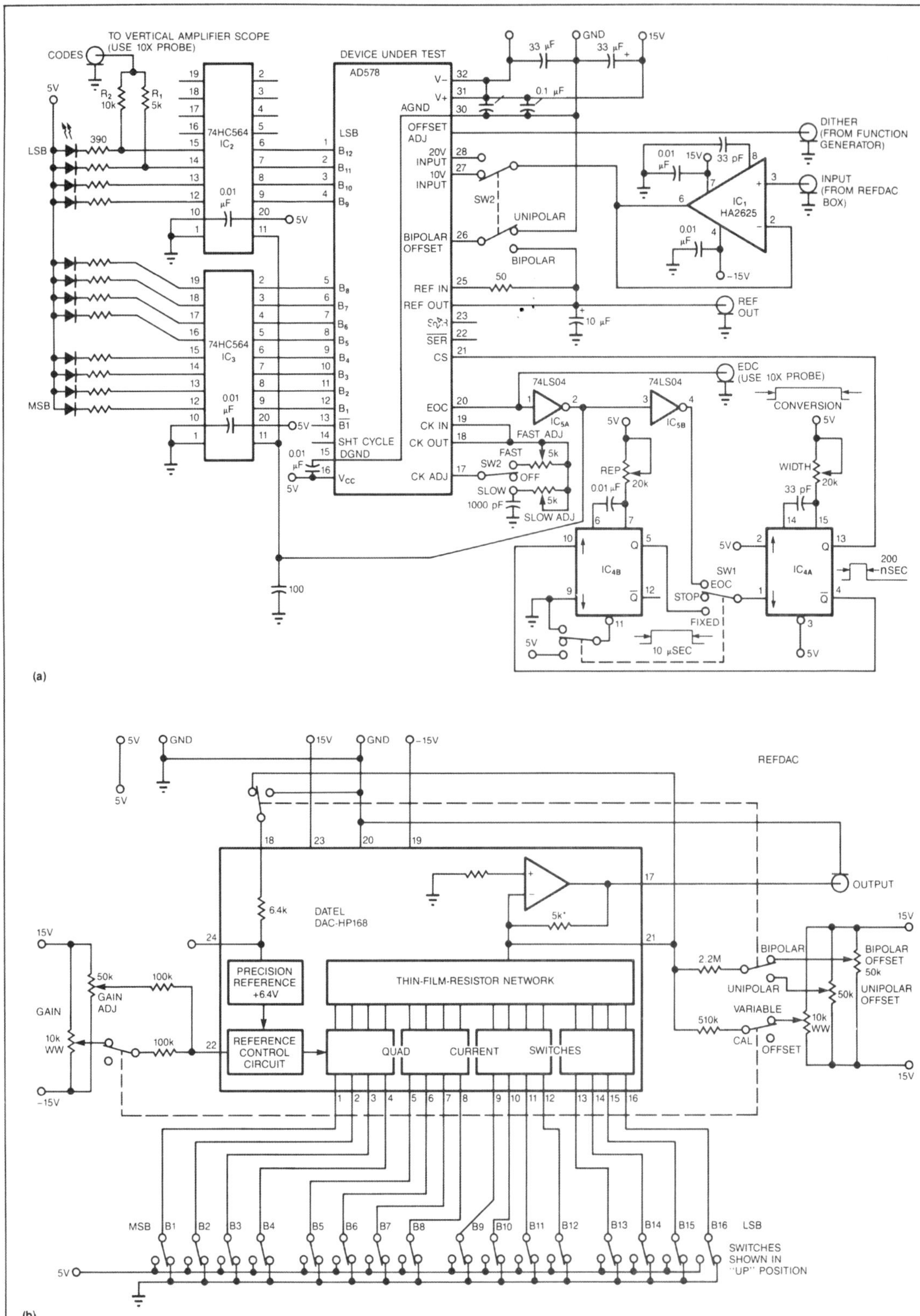

Figure 37.11 The crossplot test fixture consists of a buffer amplifier, data registers, and a dual one-shot (a), and a reference DAC (b)

Choose the right op-amp

Many different op-amps could buffer the reference DAC adequately, but you can't easily find them just by looking at data-sheet specs, because very few data sheets list the parts' settling time to 0.01%. For testing the 3-μs AD578, the Harris HA-2625 op-amp works well as long as you apply about 33 pF of external compensation and bypass the supplies. For faster ADCs, you may need to use a booster configuration: an op-amp and a high-speed buffer in a feedback loop. The buffer, an LH0033 or LH0053, for example, will absorb the transients from the ADC. Because the buffer is inside the op-amp's feedback loop, signal accuracy is maintained at the ADC input. For ADCs that don't have an offset- or zero-adjust input, you'll also need to sum the Dither input into the input buffer amplifier. This summing is not difficult, but your choice of amplifier is critical.

Because intermediate codes appear at the output during the conversion, octal registers IC_2 and IC_3 clock the valid data only at the end of each conversion. When testing an ADC with complementary-binary-output coding, you should replace IC_2 and IC_3 with a 74HC574, which is a noninverting octal register. The end-of-conversion signal (EOC) also serves to trigger a one-shot, IC_{4A}, which initiates the next conversion with SW1 in the EOC mode. If you want to set the repetition rate independently of the conversion speed, you can use IC_{4B} to retrigger IC_{4A}, as long as you set SW1 to the Fixed position. A center-off toggle switch permits you to stop the conversion process.

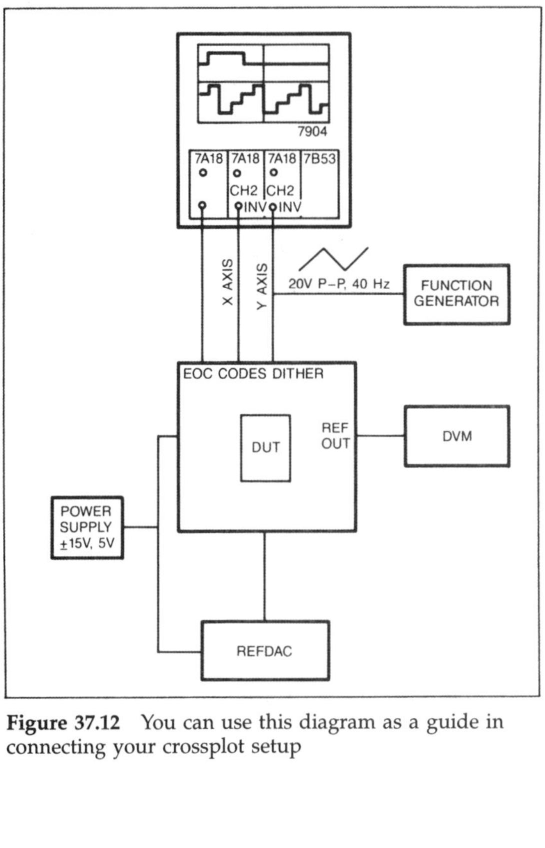

Figure 37.12 You can use this diagram as a guide in connecting your crossplot setup

The LEDs driven by the output registers help you make sure you're actually testing for the code you dialed in. If the code is the same, the illuminated pattern will match that of the code you selected with the switches on the reference DAC's input. R_1 and R_2 decode the last two LSBs. You can think of R_1 and R_2 as a single-pair R-2R ladder. You can test more bits by using an R-2R-4R set of resistors, which will produce an 8-step staircase on the oscilloscope. Alternatively, you can test for 10-bit performance by moving R_1 and R_2 up to the next two most significant bits.

Figure 37.12 shows how to connect the test fixture. With an oscilloscope such as the Tektronix 7904, which accepts as many as four plug-in amplifiers, you can simultaneously look at the codes in X-Y mode and at the EOC signal, which displays conversion speed. To do so, you need to configure one of the vertical amplifiers as a timebase amplifier. You set the vertical amplifier control to Alternate and set the timebase mode to Chop.

To make the crossplot results easy to interpret, you need to calibrate the scope display so that each code width fits into one division and all the transitions line up on the vertical lines. You simultaneously turn the OFFSET ADJ pot of the reference DAC and the X-amplifier CAL ADJ knob on the scope. It's best to ignore the bottom codes for this adjustment, because these codes are the most likely to have differential-linearity errors. Instead, you calibrate the scope so that each third high-side transition is separated from the next by exactly four divisions (Figure 37.13).

Next, you need to choose the place on the scope display where the code under test will fall when there are no errors in the system. The typical position is one division to the right of center. To choose that position, you need to make adjustments for system offset and gain. First, you switch the reference DAC inputs to zero; then, by using the OFFSET ADJ pot, you place the very first transition on the first vertical line to the right of center. This action places the zero code in the selected box (Figure 37.14). You must make sure that both the crossplot fixture and the reference DAC are set to the same range, either the unipolar (10 V) or the bipolar (20 V) range. If the image on the CRT is upside down and/or backwards, you should invert the x and/or the y input as necessary.

Similarly, you adjust the test fixture for gain by switching the first 12 reference DAC inputs to 1 and setting the GAIN ADJ pot so that the last transition falls directly on the center line (Figure

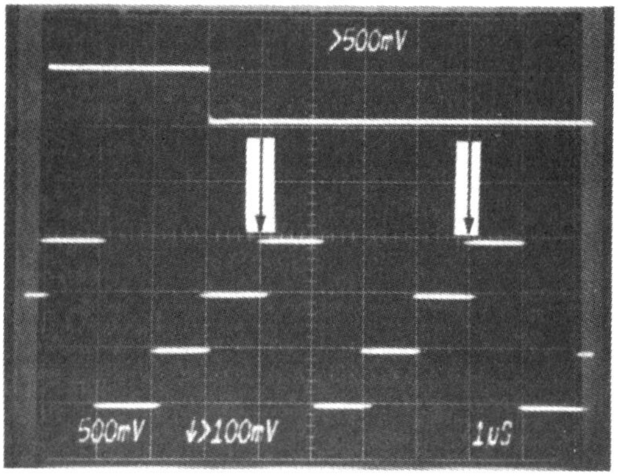

Figure 37.13 To make the crossplot results easy to interpret, calibrate the scope so that each third high-side transition (see arrows) is separated from the next by exactly four divisions. This way, each code width will fit into one division, and all the transitions will line up on the vertical lines

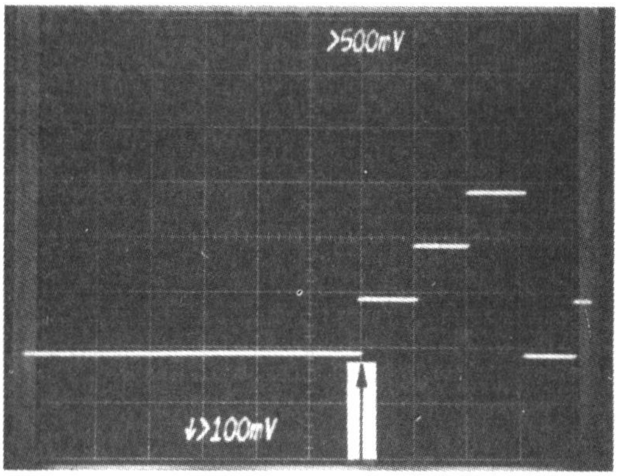

Figure 37.14 To test an ADC's digital output codes, you need to choose the place on the scope display where correct codes will fall. For example, if you choose the scope division just to the right of center, you must switch the reference DAC inputs to zero, and, with the OFFSET ADJ pot, you place the first transition on the first vertical line to the right of center

37.15). This procedure places + full scale in the designated division just to the right of center.

Note that the crossplot test method uses 'end-point' linearity calibration, so there's never a linearity error at – full scale or + full scale, which are being normalized. The normalization process rotates the actual transfer-function curve so that + full scale and – full scale are on the ideal straight line. If the ADC exhibits gain-linearity errors, those errors will now occur at ½ full scale. A limitation of this method of calibration is that you can't clearly distinguish gain errors from integral-linearity errors.

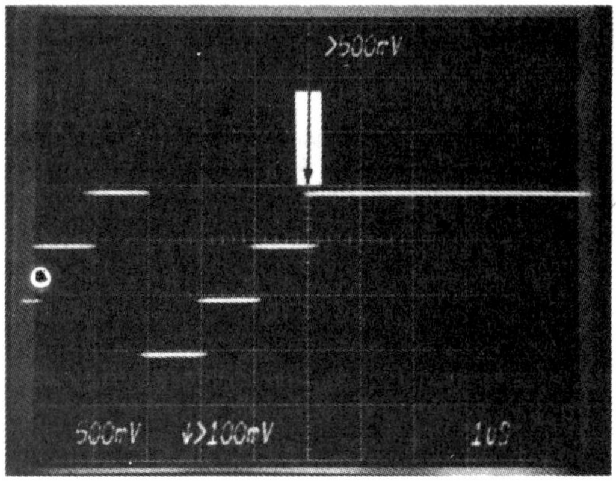

Figure 37.15 To adjust the test fixture for gain, you must switch the reference DAC's first 12 inputs to all ones and use the gain-adjustment pot on the reference DAC to line up the last transition on the center of the scope display

The ADC is now ready to be tested. Whatever code is set on the reference DAC's input switches should be in the designated location on the oscilloscope display; that is, it should be one division to the right of center. Any deviation is measured as linearity error. Because the scope is calibrated for a horizontal scale of 1 LSB/div, the errors are measured directly in LSBs.

You can test every ADC output code if you wish, but that's generally considered overkill. The test codes for a 12-bit ADC (shown in **Table 37.1**) are usually accepted as a thorough test-code set; they test the ADC in its operating range for decimal inputs from 1 to 4095. For those test codes whose two least significant bits are zero, the staircase will be at its lowest level in the division just to the right of center on the scope's CRT. Any code whose last two bits are '01' will display the first staircase level in that scope division. A code whose last two bits are '10' will display the second staircase level in that division, and one whose last two bits are '11' will display the third and top level there.

An overview of the crossplot test

An ideal transfer function for a 3-bit ADC is shown in Figure 37.16. Note that there is no unique analog value for each digital code and that the ADC output is valid over a range of input signals. The spread of the input signal for a given output code is called the 'width' of the code and is ideally 1 LSB.

The crossplot test procedure allows you to determine the actual transfer function of your device. By summing a dc signal with a low-frequency asynchronous triangular or sine-wave

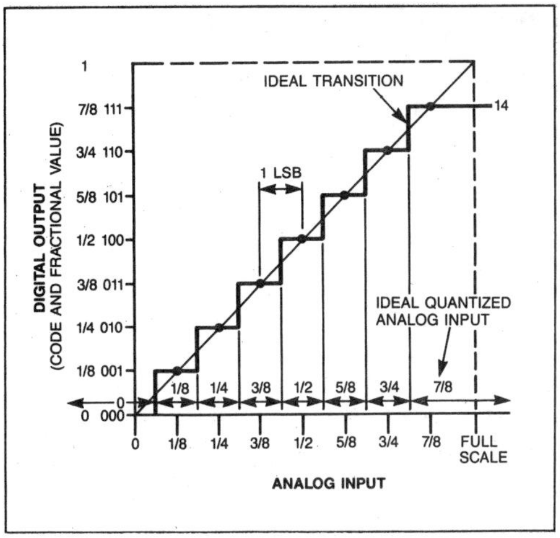

Figure 37.16 The ideal transfer function of an ideal 3-bit ADC has even code widths that are equal to 1 LSB

signal and applying the result to the ADC under test, you 'dither' the ADC output through several codes; either side of the code represents the dc voltage.

The least-significant ADC output bits are decoded by a simple resistive ladder network and plotted vertically on the scope; the ac analog input is plotted horizontally. The result is a repetitive staircase waveform, which is in essence the real transfer function. For testing most ADCs, it is sufficient to decode the two LSBs of the output, thereby obtaining a 4-step waveform.

Before performing the crossplot test, you must calibrate your fixture to the scope display so that you have a known reference point from which to recognize errors. You should calibrate the scope so that one horizontal position represents 1 LSB, and so that code transitions occur on the vertical lines of the display.

The dc signal is provided by a front-end DAC that's controlled by a set of switches. You can select the test for a particular code by setting the switches to the same code. By 'dialing in' the codes, you can test any or all of the 2^n possible output codes of your n-bit ADC.

Distinguish ADC resolution from accuracy

It's important not to confuse resolution with accuracy when selecting an ADC. The resolution (or number of bits) that an ADC has is a measure of the smallest increment of input voltage it can detect that's represented by a change in the LSB of the digital output code. The accuracy, also specified in LSBs, can be lower or higher than the resolution.

The ADC's 'accuracy' is determined by the integral and differential linearity of its digital codes. Integral nonlinearity is a measure of how many LSBs an output code is from the theoretically perfect transfer function of the converter. In other words, integral nonlinearity is the deviation of the real transfer function from the ideal straight line.

Differential nonlinearity is the minimum and maximum range of input voltage that causes the codes to change by one count; it's the deviation of the code width from the ideal 1 LSB. For an ADC to have 'no missing codes' operation (which implies that no code jumps will be greater than 1 LSB), its differential non-linearity must be lower than the resolution (1 LSB), and its integral linearity must be less than ½ LSB, because a ½ LSB error on each of the binary weighted output bits would add up to 1 LSB (that is, 1 LSB = ½ + ¼ +, etc).

Most ADCs that have 12 bits or less of resolution have linearity specifications that match the resolution, but it's common to find 16-bit ADCs with 14-bit accuracy, which means that the output data can be off by as much as 4 LSBs. On the other hand, some ADCs may have higher accuracy than resolution. For example, in a 12-bit ADC with 14-bit accuracy, the output code will have no more than a ⅛ LSB error, and the code width will be between ⅞ and 1⅛ LSB.

The absolute accuracy of the data from an ADC converter depends on many factors. An ADC's linearity is the most important of these, because the only adjustments you can make to correct it are software adjustments. However, to determine the ADC's worst-case error – and therefore its overall accuracy – you need to make a more complete list of error sources, including not only linearity, but also linearity drift, gain, gain drift, offset, offset drift, gain PSRR, and noise.

To determine the accuracy of the overall system, you must also consider error sources from the other components around the ADC, such as the input preamplifier, the multiplexer, the sample/hold amplifier, and the external voltage reference, if you use one.

References

1. Sheingold, Daniel H, ed (Analog Devices Inc), *Analog-Digital Conversion Handbook*, Prentice-Hall Inc, Englewood Cliffs, NJ, 1986.

2. Jung, W, *IC Converter Cookbook*. Howard W Sams Inc, Indianapolis, IN 1978.

38 Build a single-shot recorder to catch fast transients

Hand-knitted digital storage scope (almost)

For a cost of 23 ICs, most of them standard 74HCT, you can build a transient recorder that will interface with any old scope to turn it into a digital storage scope.

Ken Deevy, Dan Sheehan, and Mike Byrne
Analog Devices Inc

Don't tie up expensive equipment trying to capture transients that occur infrequently. If you build a low-cost transient recorder or event sampler, you can dedicate it to capturing single-shot events. Typical applications for transient recorders include monitoring powermains transients, evaluating power supplies, and capturing pressure and vacuum-line transients in medical equipment.

To build a transient recorder or burst-mode event sampler, you need a high-speed A/D converter, a wide-band track/hold amplifier, and an antialiasing filter. The A/D converter must have a sampling rate of at least twice the bandwidth to satisfy the Nyquist criterion. In practice, you should oversample the input signal. At 2 × oversampling (a sampling frequency of twice the input bandwidth), you'll need to use a filter with an infinite roll-off rate to avoid aliasing effects. At 3 × oversampling, the roll-off requirement drops to 50 dB/octave in an 8-bit system. With an oversampling ratio of 10:1, the filter roll-off need be only about 16 dB/octave. (See 'Oversampling reduces antialiasing requirements'.)

High-speed sampling A/D converter chips routinely include track/hold amplifiers on the same chip. The AD7821 is an example of this trend. It combines a 100-kHz track/hold amplifier with a 1M-sample/s 8-bit A/D converter. Because the A/D conversion rate is 10 times the input bandwidth, you don't have to design a complex antialiasing filter. In fact, if the input signal exhibits only a low-power spectral content at and above 500 kHz, you can eliminate the filter altogether.

The AD7821 uses a half-flash conversion technique to perform an 8-bit conversion in 660 ns. A requirement of a 350-ns signal-acquisition period between conversions results in a maximum acquisition rate of 1M samples/s. You can operate the A/D converter with a single or dual supply for either unipolar or bipolar inputs.

Capture single-shot waveforms

One of the difficulties in capturing single-shot events is the speed at which the transient recorder circuit responds once the input signal has crossed a predetermined trigger point. If the recorder circuit responds too slowly, it can miss fast transients altogether. Therefore, to accurately capture fast events, you need a high-speed A/D converter and a wide-bandwidth track/hold amplifier. For example, an 8-bit A/D converter that has a 1-μs conversion time can capture 1-μs transients only if it's not preceded by a track/hold amplifier. If you match this A/D converter with a track/hold amplifier that has a 100-kHz bandwidth, the ADC can recover 6-μs-wide 5 V transients.

To simplify fault detection or take corrective measures, you need a transient recorder that can grab pretransient information. You can use this pretransient data to learn timing relationships between the transient and another waveform. Additionally, your recorder should be able to react to both positive and negative transients.

Another important criterion for transient recorders is cost. Although you could use a digital storage oscilloscope (DSO) to capture frequently occurring or very fast transients, dedicating a DSO to capturing random events would be an expensive proposition.

Transient recorder

A block diagram of a transient recorder (Figure 38.1) shows the minimum hardware you'll need

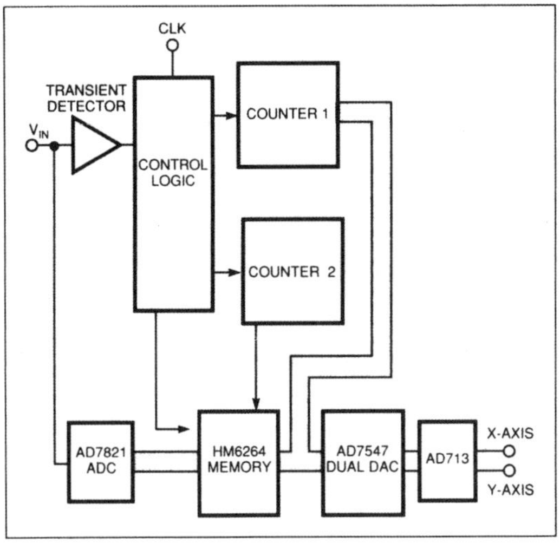

Figure 38.1 If you think of a transient recorder as merely an ADC, look again. This recorder contains several purely digital blocks as well as the DAC that drives the display

to build a high-speed transient recorder with playback. For simplicity, the design uses a clock with an even mark/space ratio. The clock's 50% duty cycle limits the acquisition rate to 660 k samples/s rather than the A/D converter's 1-M sample/s maximum rate. (This simplification reduces the oversampling ratio to 6.6:1.) A memory chip stores the digitized data for later playback on an X-Y plotter or oscilloscope via a dual 12-bit D/A converter and a quad op-amp. One half of the samples are pretransient information; the other half are transient data.

A more detailed schematic (Figure 38.2) shows that two counters, IC_1 and IC_2, control where the circuit stores pretransient and transient data. The counters also clock out data to either the oscilloscope or the X-Y plotter. You can use the fast clock input, CLK IN_1, for the clock source when you're using the circuit in the record mode or displaying stored data on an oscilloscope. The design also provides a slower clock input, CLK IN_2, to print data on an X-Y plotter.

The transient recorder operates in two basic modes: record and playback. You select the record mode by placing switch S_1 in the record position. (IC_{18A} and IC_{18B} provide debouncing for this switch.) Having the MODE output of IC_{13D} low makes one input of both IC_{15D} and IC_{14D} low. Hence the clock inputs of IC_{9A} and IC_{9B} (pins 10 and 2, respectively) are disabled, ensuring that the $1\overline{Q}$ and $2\overline{Q}$ outputs of IC_{9A} and IC_{9B} are high. Besides disabling the chip-select inputs of the D/A converter, $\overline{CSA}$ and $\overline{CSB}$, the circuit disables the output enable signals of IC_3, IC_4, and IC_5 (the HM6264 memory chip), ensuring that the playback portion of the transient recorder is turned off.

CLK IN_1 serves as the clock source for the counters via IC_{18C}, IC_{-14B}, IC_{15A}, and IC_{15C}. While the MODE signal is low, CLK is the clock input for both counters and provides the $\overline{RD}$ (convert) signal for the A/D converter, IC_6. At the same time, IC_6's $\overline{CS}$ input is active, ensuring that the device is selected. After a reset from S_3 initializes the circuit, counter 2 begins counting. IC_{17} and IC_{23B} hold the reset (CLR) input of counter 1 high from power-up and keep counter 1 in a reset condition until the circuit detects a transient.

You configure the A/D converter by tying its MODE input (pin 7) to GND. (Note that the MODE pin of the AD7821 shown bears no relation to the signal labeled MODE in the circuit diagram.) When the CLK signal toggles its $\overline{RD}$ input, the A/D converter executes continuous conversions of the input signal, V_{IN}. Counter 2 provides the memory addresses for the A/D conversion results. Data transfers from the digital outputs of IC_6 to IC_5 employ the $\overline{INT}$ output of IC_6 to drive the $\overline{WE}$ input of IC_5.

The circuit automatically loads the first conversion result after reset into location 0 of memory and the second result into location 1. After transferring the result of the 4096th conversion to memory location 4095, the counter resets and stores the next conversion result in location 0. The memory will therefore always contain the most recent 4096 samples of the input waveform.

Detect fast transients

You apply the input signal to V_{IN}. This terminal connects directly to two TL311 comparators and the analog input of the A/D converter. Comparator IC_{19} detects positive transients, and IC_{20} detects negative ones. To set the threshold level for a positive-going signal, adjust resistor R_5; adjust R_6 for negative-going transients. Wire the outputs of both comparators together to ensure that they produce a rising edge to the clock input of IC_8 when either a negative or a positive transient occurs.

Once the circuit detects a rising edge at pin 11 of IC_8, it illuminates an LED, D_1. At the same time, it releases counter 1 from its reset condition by taking RS_1 low. Now the circuit clocks both counter 1 and counter 2 as A/D conversions continue. Counter 2 counts up from the value it held before the transient was detected. The memory locations determined by the output of counter 2 store the transient data while overwriting the oldest 2048 samples of pretransient data already stored in memory. Counter 1 counts off the 2048 clock states that correspond with the samples.

Because the output of IC_{16B} is always high in the record mode, when counter 1 reaches 2047, all inputs to IC_{10} and IC_{11} are high and both IC's outputs go low. As a result, the output of IC_{12A} goes high, causing the output of IC_{14A} to go low via IC_{13A} and IC_{12B}. This DIS REC CLK signal gates off CLK IN_1 from the rest of the circuit at IC_{14B}. The output of IC_{18C} ensures that the CLK signal is held low, stopping both counters and the A/D converter.

At the end of the transient-record cycle, the memory will contain 4096 samples of the input waveform. One half of these samples are transient data, the other half represent pretransient information. Whatever value is in counter 2 will be the last memory location for the transient data. The next memory location will hold the first of the 2048 words of pretransient data. Now when you start the playback mode, the first output from the counter will correspond to the memory location of the first pretransient sample. (To alter the ratio of transient to pretransient samples, simply alter the connections from counter 1 to IC_{10} and IC_{11}.)

To accurately convert the input waveform to stored data, you must pay close attention to the

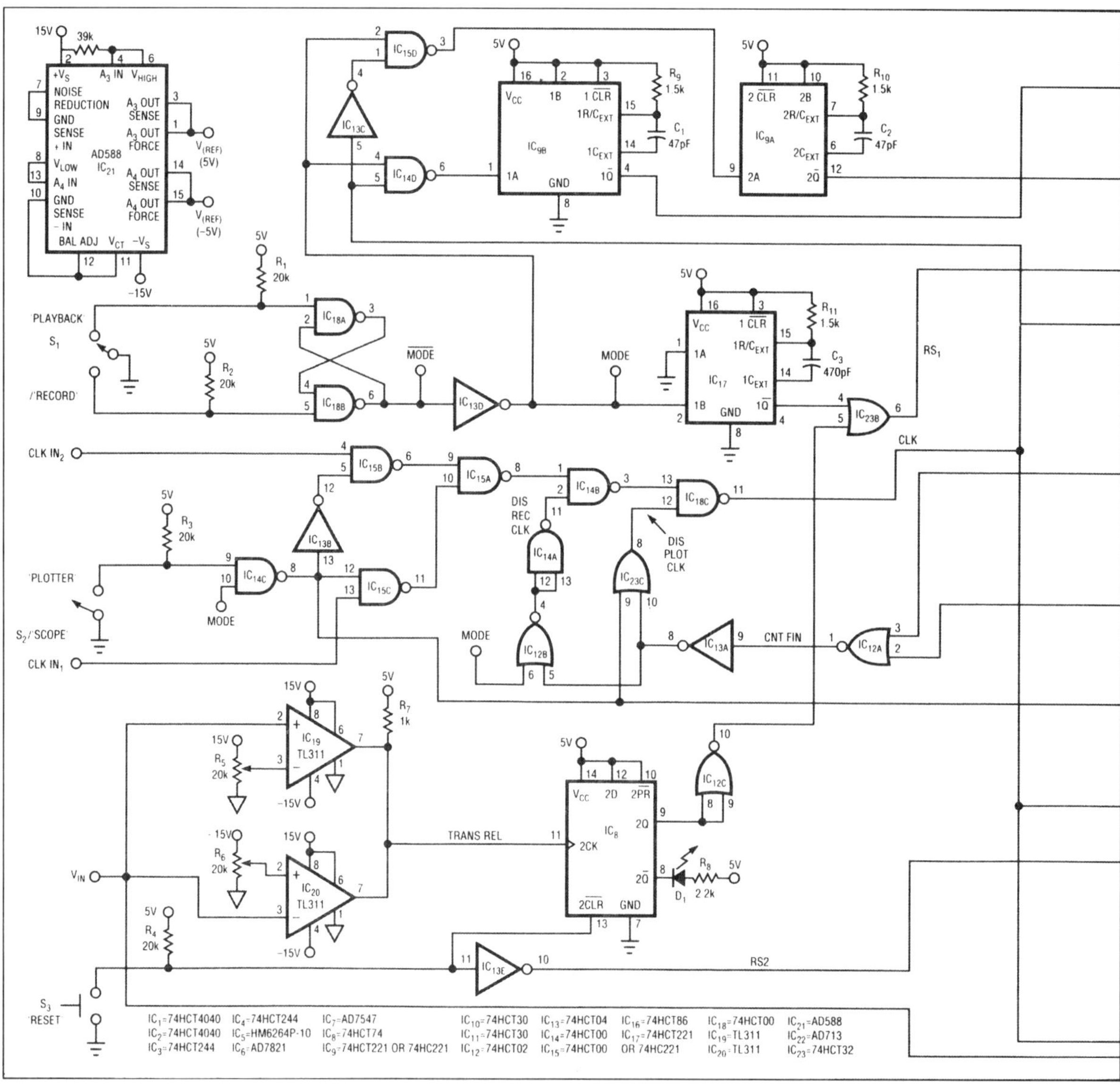

Figure 38.2 At the component level, the basic transient recorder is a circuit of moderate complexity, requiring 23 ICs

circuit. Use a precision reference, IC_{21}, to generate 5 V and – 5 V references for the $V_{REF\,+}$ and $V_{REF\,-}$ inputs of the A/D converter. Make sure that you properly decouple these reference voltages along with the V_{DD} and V_{SS} lines of the A/D converter. Connect IC_6's GND pin to the star ground of the system (the point in the circuit at which you connect the analog and digital grounds). Make sure that the conductor between the A/D converter and the star ground is as wide as circuit-board layout constraints allow. Further, ensure that the $\overline{WR}$/RDY line is pulled high via R_{19} to avoid noise pickup on this pin.

Play back captured signals

Once you've captured that bothersome transient, you can play it back at any convenient time; the recorder will retain the information as long the power remains on, or until you depress the reset button. Select the playback mode with S_1. Playback takes the $\overline{MODE}$ signal low, activates the $\overline{WR}$ input to IC_7, and deselects IC_6 by taking its $\overline{CS}$ high. You can display the transient on either an analog oscilloscope or an X-Y plotter, depending on the position of S_2. Make sure to select the oscilloscope or the plotter before starting playback.

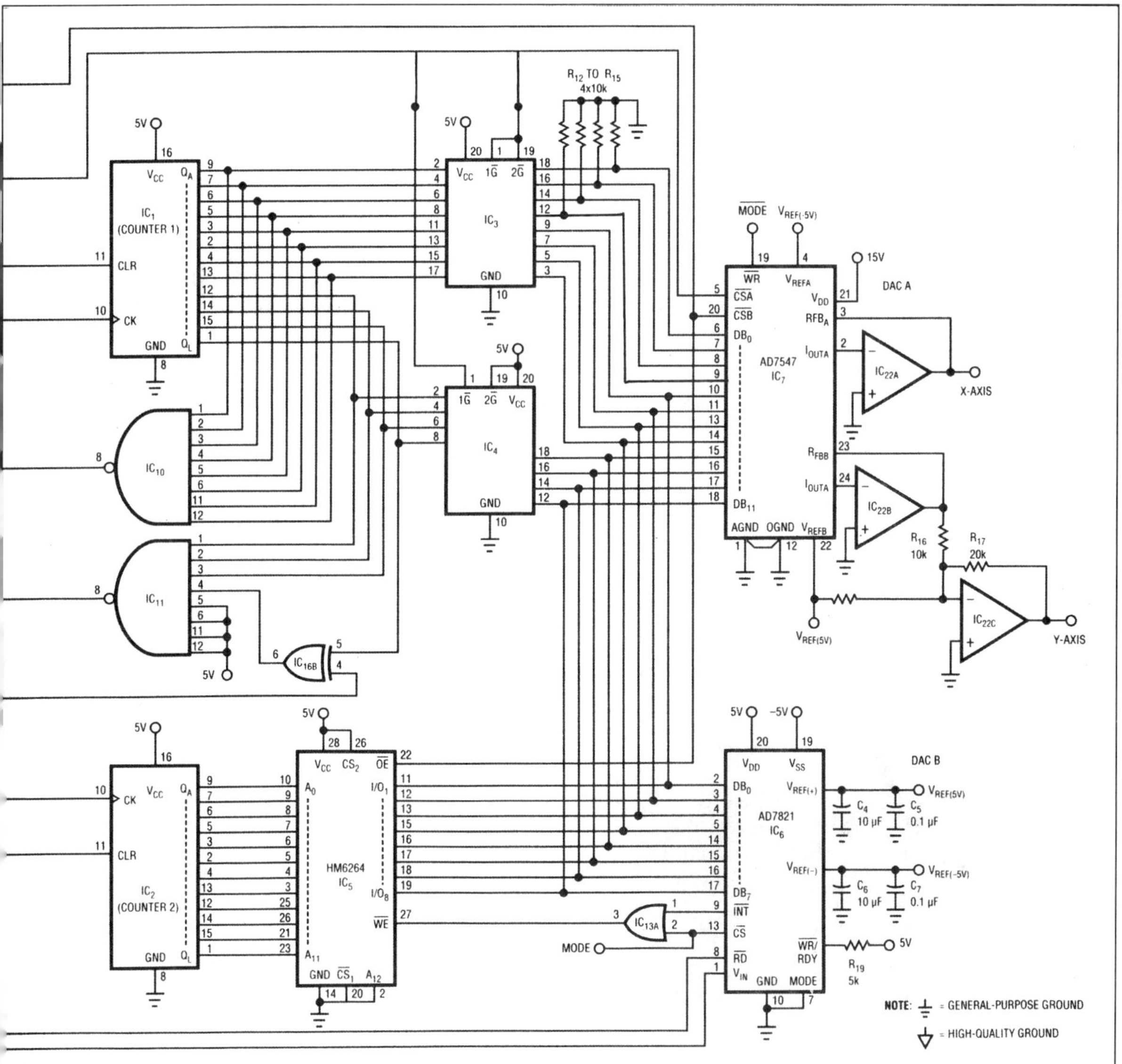

If you decide to display the transient on an oscilloscope, the clock source for the circuit is the same as in the record mode. If you use a plotter for playback, the clock frequency is much lower and is applied via the CLK IN $_2$ input. CLK (from either CLK IN $_1$ or CLK IN $_2$) passes through gates IC_{15D} and IC_{14D} because the $\overline{\text{MODE}}$ signal is high. IC_{9A} and IC_{9B} generate the $\overline{\text{CSA}}$ and $\overline{\text{CSB}}$ pulses for IC_7 from this CLK signal.

IC_{9A} drives the $\overline{\text{CSA}}$ input of IC_7 as well as the enable signals for IC_3 and IC_4. When you choose the playback mode, counter 1 resets and starts counting from 0 to 4095. The counter's output is the digital input code to DAC A of IC_7. This DAC drives the X axis of either the oscilloscope or the plotter. DAC A produces a unipolar output range from 0 to 5 V, with a resolution of 4096 steps.

The output of IC_{9B} drives the $\overline{\text{CSB}}$ input of IC_7 and also sets the logic level on IC_5's output-enable line, $\overline{\text{OE}}$. This action latches the data from memory into DAC B, which drives the Y axis of the oscilloscope or plotter. By using dual supplies, you can set DAC B for a bipolar output range to reconstruct both positive and negative transients.

Counter 2 starts its count from the point at which it stopped at the end of the record mode; the first memory output word to IC_7 is the oldest sample in memory. The scan will then proceed through the 2048 samples of pretransient information and the 2048 samples of transient information. The output of each sample from memory to the Y axis, via DAC B, corresponds to the output of a count value from counter 1 to the X axis via DAC A. In this way, the circuit reconstructs the pretransient and transient waveforms.

For oscilloscope display of waveforms, place S_2 in the scope position. Doing so locks out CLK IN_2 from the rest of the circuit but allows CLK IN_1 to operate as clock signal for the circuit. Unlike the plotter display option, where counter 1 runs through once and then stops, CLK runs continuously. CNT FIN does go high when counter 1 reaches a count of 4095, but because the output of IC_{14C} is high, the DIS PLOT CLK signal does not go low. You can see the typical oscilloscope waveform display in Figure 38.3(a).

You display the stored waveform on an X-Y plotter by placing S_2 in the plotter position. Doing so locks out the CLK IN_1 input from the rest of the circuit and permits CLK IN_2 to generate the clock signal for the circuit. IC_{16B}, IC_{10}, IC_{11}, and IC_{12A} function in a manner similar to the record mode to generate a high CNT FIN signal. But this time, IC_{10} and IC_{11} go low when counter 1 reaches a count of 4095. IC_{13A} goes low, and, because the output of IC_{14C} is already low, the DIS PLOT CLK signal goes low, turning off CLK IN_2 at IC_{18C} and holding the CLK signal high. Figure 38.3(b) shows a captured transient displayed using a plotter as the display method.

Record-mode timing and clock waveforms

The timing diagrams in Figure 38.4 show the logic relationships for the record mode. The MODE signal (not shown) is low and the DIS REC CLK signal is high. The RS_2 signal goes high when the recorder receives a reset command via S_3 resetting counter 2. The next falling edge of the CLK signal clocks out an address for IC_5 from counter 2. A conversion is also initiated on this falling CLK edge, and, within 700 ns, the $\overline{\text{INT}}$ signal of IC_6 goes low, activating the $\overline{\text{WE}}$ input of IC_5. The rising edge of CLK resets the $\overline{\text{INT}}$ line 50 ns later.

When the circuit detects a transient, the TRANS REC signal goes high, causing the RS_1 line to go low and releasing counter 1 from its reset state.

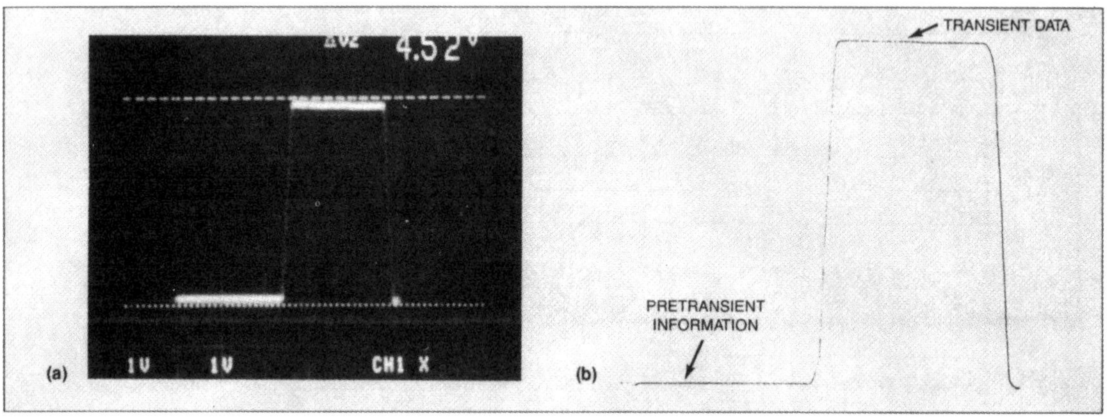

Figure 38.3 The recorder can display captured data on a scope, (a) or on an X-Y pen plotter, (b). The first half of each display shows data acquired before the triggering event

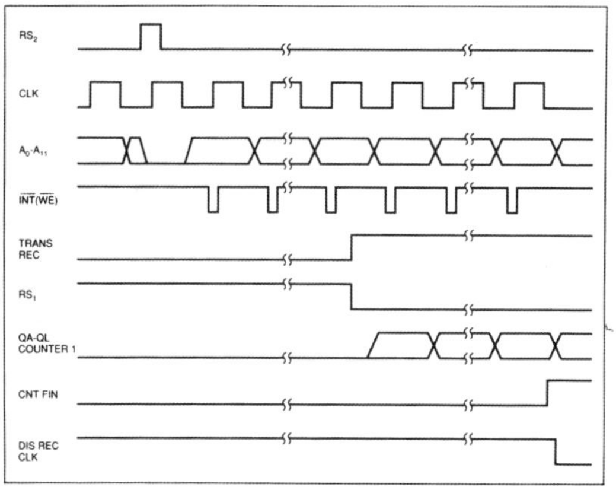

Figure 38.4 The timing relationships in the record mode show that recording starts on the first falling clock edge after the reset line goes high.

The next falling edge of CLK clocks out the outputs from counter 1. When the count output from counter 1 reaches 2047, the CNT FIN signal goes high and causes the DIS REC CLK signal to go low, shutting off the CLK signal.

In the record mode in the waveforms shown, the 50/50 mark/space ratio of the CLK signal limits the clock frequency to 660 kHz. You need a CLK-low time of 750 ns for the A/D converter to perform its conversion correctly and latch the data into IC_5. However, the CLK-high time can be as short as 350 ns, the time required between conversions by the AD7821. Therefore, if the input to CLK IN_1 has a low time of 750 ns and a high time of 350 ns, the circuit can make one conversion every 1100 ns – equivalent to approximately 900 k samples/s.

Playback to a scope

During playback to an oscilloscope (Figure 38.5(a)), the MODE signal, the $\overline{WE}$ input of IC_5, and the DIS REC CLK signal are high. When you place S_1 in the playback mode, RS_1 goes high, resetting counter 1. The CLK signal generates a $\overline{CSA}$ signal for IC_7 on its rising edge and a $\overline{CSB}$ signal on its falling edge. The falling edge of the CLK signal clocks data from counter 1, and the rising edge of $\overline{CSA}$ updates the X axis. The falling edge of $\overline{OE}$ outputs stored data from memory, and the rising edge of $\overline{CSB}$ updates the Y axis. The CLK signal runs continuously when the circuit is in the scope-playback mode.

The timing diagram of Figure 38.5(b) shows operation of the circuit for playback on a plotter. Once again, MODE, the $\overline{WE}$ input of IC_5, and the DIS REC CLK signals are high. The circuit generates $\overline{CSA}$ and $\overline{CSB}$ to update the X and Y axes. Compared with scope playback, the difference in the circuit's operation is that when the output count from counter 1 reaches 4095 and the CNT FIN signal goes high, the DIS PLOT CLK signal goes low, forcing the CLK signal into a high state.

Burst-mode event sampling places requirements on an A/D converter similar to those for transient recording. In burst-mode sampling, the recorder looks at the input waveform infrequently, but when it does, it must acquire a large number of samples in a short time. With slower microprocessors or microcontrollers, you'll find that because of instruction- and bus-timing constraints, you can't achieve anything like the A/D converter's maximum throughput.

You can overcome timing limitations in a burst-mode sampler by using a DMA controller to initiate A/D conversions and transfer conversion data to memory. Doing so lets you run the A/D converter at or near its maximum sample rate, permitting high oversampling ratios and the acquisition of short transients.

Building a burst-mode sampler is a simple matter with the popular 8052 microcontroller (Figure 38.6). Although the 8052 does not support hardware DMA, it does support what is termed 'fake DMA'. However, expect the response time to DMA requests to be much slower than is possible with microcontrollers that support genuine DMA.

The HM6264P memory chip, IC_3, stores the control program for IC_1. The first part of the control program is the initialization routine. This routine (Listing 38.1) sets up the sense of the DACKO line of the 8237, IC_2, to be active high. It also loads the starting data address into IC_2 for the first conversion results. IC_1 initializes the

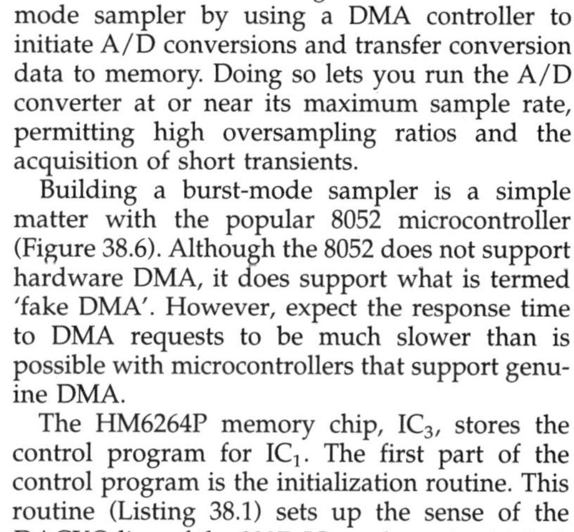

Figure 38.5 The timing for waveform playback is almost the same in the scope mode (a) as in the plotter mode (b). The difference is that the scope playback runs continuously, whereas the circuit produces a plotter output only once.

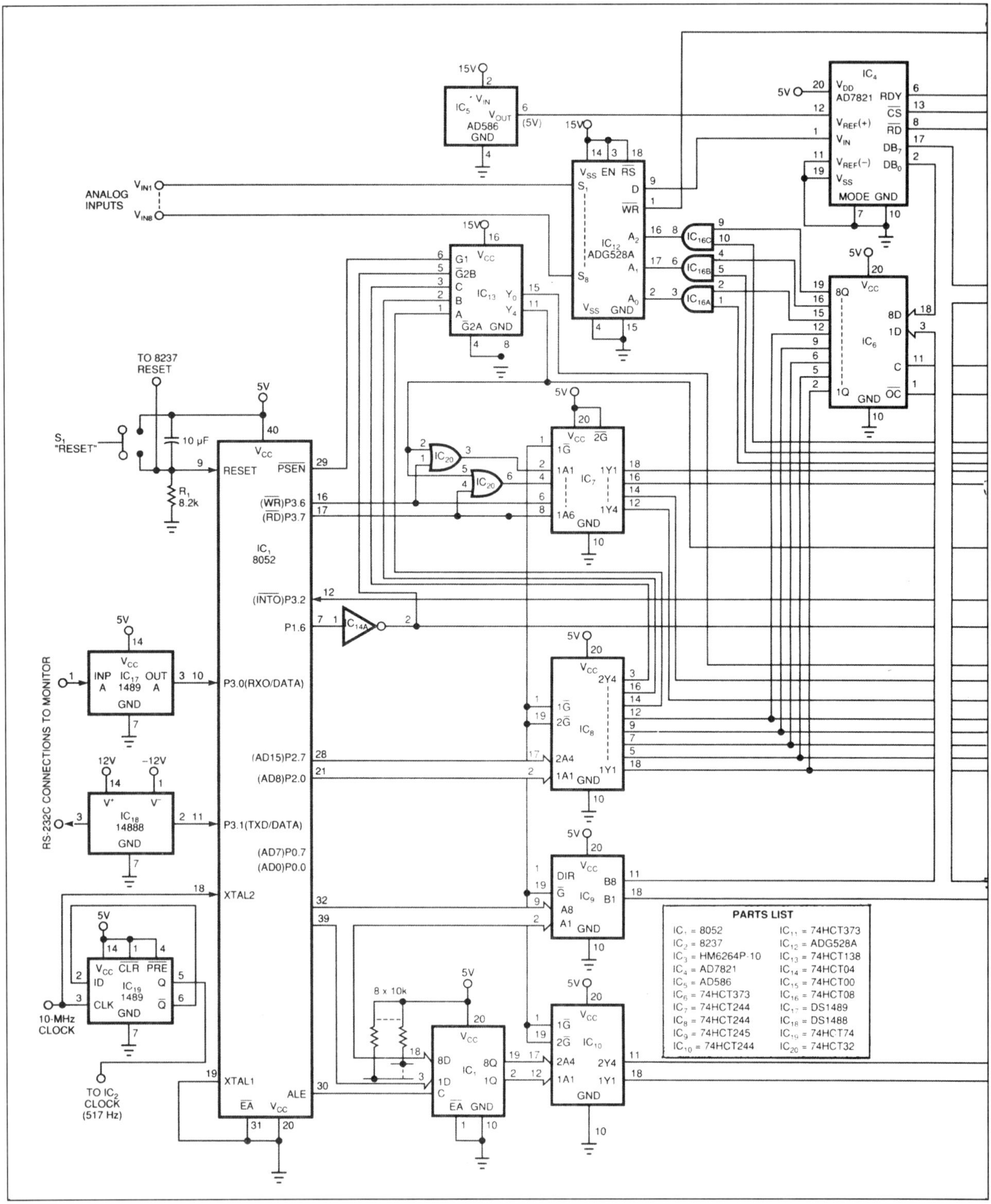

Figure 38.6 A burst-mode sampler using an 8052 microcontroller is roughly equal in complexity to the transient recorder shown earlier

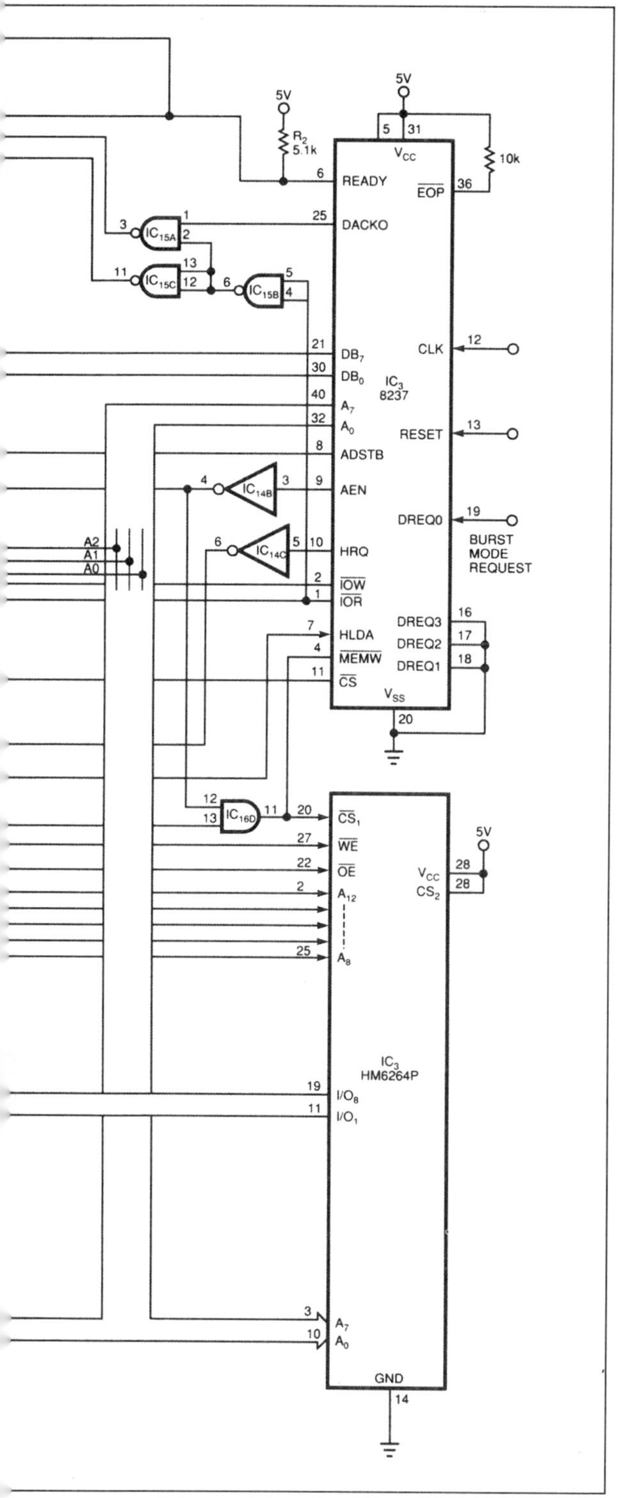

counting register to control the number of conversions before IC_2 returns control to IC_1. The program must also set up IC_1 for 'fake DMA'.

Once you've run the initialization program, IC_2 is ready to take control when requested to do so. Although IC_2 has four interrupt-request lines, this circuit uses only one, DREQO. An external command signal drives this interrupt line high, telling IC_2 to take control of the circuit and start the A/D converter sampling the input waveform.

After IC_2 receives the DREQO request (Figure 38.7), its HRQ line goes high and feeds IC_{14C}, which takes the $\overline{INTO}$ line of IC_1 low. IC_1 responds to this 'fake DMA' request by bringing its P1.6 line low and the output of IC_{14A} high, selecting inputs of IC_7, IC_8, IC_9, and IC_{10}. When the output of IC_{14A} goes high, it shuts off IC_1's address and data lines from the rest of the circuit and deselects the output's address decoder, IC_{13}. The inverted P1.6 line also feeds the HLDA input of IC_2, acknowledging IC_2's request for control. IC_2 then takes control of the address and data bus and the sampling of the input waveform.

To reduce pin count, IC_2 multiplexes the eight higher-order address bits on the data lines. You need an external device to latch these address bits. The address strobe signal, ADSTB, takes AEN high and switches the $\overline{OC}$ line of IC_6 low. ADSTB drives the C input of IC_6 to latch the higher address lines to the outputs of IC_6. The inverted AEN line also drives one input of IC_{16D}. The decoded output, Y_0, of IC_{13} controls the other input of this gate. Therefore, either a high on AEN or a low on the decoder output selects IC_3. You need this control logic because both IC_2 and IC_1 must be able to access IC_3.

The DACK0 line goes high at about the same time that ADSTB latches the address and drives one input of IC_{15A}. IC_{15A} and IC_{15B} ensure that the $\overline{CS}$ line of IC_4 goes low only when an input/output read operation of IC_2 occurs. IC_{15C} provides the correct polarity for the $\overline{RD}$ input and equalizes the delay paths for the $\overline{CS}$ and $\overline{RD}$ lines, ensuring that the circuit obeys the $\overline{CS}$-to-$\overline{RD}$ setup time.

Once IC_4 receives a $\overline{CS}$ signal, it acknowledges receipt of the signal by bringing its RDY line low, placing the controller, IC_2, into a wait state for as long as its READY input is low. When the device completes a conversion, the RDY line goes high, releasing IC_2 from its wait state. Because IC_4's RDY output is an open-drain output, you need to install an external pullup resistor, R_2.

When the circuit releases IC_2 from its wait state, data from IC_4 is valid. The address lines of IC_2 determine where data loads into memory. IC_2 performs all of these operations automatically because a memory write accompanies each input/output read. Depending on the value loaded into the counting register, IC_2 will con-

Listing 38.1 Initialization routine

```
10     XBY(8008H) = 80H                   : SETS DACK SENSE ACTIVE HIGH
20     XBY(800FH) = 0EH                   : CLEARS DREQ0 MASK REGISTER
30     XBY(800BH) = 94H                   : SETS MODE REGISTER
40     XBY(800CH) = 00H                   : CLEARS FIRST/LAST FLIP-FLOP
                                          :  (ONLY REQUIRED IF 8237 IS
                                          :  NOT RESET BETWEEN DMA REQUESTS)
50     XBY(8000H) = 00H                   : LOADS LOWER BYTE OF STARTING DATA
                                          :  ADDRESS TO BASE AND CURRENT ADDRESS
60     XBY(8000H) = 08H                   : LOADS HIGHER BYTE OF STARTING DATA
                                          :  ADDRESS TO BASE AND CURRENT ADDRESS
70     XBY(8001H) = 00H                   : LOADS LOWER BYTE OF COUNTING NUMBER
                                          :  TO COUNT REGISTER
80     XBY(8001H) = 02H                   : LOADS HIGHER BYTE OF COUNTING NUMBER
                                          :  TO COUNT REGISTER
90     DBY(38) = DBY(38).OR.02H
100    IE = IE.OR.81H
110    GOTO 10
```

tinue to issue read commands to IC_4 until the circuit completes the required number of conversions. IC_2 automatically increments the memory address after every write operation.

The multiplexer, IC_{12}, accommodates eight input channels. The three upper and three lowest address lines of IC_2, gated through IC_{16A}, IC_{16B}, and IC_{16C}, select the input channel. If the three upper address lines are set to all 1s, IC_4 will convert each channel in sequence, and the conversion results will be stored in consecutive memory locations. For example, if the first conversion takes place on the channel-1 input voltage, V_{IN1}, and the result is stored in location M of IC_3, the next conversion will take place on V_{IN2}, and the result will be stored in location M + 1. If the three uppermost address bits are set to 011, the circuit will sequence through channels 1 to 4 only.

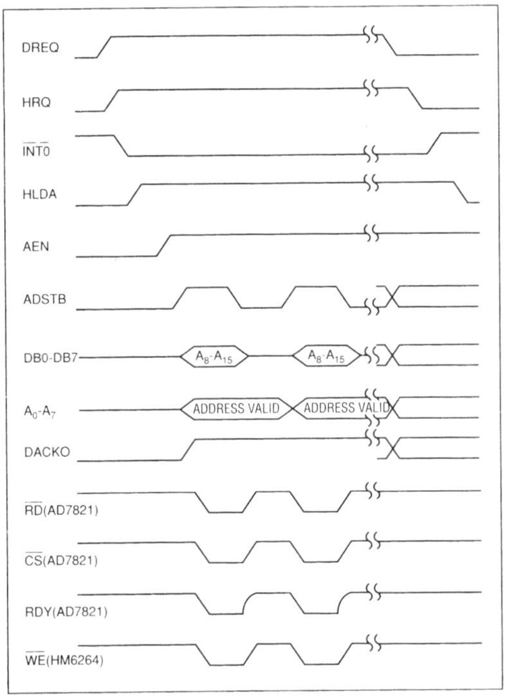

Figure 38.7 Although the 8052 does not support true DMA, you can create a 'fake DMA' mode, which, though not as fast as real DMA, lets you transfer blocks of data directly to memory

Ready or not

The RDY line of IC_4 drives the $\overline{WR}$ input of IC_{12}, loading the address for the next channel to be converted into the multiplexer. When you have only one input channel to convert, you can use an alternate design: remove IC_{16A}, IC_{16B}, and IC_{16C}, and drive the A0, A1, and A2 inputs of IC_{12} directly from the three uppermost address lines. Using this scheme, the program chooses the input channel.

IC_1 selects the device it talks to using a 1-of-8 address decoder, IC_{13}. The outputs of IC_{13} provides signals for IC_{12}'s $\overline{WR}$ line, IC_3's CS input, and IC_2's $\overline{CS}$ input. One of the outputs also gates the P3.7 ($\overline{RD}$) and the P3.6 outputs from IC_1 to drive the $\overline{IOW}$ and $\overline{IOR}$ inputs of IC_2. The upper three address lines of IC_1 select the required device. The lower address lines are multiplexed with the data lines in a manner similar to the way IC_2's address and data lines are multiplexed. IC_{11} demultiplexes the lower eight address lines. IC_1's ALE signal latches these address lines. The 3-state buffers, IC_7, IC_8, and IC_{10}, isolate IC_1 outputs from the address bus when IC_2 takes control. You must use these buffers because IC_1 can't place its address and data buses into a high-impedance state when IC_2 takes control of the circuit. IC_9 also acts as a buffer but is bidirectional because IC_1 must read data from and write data to memory.

IC_1 uses a 10-MHz input-clock frequency. A 74HCT74 counter (IC_{19}) divides down this clock to form the clock input to IC_2. The standard 8237 operates from a 3-MHz maximum clock frequency, so you can divide the 10-MHz clock by 4 to provide IC_2's clock. You'll have a resultant acquisition rate of 608k samples/s. A faster version of the 8237, the 8237–5, operates from a 5-MHz input clock, allowing you to divide the clock frequency by 2 and enabling the circuit to take 812k samples/s. If you were to use IC_1 on its own to control the sampling of the input waveform, the best acquisition rate you could obtain would be approximately 100k samples/s.

The entire circuit operates from 15V and 5V supplies. If there isn't a 5V supply in your system, you can add a regulator to generate 5V. In addition, plan to use a precision 5V reference (IC_5) for the A/D converter, allowing an input range of 0 to 5V. To obtain accurate conversion results, you must obey the same guidelines regarding decoupling and grounding as apply to the transient-recorder circuit.

You can use the same design (Figure 38.6) with slow- and medium-speed microprocessors that support DMA requests. With these microprocessors, you'll find the DMA response time will be much faster than the response of the 8052's 'fake DMA'. Because microprocessors that support genuine DMA will 3-state their address and data lines during a DMA transfer, you can eliminate the 3-state driver chips.

Oversampling reduces antialiasing requirements

In the spectrum of a periodically sampled waveform, the spectrum of the (unsampled) input signal repeats around harmonics of the sampling frequency. Any frequency contained in the input signal is repeated above and below each harmonic of the sampling frequency. Therefore, in the spectrum of the sampled signal, the band between 0 and f_{IN} (the input spectrum), appears – among other places – between $f_s - f_{IN}$ and f_s, where f_s is the sampling frequency. Though you may be under the impression that the input-signal bandwidth is 100kHz, if the sampling frequency is 1M sample/s, a signal at 991kHz in the input spectrum would appear as a 9-kHz 'alias' component in the sampled signal's spectrum.

The purpose of an antialiasing filter is to remove or at least attenuate any noise or spurious signals that could be aliased back into the bandwidth of interest. Figure 38.8 shows the frequency response of an antialiasing filter for a generalized A/D converter. You determine the filter roll-off by drawing a straight line between the highest signal frequency of interest, f_{IN}, and the stopband attenuation frequency, $f_s - f_{IN}$. As the ratio of f_s to f_{IN} increases (that is, as the oversampling ratio increases) the slope of the line decreases.

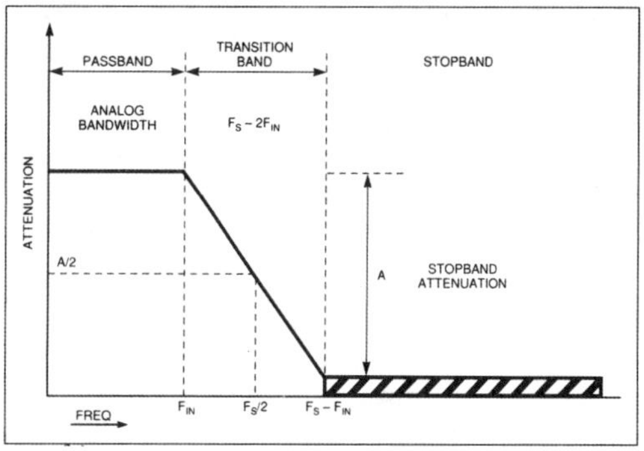

Figure 38.8 The antialiasing filter that precedes the transient recorder's ADC can be simple or complex depending on the degree of oversampling. When the sampling frequency is 10 × the highest frequency of interest, the filter has three octaves to roll in its attenuation. A simple 3-pole filter has 18dB/octave roll-off. An 8-bit ADC needs slightly more than 48dB of attenuation. Hence, a 3-pole filter is usually sufficient for 8-bit resolution.

In an 8-bit system, an ideal ADC's signal-to-noise ratio (S/N) is slightly greater than 256:1 or 48dB. To avoid having noise limit the system performance, the ratio of the input signal to noise should exceed the approximately 48-dB limit imposed by the ADC. Here, the signal is the peak-to-peak value of the signal within the band of interest, and the noise is the square root of the sum of the squares of the amplitudes of all of the frequency components outside that band. The attenuation required on signals outside the band of interest depends on the application and the expected magnitude of the out-of-band signals. In most cases, the magnitude of these signals is much lower than that of the desired signal.

Usually, 8-bit systems require 50dB of attenuation for signals that can be aliased into the band of interest. Even if 50dB is not the desired number, the following calculations show the kind of reduction in antialiasing filter requirements brought about by oversampling. With 2 × oversampling (that is with $f_s = 2{\cdot}f_{IN}$), f_s and f_{IN} are at the same point and the filter has to have infinite roll-off to attenuate signals at $f_s - f_{IN}$. With $f_s = 3{\cdot}f_{IN}$ (3 × oversampling), the filter's attenuation must drop from 0dB at f_{IN} to 50dB at $2.f_{IN}$. In other words, the slope of the attenuation vs frequency curve must be 50dB/octave; the filter (if it has a Butterworth characteristic) must have more than eight poles.

With 10 × oversampling, there are three octaves for the attenuation to drop from 0 to 50dB. The required slope is a little more than 16dB/octave; a 3-pole Butterworth filter will do the job.

The above analysis of the antialiasing filter holds true regardless of the type of ADC that follows the filter. No matter what the conversion technique, oversampling reduces the antialiasing filter requirements. Oversampling also reduces the ADC noise within the signal bandwidth because it spreads the quantization noise over a wider bandwidth. Oversampling has recently gained considerable popularity in connection with sigma-delta ADCs. In the case of these converters, the advantages of oversampling are much greater than with successive-approximation or flash ADCs because noise shaping produces dramatic improvements in noise performance as the oversampling ratio increases.

The relationship between antialiasing-filter performance and oversampling is, however, exactly the same for an oversampled sigma-delta modulator as for a half-flash or a successive-approximation ADC. A sigma-delta ADC and a half-flash ADC with the same oversampling ratio place the same requirements on the antialiasing filter.

The disadvantage of the sigma-delta process for transient recording is the pipelining or averaging technique inherent in sigma-delta converters. Because of the pipelining, a step change requires a significant time (the settling time of the ADC's digital filter) to ripple through to the output. Therefore, there is a delay before a sigma-delta converter produces an output that represents an input change. Between the time the input changes and the sigma-delta converter's output reflects the change, the ADC's output does not accurately represent the converter's input. Such performance is not appropriate for transient recorders of the type discussed here.

39 Op-amp distortion measurement bypasses test-equipment limitations

Measuring the immeasurable

With op-amps now boasting distortion levels down to around 0.0001% (– 120 dB), the ideal of having test equipment an order of magnitude better than the item under test is no longer possible. This article discusses the use of an auxiliary amplifier to measure the distortion of the amplifier under test.

Jerald Graeme
Burr-Brown Corp

Until recently, distortion performance was not important in most op-amp applications. Now, common use of the fast Fourier transform (FFT) extends the importance of op-amps' distortion beyond audio applications into general signal processing. Any distortion introduced by an amplifier produces erroneous Fourier components. To predict these errors components, you must first characterize your op-amp's distortion. But op-amp's distortion performance often surpasses that of available test equipment, defying characterization. Making the amplifier-under-test part of the test system solves this characterization problem. This solution works exclusively with feedback amplifiers, op-amps included.

Feedback reduces an amplifier's distortion – at its output – to minuscule levels. Feedback also separates the amplifier's distortion from the test signal. This separated distortion is none other than the error signal fed back to the op-amp's inputs. Once separated, the amplifier-distortion signal is insensitive to any distortion in the incoming test signal. Also, the separated signal has a reduced magnitude that reduces the dynamic range your test equipment has to handle.

Three distortion-measurement methods capitalize on the signal-separating action of op-amp feedback. In the first method, you measure the separated signal directly. This method circumvents test-equipment limitations. In the second method, selectively amplifying the amplifier's distortion raises this error signal above the threshold of the test equipment. Finally, the third method removes the test signal yet avoids measuring any effects of the added amplifier. This method bootstraps the op-amp's power supplies on the test signal itself to remove the test signal from the measurement. This article covers direct measurement and selective amplification; the next article covers selective amplification and bootstrapping.

Each approach greatly improves distortion resolution but also has specific constraints. Signal separation adds an amplifier to the test system; selective amplification reduces the measurement bandwidth; and bootstrapping requires using a signal to drive the reference point of the op-amp's power supplies.

Noise seldom limits op-amp-distortion measurement. Only in very low-distortion op-amps does noise impose a limit on distortion analyzers. Amplifier noise is almost never a problem for spectrum analyzers because they are highly insensitive to noise.

Translate to ground

First consider how feedback separates the amplifier-distortion products from the test signal. This consideration is fundamental to each of the measurement circuits that follow. You can visualize the signal separation most easily with a voltage follower (Figure 39.1). In Figure 39.1, input signal V_{IN} drives the op-amp's input to produce output signal V_{OUT}, and a simple loop equation shows that

$$V_{OUT} = V_{IN} - V_{ERR},$$

where V_{ERR} is the differential-input error signal of the op-amp.

As trivial as this equation seems, it holds the answer to measuring op-amp distortion with high

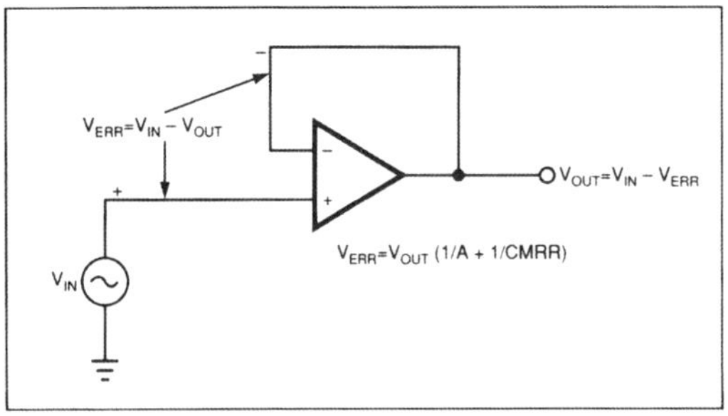

Figure 39.1 Feedback separates an op-amp's distortion products from the test signal by developing a signal V_{ERR} equal to the difference between input and output signals

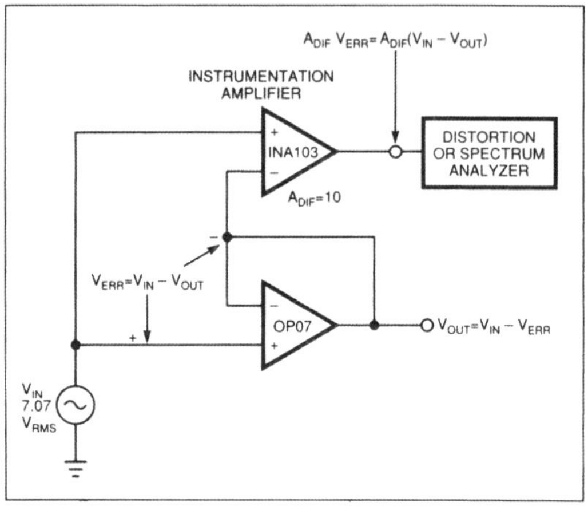

Figure 39.2 Directly implementing Figure 39.1's signal separation requires adding an instrumentation amplifier to provide a ground-referenced signal

resolution. The equation states that the output signal, V_{OUT}, is a replica of the input signal, V_{IN}, except V_{OUT} does not include the input error signal V_{ERR}. Thus, any distortion the amplifier introduces is in V_{ERR}.

Measuring V_{ERR} instead of V_{OUT} removes any effects of signal-generator distortion and reduces the dynamic range required of your test equipment. The op-amp's open-loop gain and common-mode rejection attenuate whatever test signal remains in V_{ERR}.

Distortion measurement with Figure 39.1's setup requires additional processing of the signal V_{ERR}. Signal V_{ERR} rides on the input signal V_{IN}. Consequently, any ground-referenced measurement of V_{ERR} still includes the test signal V_{IN}.

The instrumentation amplifier in Figure 39.2 references V_{ERR} to ground and increases the signal level presented to the analyzer. Finding a low-distortion instrumentation amplifier is easier than producing a better signal generator and a better signal analyzer. This instrumentation-amplifier alternative serves the measurement of intermediate levels of distortion in feedback amplifiers.

After measuring distortion using the setup in Figure 39.2, you must adjust your results. These adjustments transform the distortion percentage measured in V_{ERR}, THD + N_M, to the equivalent percentage present in V_{OUT}, THD + N_0. (THD + N_M is the measured value and THD + N_0 is the corresponding output distortion and noise.)

$$\text{THD} + N_0 = (V_{ERR}/V_{OUT})\text{THD} + N_M.$$

When using a spectrum analyzer, adjusting the THD result as you calculate it is the easiest way to go. Taking this tack necessitates two changes. THD expresses distortion as the ratio of the rms sum of the distortion products to the signal fundamental:

$$\text{THD} = \sqrt{(V_2^2 + V_3^2 + V_4^2 + \ldots)} \times (100\%)/V_1.$$

Here, V_1 represents the fundamental component of the signal, and V_2, V_3, V_4 and so forth represent the distortion components. For the measurement shown in Figure 39.2, the magnitude of V_{OUT} substitutes for the fundamental V_1 to correct for the smaller fundamental signal present in V_{ERR}. Also, the harmonic amplitudes measured require adjusting to account for the gain they receive from the instrumentation amplifier. For this adjustment, divide the overall THD equation by the instrumentation amplifier's differential gain, A_{DIF}.

$$\text{THD}_{OUT} = \sqrt{(V_2^2 + V_3^2 + V_4^2 + \ldots)} \times 100\%/A_{DIF}V_{OUT}.$$

For the unity-gain amplifier under test, subtraction obviously separates the op-amp's distortion from the test signal. However, this condition is a coincidence unique to the voltage follower. In other op-amp configurations, the signal translation of V_{ERR} does not subtract the op-amp's output from the input signal.

Figure 39.3 shows the generalized, noninverting, feedback configuration along with the equations relating V_{ERR} to V_{OUT}. Here, a feedback network attenuates the effect of V_{OUT} on V_{ERR}. Thus, the amplifier-distortion products reflected in V_{ERR} are smaller than those in V_{OUT}.

As before, you must separate the V_{ERR} signal from the common-mode test signal in Figure 39.3's circuit. The first method for this separation is translation to a ground-referenced signal (Figure 39.4). Distortion measurement with this configuration is easiest to see by considering the circuit to be an extension of Figure 39.2's voltage follower. In Figure 39.4, the voltage-divider action of the feedback network presents a signal

$V_{OUT}R_1/(R_1 + R_2)$ to the amplifier's inverting input. For the voltage follower, this signal was the full V_{OUT}. Now, the feedback signal is attenuated, and a simple loop equation shows that for Figure 39.4.

$$V_{ERR} = V_{IN} - (V_{OUT}R_1/(R_1 + R_2)).$$

Figure 39.4's circuit amplifies V_{IN} and its distortion in producing V_{OUT}. Thus, subtracting V_{OUT} from V_{IN}, as with Figure 39.2, would not remove the generator's distortion for Figure 39.4. However, subtracting an appropriately attenuated V_{OUT} from V_{IN} does remove this distortion. Figure 39.4 has a gain of $(R_1 + R_2)/R_1$. Then, feedback attenuates V_{OUT} by the *inverse* of this gain or $R_1/(R_1 + R_2)$.

For a distortion-analyzer measurement like that shown in Figure 39.4, first compensate the result for the smaller fundamental measured through V_{ERR}. Multiply the measured $THD + N_M$ result by V_{ERR}/V_{OUT} as before. This calculation yields the input $THD + N_{IN}$ result, which you then multiply by the $1/\beta = (R_1 + R_2)/R_1$ of the op-amp's configuration.

$$THD + N_O = \frac{V_{ERR}(R_1 + R_2)}{V_{OUT}R_1} THD + N_M$$

$$= \frac{R_1 + R_2}{R_1} THD + N_{IN}$$

You must also adjust your results for spectrum-analyzer measurements using Figure 39.4's setup. Once again, you discard the measured fundamental because it does not represent the output signal. Then, substitute V_{OUT} for fundamental amplitude V_1 in the THD equation and divide this equation by A_{DIF} to remove the effect of the instrumentation amplifier's gain. For Figure 39.4, also make a gain adjustment for the circuit's gain of $1/\beta = (R_1 + R_2)/R_1$.

$$THD_{OUT} = \frac{(R_1 + R_2)\sqrt{(V_2^2 + V_3^2 + V_4^2 + \ldots)}}{R_1 A_{DIF} V_{OUT}} (100\%)$$

Signal separation extends to inverting case

For the generalized inverting amplifier, distortion resolution is even greater (Figure 39.5) than for the noninverting amplifier. The most significant improvement with inverting circuits actually results from removing the instrumentation amplifier of Figure 39.4. The inverting configuration of Figure 39.5 removes common-mode voltage from the op-amp's input and avoids the added amplifier along with the added amplifier's distortion.

For the inverting circuit in Figure 39.5, the relationship between input and output distortion is not as obvious as with Figure 39.4's circuit. Previously, the feedback network relayed a large signal to the amplifier's input. But inverting circuits keep this input near zero voltage, balancing V_{IN} and V_{OUT} at the amplifier's input. Both signals influence the voltage at the inverting input through the feedback network. To find the result, consider the two signals separately using superposition. This exercises the feedback network as a voltage divider driven from each end. Then, the amplifier input signal is

$$V_{ERR} = \frac{V_{IN} R_2}{R_1 + R_2} - \frac{V_{OUT} R_1}{R_1 + R_2}.$$

Although not immediately obvious, the distortion introduced by V_{IN} still cancels in this V_{ERR}

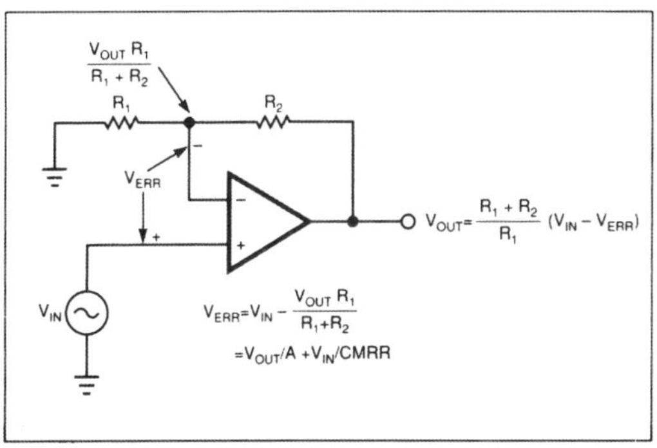

Figure 39.3 In the generalized feedback case, the output signal, along with its distortion, reflects to V_{ERR} through an attenuation

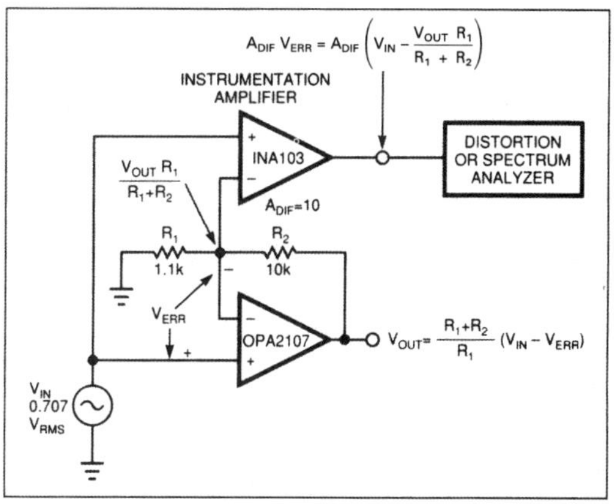

Figure 39.4 The amplifier's distortion products, included in V_{ERR}, remain separated from the test signal in measurements of the generalized noninverting circuit

signal. The above equation shows that signal V_{IN} influences V_{ERR} directly in the first term of the equation and then indirectly through feedback in the second term.

In the direct path, V_{IN} contributes to V_{ERR} through an attenuation of $-R_2/(R_1 + R_2)$. Added to this contribution is the V_{IN} component transmitted through V_{OUT}. In this path, V_{IN} and its distortion products first receive a forward gain of $-R_2/R_1$ to produce V_{OUT}. Feedback then attenuates V_{OUT} by a factor of $-R_1/(R_1 + R_2)$. The total gain of this path is the product of the forward gain and the feedback attenuation, or $R_2/(R_1 + R_2)$. This product has the same magnitude as the attenuation of the direct path above, but these two gains have opposite polarities. Thus, the direct and feedback distortion effects of V_{IN} again cancel in the V_{ERR} signal.

Figure 39.5's measured results require two adjustments to account for the THD_{OUT} of the amplifier's configuration. These adjustments follow directly from the Figure 39.4 results and use the same equations. One adjustment accounts for the smaller fundamental actually measured and the other corrects for the $1/\beta$ gain that the harmonics included in the measurement don't receive. For distortion-analyzer measurements using Figure 39.5's setup,

$$THD + N_0 = \frac{V_{ERR}\ (R_1 + R_2)}{V_{OUT}\ R_1}\ THD + N_M$$

$$= \frac{(R_1 + R_2)}{R_1}\ THD + N_{IN}$$

And for spectrum-analyzer results,

$$THD_O = \frac{(R_1 + R_2)\ \sqrt{(V_2^2 + V_3^2 + V_4^2 + \ldots)}}{R_1\ V_{OUT}} \times (100\%).$$

References

1. Pryce, Dave, 'Audio DACs Push CD Players to Higher Performance', *EDN*, December 7, 1989.
2. The Institute of High Fidelity Inc, 'Standard Methods of Measurement for Audio Amplifiers', IHF-A-202, 1978.

40 Advanced techniques tackle advanced op-amps' extremely low distortion

Still measuring the immeasurable

This article, a companion to the previous one, describes other distortion-measuring techniques, applicable to the latest generation of extremely low-distortion amplifiers.

Jerald Graeme
Burr-Brown Corp

Selective amplification offers an alternative to the added amplifier described in the previous article. This alternative moderates, rather than negates, the limitations of signal generators and signal analyzers. In addition to separating distortion and test signals, selective amplification makes the amplifier distortion signal dominant in the measurement; however, it also reduces measurement bandwidth.

As with signal separation, the selective-amplification approach is easiest to understand starting with a voltage-follower connection. Figure 40.1 shows a bootstrapped feedback network added to a voltage follower. In Figure 40.1, the common-mode rejection of the amplifier-under-test replaces the instrumentation amplifier used before. However, taking this tack moves the measurement back to the amplifier's output.

Resistors R_1 and R_2 form a feedback network that produces gain for V_{ERR} but not for V_{IN}. Signal V_{ERR}, which includes the amplifier's distortion products, appears across resistor R_1. There, this signal produces a feedback current that goes to resistor R_2. This operation develops an error-signal gain, $A_{ERR} = 1 + R_2/R_1$, for V_{ERR} alone.

Input signal V_{IN} does not experience this amplification because R_1 is bootstrapped rather than grounded. The resulting output signal for Figure 40.1 is

$$V_{OUT} = V_{IN} - ((R_1 + R_2)/R_1 V_{ERR}).$$

To V_{IN}, the circuit remains a voltage follower. The amplifier's output follows V_{IN} except for the difference produced by the amplified distortion signal. This difference is small as long as high loop gain keeps error signal V_{ERR} low.

Distortion above the measurement floor

Similarly, signal V_{IN} directly varies the voltage at the amplifier's noninverting input in Figure 40.1. Feedback forces the amplifier's inverting input to also follow this signal.

The selective amplification in Figure 40.1 raises the relative magnitudes of the V_{ERR} distortion products for increased resolution. But the signal generator's distortion now remains in the signal measured. However, this distortion signal is not amplified and its relative significance diminishes in proportion to the gain V_{ERR} receives. Similarly, this gain moderates the dynamic range demands on the signal analyzer. Thus, selective amplification raises the amplifier-distortion signal above the measurement floor of your instruments.

Following the measurement, a THD calculation removes the effect of the selective gain. Divide the measured distortion by the distortion gain of $(R_1 + R_2)/R_1$. For Figure 40.1 the output-referred distortion for a voltage follower is:

$$\mathrm{THD} + \mathrm{N_O} = \frac{R_1}{(R_1 + R_2)} \mathrm{THD} + \mathrm{N_M}$$

$$= \mathrm{THD} + \mathrm{N_{IN}}$$

OR

$$\mathrm{THD_{OUT}} = \frac{R_1 \sqrt{(V_2^2 + V_3^2 + V_4^2 + \ldots)}}{(R_1 + R_2)\, V_1} \times (100\%)$$

At first blush, you'd think that maximizing the selective gain would achieve the greatest measurement accuracy. However, measurement bandwidth declines because of feedback-factor reduction as this gain increases[1]. Because the op-amp is now part of the measurement system, the amplifier's bandwidth limits resolution of higher-order

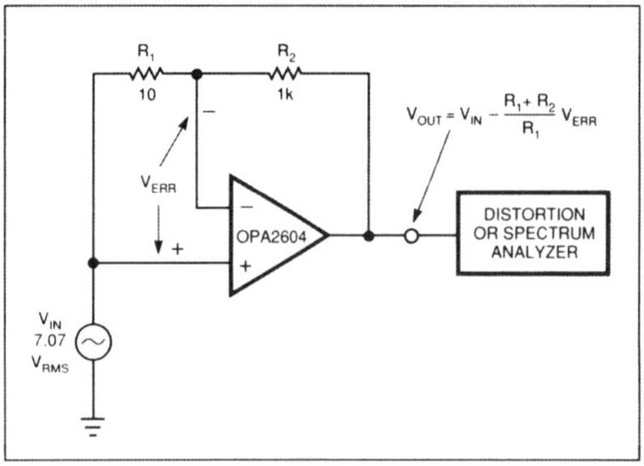

Figure 40.1 Selectively amplifying V_{ERR} magnifies the amplifier's distortion signal for direct measurement at the output of the amplifier tested

distortion harmonics. Thus, you should choose the selective gain for Figure 40.1's setup to be as large as possible within your bandwidth constraints. Note that the low-value feedback resistors avoid adding noise.

Generalizing selective gain

The selective-gain approach of Figure 40.1 extends to generalized noninverting and inverting op-amp configurations. The generalized noninverting version in Figure 40.2 has R_3's added gain for selectively amplifying distortion products. Resistors R_1 and R_2 set the normal closed-loop gain presented to V_{IN}. As usual, this gain is simply $A_{CL} = 1 + (R_2/R_1)$. V_{ERR} experiences greater gain because it develops a feedback current through R_3, as well as through R_1. The resulting error-signal gain relates the parallel combination of resistors R_1 and R_3 and is $A_{ERR} = 1 + R_2/(R_1 \parallel R_3)$. The proper choice for R_3 makes V_{ERR}'s distortion dominant at the amplifier's output.

The distortion measurement's resolution remains unchanged between Figure 40.1 and Figure 40.2. These circuits differ by the closed-loop gain, A_{CL}, supplied to V_{IN} and its distortion, but practical limits equalize the results. As A_{CL} increases, the magnitude of V_{IN} diminishes to maintain a given output-signal level. Thus, the magnitude of the input-signal distortion decreases by the same amount that its gain increases. The resulting output distortion arising from V_{IN} is then, unchanged in magnitude from that of Figure 40.1. Adding R_3 keeps this distortion in the background by ensuring sufficient additional gain for the distortion products of V_{ERR}.

Dynamic-range constraints of the signal analyzer are also independent of A_{CL} in Figure 40.2. The relative levels of the fundamental signal and the distortion signals determine this range. For a given test condition, the output level is fixed and is essentially the level of the fundamental signal. To reduce dynamic-range requirements, raise the level of the distortion signal by amplifying V_{ERR}. This amplification results from either the intended closed-loop gain of the circuit or from this gain in conjunction with the selective gain R_3 provides. However the gain occurs, it raises the relative proportion of V_{ERR} in the output signal. As long as V_{ERR} receives sufficient gain, you can easily distinguish amplifier-distortion products in the output signal.

However, you must limit the gain you choose for V_{ERR}, or amplifier-response roll-off will restrict measurement of higher-frequency harmonics. The gain applied to V_{ERR}, not that applied to V_{IN}, sets the amplifier's bandwidth. To determine your measurement's bandwidth, calculate the feedback factor, considering R_3 to be grounded rather than bootstrapped. Then, for the circuit in Figure 40.2,

$$\beta = (R_1 \parallel R_3)/(R_1 \parallel R_3 + R_2).$$

This feedback factor defines a measurement-bandwidth limit of βf_C, where f_c is the unity-gain crossover frequency of the op-amp. Beyond this limit, higher-order harmonics are attenuated in the measurement. Thus, again, you should consider a balance between test-equipment error suppression and higher-frequency resolution when choosing R_3.

Accounting for gain differences

However, determining the output-referred distortion still requires separating the A_{CL} and A_{ERR} effects on the circuit in Figure 40.2. Selectively amplifying the distortion signal makes its effect dominant in the measurement. You must again

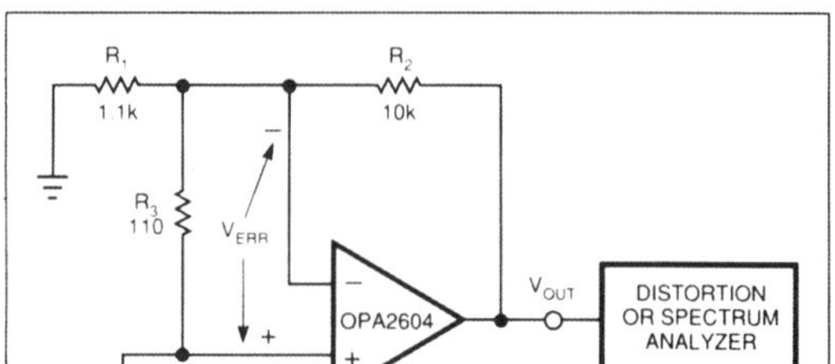

Figure 40.2 Adding R_3 extends selectively amplifying V_{ERR} to measuring the distortion of a generalized noninverting amplifier

adjust the measured distortion to account for the difference in signal and distortion gains. To adjust the measurement result, remove the selective gain that the amplifier distortion receives. In Figure 40.2, resistors R_1 and R_2 supply a gain of $A_{CL} = 1 + (R_2/R_1)$ to both V_{IN} and V_{ERR}. R_3 supplies additional gain to V_{ERR}. This added gain $(1 + R_2/R_3)$ amplifies only the distortion signal. To compensate, divide the measured distortion result by this added gain.

$$THD + N_O = \frac{R_3}{(R_3 + R_2)} THD + N_M$$

$$= \frac{(R_1 + R_2)}{R_1} THD + N_{IN}$$

OR

$$THD_{OUT} = \frac{R_3 \sqrt{(V_2^2 + V_3^2 + V_4^2 + \ldots)}}{(R_3 + R_2)\, V_1} \times (100\%)$$

The selective amplification in Figure 40.2 translates directly for inverting op-amp configurations. To convert Figure 40.2 to an inverting amplifier, simply switch the circuit connections to the common return and the input signal (Figure 40.3). This switch returns the op-amp's noninverting input and R_3 to ground and causes V_{IN} to drive R_1. As before, resistors R_1 and R_2 set the gain, A_{CL}, presented to V_{IN}, and resistor R_3 boosts this gain to a higher level, A_{ERR}, for V_{ERR}. This higher gain determines the feedback factor and resulting measurement's bandwidth. The circuit in Figure 40.3 retains a measurement resolution that is independent of A_{CL}. Only gain A_{ERR} affects this

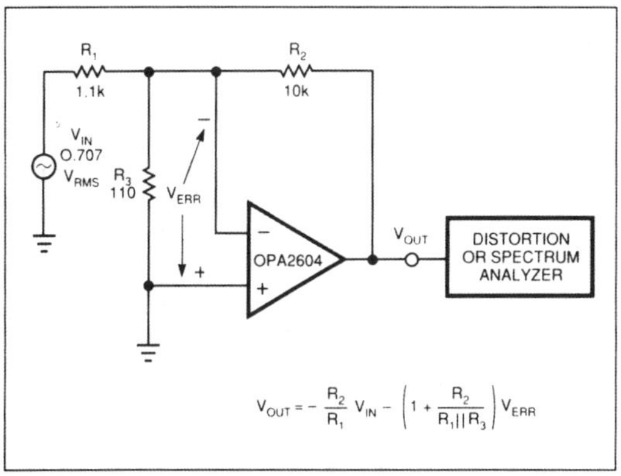

Figure 40.3 Selective amplification of an inverting amplifier follows directly from the noninverting case of Figure 40.2

resolution. Finally, convert the measured THD_M of Figure 40.3 to THD_{OUT} or THD_{IN} using the equations for Figure 40.2.

The only way Figure 40.3 differs from Figure 40.2 is in the common-mode input signal of the amplifier. In the noninverting circuit in Figure 40.2 input signal V_{IN} is a common-mode signal to the amplifier's inputs, and it exercises nonlinearities in the amplifier's CMRR. The inverting circuit in Figure 40.3 removes the common-mode signal from the amplifier's inputs. Then, only the gain nonlinearity of the amplifier influences the amplifier's distortion. This difference permits you to separate gain and CMRR distortion effects.

Combining the two methods

Selective amplification in the inverting case offers another alternative. Both selective amplification and signal separation work in inverting circuits. However, the combination places greater demands on measurement bandwidth. Selective amplification obviates the instrumentation amplifier used before. To eliminate the instrumentation amplifier, Figure 40.1 and 40.2 move the signal measurement to the op-amp's output. There, signal separation is compromised because the full test signal remains in the measurement.

This compromise is unnecessary for inverting configurations. As mentioned before, inverting configurations do not require the instrumentation amplifier for the signal-separation measurement. Thus, with inverting configurations you need not move the measurement to the amplifier's output. Instead, signal separation and selective amplification combine at the amplifier's input (Figure 40.4). There, R_3 develops a feedback current with V_{ERR} just as before.

However, the circuit in Figure 40.4 does not rely on R_2 to convert this feedback current to an amplified output error. Instead, a second resistor, R_4, added at the amplifier's input, does this job. The feedback current produced in R_3 conducts through R_4 to produce the desired amplification right at the amplifier's input. At the top of R_4, the signal is $-(1 + R_4/R_3)V_{ERR}$. This amplified error signal remains free of the large test signal present in the amplifier's output. As before, this separated error signal permits measurements free from signal-generator distortion and eliminates large dynamic-range requirements.

The distortion measured in Figure 40.4 requires three adjustments for converting it to output-referred distortion. First, compensate the difference in measured and actual fundamental signals as in previous signal-separation measurements. Then, make two gain adjustments. The measured signal receives a measurement gain of $(1 + R_4/R_3)$ but does not receive the circuit closed-loop gain of $A_{CL} = (1 + R_2/R_1)$. To compensate, divide the

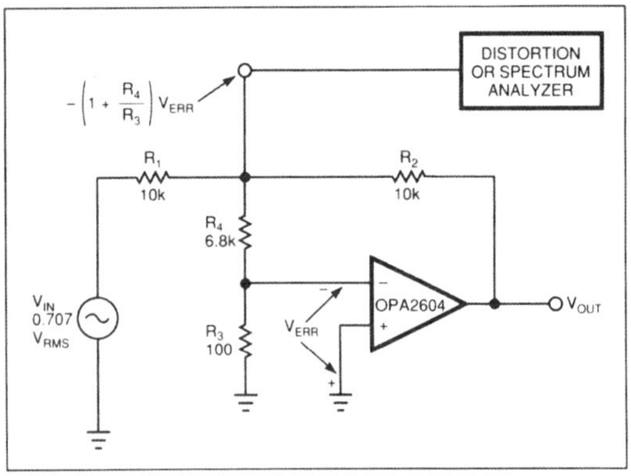

Figure 40.4 Adding R_4 combines selective amplification with signal separation for the inverting amplifier

measured distortion by the measurement gain and multiply it by A_{CL}.

$$\text{THD} + \text{N}_O = \frac{R_3\ (R_1 + R_2)\ V_{ERR}}{R_1\ (R_3 + R_4)\ V_{OUT}}\ \text{THD} + \text{N}_M$$

$$= \frac{(R_1 + R_2)}{R_1}\ \text{THD} + \text{N}_{IN}$$

OR

$$\text{THD}_{OUT} = \frac{R_3\ (R_1 + R_2)\ \sqrt{(V_2^2 + V_3^2 + V_4^2 + \ldots)}}{R_1\ (R_3 + R_4)\ V_{OUT}} \times (100\%)$$

Figure 40.4 introduces an added attenuation to the circuit's feedback factor, restricting measurement bandwidth. In addition to the normal feedback attenuation of R_1 and R_2, a second feedback attenuation results from R_3 and R_4. R_3 and R_4 also produce a loading effect on the attenuation of R_1 and R_2. The net Figure 40.4 feedback factor is

$$\beta = \frac{R_1R_3}{R_1R_2 + R_1R_3 + R_1R_4 + R_2R_4 + R_2R_3}$$

The relationship of bandwidth to gain bandwidth, BW = βGBW, then determines the bandwidth for the circuit in Figure 40.4. For the specific components of Figure 40.4, $\beta = 0.0043$ and GBW = 10 MHz for BW = 43 kHz.

Because of the low feedback factor, this measurement's bandwidth is below the 80 kHz desired for audio applications. Other choices for R_3 and R_4 offer higher feedback factors to improve bandwidth, but such choices lower the selective gain. With less gain, the distortion signal's level is closer to the test equipment's measurement floor. Because of this compromise, you should use the circuit in Figure 40.4 only where signal-generator distortion must be separated from the test signal. In other cases, the basic selective-gain configuration offers a better compromise.

The input capacitance of the signal analyzer, C_1, alters the feedback factor in Figure 40.4. This capacitance bypasses R_1 and can cause gain peaking or even oscillation. Such problems occur only if the break frequency of the bypass, $1/2\ \pi\ R_1C_1$, is within the amplifier's closed-loop bandwidth. In this case, add a compensating capacitor in parallel with R_2 to roll off the gain peaking. Choose this capacitor to break with R_2 at the same frequency that C_1 breaks with $R_1 \parallel (R_3 + R_4)$. Then, the feedback-divider action of the R_2 and R_1 legs remains approximately constant with frequency.

Gain variation extends resolution

Some op-amps' distortion-measurement requirements exceed test equipment's capabilities even when you use the preceding methods. When your op-amp has low distortion over wider bandwidths, or just extremely low distortion, you need variable test configurations to characterize fully its distortion-versus-frequency performance.

First, low distortion levels automatically rule out the basic signal-separation approach of the previous article because that approach requires an instrumentation amplifier of even lower distortion than the op-amp under test. Instead, use selective amplification, which places measurement bandwidth and measurement resolution in competition. You must maintain measurement bandwidth to around 80 kHz to resolve harmonics important to the audio range. This bandwidth limits the selective amplification to a gain of $1/\beta$ = GBW/80 kHz.

However, your setup need not maintain full bandwidth at every test frequency. The amplitude of distortion harmonics drops as their frequencies get further away from the fundamental's frequency. Because of this decline, a measurement bandwidth that spans only five or six harmonics is sufficient. A smaller measurement bandwidth permits the use of higher selective gains to better resolve the lower distortion levels encountered at lower frequencies. Higher test frequencies require the full bandwidth, but they also cause correspondingly higher amplifier distortion. Accordingly, higher test frequencies require less selective gain, extending measurement bandwidth. Thus, the gain/bandwidth compromise of selective amplification yields to distortion measurement with varied gains.

The OPA627, for example, requires three selective-gain steps, each step providing a different gain/bandwidth combination. Figure 40.5 details this gain variation, which revolves around the THD_{IN}-vs-frequency plot. As this plot shows, op-amp distortion typically rises at higher frequencies, where measurement bandwidth is most needed.

Power-supply bootstrapping for noninverters

Signal separation is a complete solution only for inverting configurations. Using a power-supply bootstrap avoids limitations in noninverting solutions. Given care to avoid ground loops, the bootstrapping approach separates the common-mode signal, extending signal separation to the noninverting case.

To permit an optimal analyzer connection, power-supply bootstrapping moves the circuit's common from the normal circuit ground to the op-amp's noninverting input (Figure 40.6). As odd as it may seem to consider an op-amp's noninverting input to be the common, the common of a circuit is a relative point that you can define to be anywhere you choose. This connection retains the common-mode swing for the amplifier but removes that swing relative to common and, thus, removes it from the analyzer's input.

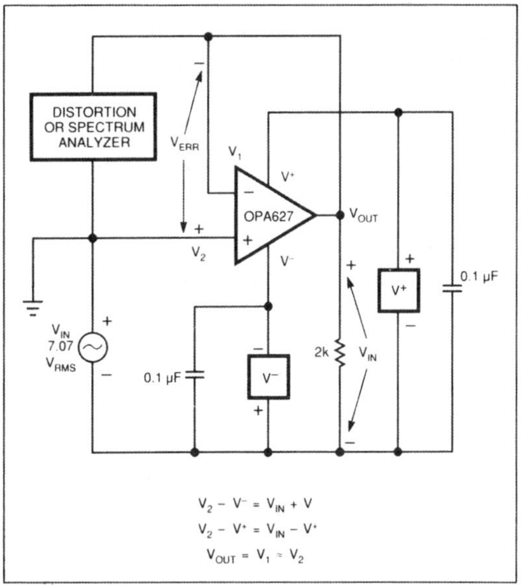

Figure 40.6 Power-supply bootstrapping permits directly measuring V_{ERR} and does not change an amplifier's internal voltage swings

Theoretical niceties aside, redefining the common introduces ground-loop errors. The effects of these errors depend on the sensitivity of the circuit to voltage drops in its connecting lines. In Figure 40.6, the element most sensitive to such voltages is the signal analyzer because it measures a small signal superimposed on a larger one. For this reason, the figure shows the signal analyzer returned to the circuit's new common. Figure 40.6 makes the power-supply connections vulnerable to line drops, but the power-supply rejection of the op-amp attenuates the resulting voltages.

For Figure 40.6, the test-equipment demands again decrease by a factor of $1/(1/A + 1/CMRR)$. However, the measurement made in Figure 40.6 requires adjustment to account for the reduced fundamental measured. For this figure, the relevant signal swing is that across the load resistor, or V_{IN}. Therefore, multiply the measured distortion by V_{ERR}/V_{IN}.

The actual adjustment made depends on the type of signal analyzer used. Measurements made with a distortion analyzer directly produce a THD + N percentage. Simply multiply this percentage by V_{ERR}/V_{IN} and Figure 40.6's output distortion plus noise is then

$$THD + N_O = \frac{V_{ERR}}{V_{IN}} THD + N_M = THD + N_{IN}.$$

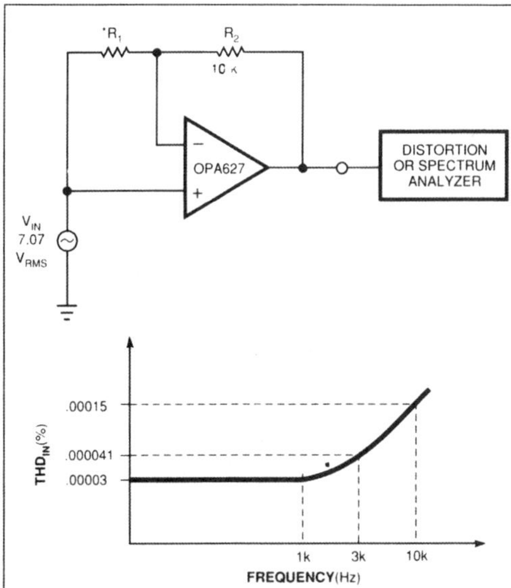

	*R₁	A	BW
20Hz TO 3 kHz	10	1001	16 kHz
4 TO 9 kHz	33	303	53 kHz
10 TO 20 kHz	100	101	160 kHz

Figure 40.5 For extremely low-distortion amplifiers, vary the selective gain to retain resolution and bandwidth

You must measure the magnitude of V_{ERR} separately because distortion-analyzer outputs do not normally indicate this magnitude.

When you measure distortion with a spectrum analyzer, no separate measurement is required. Spectrum analyzers display the magnitudes of the fundamental and harmonic signals individually. You can then calculate distortion from the fundamental THD equation. Multiply this equation by V_{ERR}/V_{IN}, where V_{ERR} is equal to and therefore replaces V_1 and V_{IN} remains in the denominator. Then, the spectrum analyzer result for Figure 40.6 is

$$THD_{OUT} = \frac{\sqrt{(V_2^2 + V_3^2 + V_4^2 + \ldots)}}{V_{IN}} (100\%).$$

Bootstrapping resolves noninverting cases

The convenience of Figure 40.6 extends to the generalized noninverting amplifier. As Figure 40.7 shows, power-supply bootstrapping again permits directly measuring V_{ERR} with a grounded signal analyzer.

Only one difference separates the measurements of the two circuits. The greater gain of Figure 40.7 results in a larger load signal V_{LOAD}. This gain also amplifies V_{ERR}, making the distortion in V_{LOAD} greater than that measured in V_{ERR}.

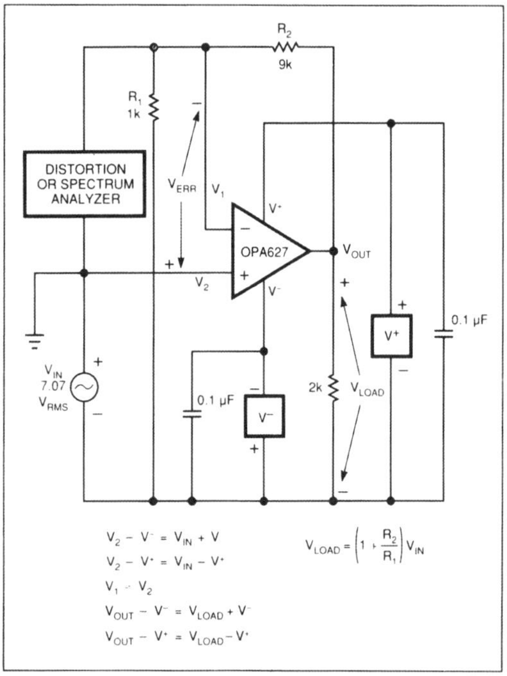

Figure 40.7 The bootstrapping of Figure 40.6's voltage follower also applies to general noninverting circuits

You can adjust the measured distortion later to compensate for the effect of this gain. Finally, the added gain further reduces the performance requirements of the test equipment. For a given level of V_{LOAD}, V_{IN} is smaller for Figure 40.7 than for Figure 40.6. Thus with Figure 40.7, V_{IN}'s reacting with the amplifier's CMRR produces a smaller V_{ERR} signal. The noninverting configuration reduces the signal measured by a factor of $V_{LOAD}/V_{ERR} = 1/(1/A + \beta/CMRR)$.

Consider Figure 40.7 with the common return first at the top and then at the bottom of the signal generator. This change makes no difference in the equations relating amplifier voltages to the V^+ and V^- supply terminals. For both configurations, the $V_2 - V^-$ and $V_2 - V^+$ equations are the same as those for Figure 40.6. Amplifier feedback forces $V_1 \approx V_2$ to again extend these equations to Figure 40.7's input. Thus, whether bootstrapped or not, the amplifier distorts the input signal.

Figure 40.7's greater gain produces a different output result than Figure 40.6. To define V_{OUT} relative to V^+ and V^-, first determine the load voltage, V_{LOAD}. You can find this voltage from the loop formed by the load resistor with resistors R_1 and R_2. The input signal, V_{IN}, appears across resistor R_1, producing a feedback current of V_{IN}/R_1. This current flows in R_2 to develop a voltage of $V_{IN}R_2/R_1$. Adding the voltages of R_1 and R_2 shows the voltage on the load to be $V_{LOAD} = (1 + R_2/R_1)V_{IN}$. This result portrays the familiar response of a noninverting op-amp configuration and is independent of Figure 40.7's redefined common. Thus, the bootstrapping does not affect the load voltage and the corresponding amplifier output current.

Similarly, the loops relating V_{OUT} to V^+ and V^- remain unchanged. With the common on either side of the signal generator, the output voltages with respect to the amplifier supply terminals are

$$V_{OUT} - V^- = V_{LOAD} + V^-$$
$$V_{OUT} - V^+ = V_{LOAD} - V^+.$$

Thus, both input- and output-signal conditions are independent of the Figure 40.7 common connection, and the bootstrapping does not change the amplifier's distortion products.

You must convert the distortion measured in Figure 40.7 to output-referred distortion. In Figure 40.7, the circuit amplifies the distortion products in V_{ERR} by $1/\beta$ to produce greater distortion signals in V_{OUT}. This effect changes the correction factor to $V_{ERR}/\beta V_{IN}$. However, because the same gain amplifies the input signal, V_{IN}, the final correction factor becomes $\beta V_{ERR}/\beta V_{IN} = V_{ERR}/V_{IN}$. Thus Figure 40.7's correction equations are the same as Figure 40.6's.

For the voltage-follower case, combining bootstrapping and selective amplification achieves even greater distortion resolution. This way, you can test the very lowest distortion amplifiers with a distortion analyzer. This particular amplifier-and-analyzer combination is the one case where noise becomes a limit to op-amp distortion measurement. And for a spectrum analyzer, the ambient noise of the test environment requires your careful attention to avoid coupling stray noise into your circuit. With either type of analyzer, selective amplification expands distortion resolution for the bootstrapped voltage follower.

The benefits of power-supply bootstrapping and selective amplification combine in Figure 40.8. In this circuit, the only signal developed at the amplifier's output is the amplified error signal:

$$V_{OUT} = -(1 + R_2/R_1)\ V_{ERR}.$$

A signal analyzer measures this amplified signal referenced to ground with no interference from the test signal. In addition, the amplified distortion signal conveniently overrides the background noise of the signal analyzer and the measurement environment. This convenience does not extend to the general noninverting case because added gain there restores the test signal to the amplifier's output.

Other characteristics of the measurement shown in Figure 40.8 follow directly from earlier results. Selective amplification reduces the measurement bandwidth from f_C to βf_C. Hence, f_C is the unity-gain bandwidth of the op amp. The feedback factor is $\beta = R_1/(R_1 + R_2)$. Figure 40.8's test-equipment requirements are the same as for the bootstrapped follower of Figure 40.6. As with that circuit, the distortion and dynamic-range requirements of the test equipment decrease by a factor of $1/(1/A + 1/CMRR)$. Because the selective amplification amplifies both the amplifier's distortion products and the background signal, V_{ERR}, it does not improve this factor. The attenuated generator distortion present in V_{ERR} gets amplified along with the amplifier distortion

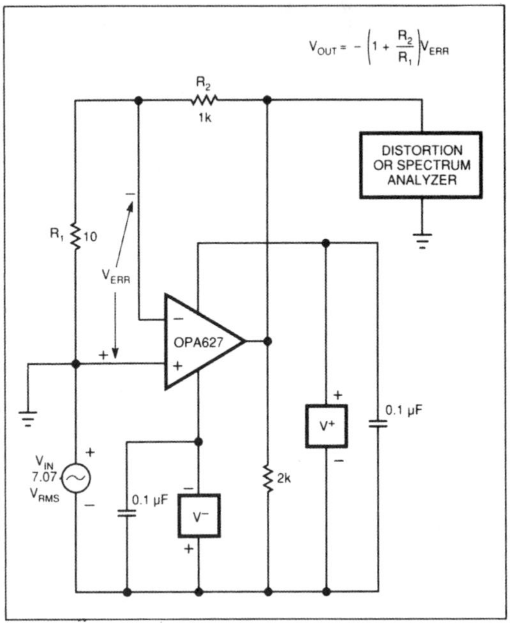

Figure 40.8 Combined bootstrapping and selective amplification expand distortion resolution for the voltage follower

products. The relative significance of generator distortion is unchanged. Similarly, the selective gain amplifies both the maximum and minimum signals to be resolved by the analyzer. Thus, the dynamic range of the measurement is also unchanged.

For the same reasons, results measured with the circuit in Figure 40.8 translate output-referred distortion with the same equations as those used in Figure 40.6 and 40.7. EDN

References

1. The Institute of High Fidelity Inc. 'Standard Methods of Measurements for Audio Amplifiers', IHF-A-202, 1978.

2. Graeme, J, 'Feedback Plots Offer Insight into Operational Amplifiers', *EDN*, January 19, 1989.

3. Pryce, Dave, 'Audio DACs Push CD Players to Higher Performance', *EDN*, December 7, 1989.

41 Characterizing high-speed digital circuits: a job for wideband scopes

High end DSOs can cope with the fastest logic families

Ever since, as a lad, I converted an ex-RAF airborne radar indicator unit type 182A into an oscilloscope I have been all too aware of bandwidth limitations. Even my trusty 475A is really no match for the current fastest generations of logic. Digital storage 'scopes on the other hand can offer much wider bandwidths at a modest price, using repetitive sampling, but only the most expensive types with a high-speed single shot capability and deep memory will cope adequately with funnies like occasional glitches and metastable states.

Lawrence B Levit and Marco L Vincelli
LeCroy Corp

With the advent of fast TTL and CMOS ICs, subnanosecond edges have become a reality – forcing you to use circuit-design and board-layout techniques originally developed for high-speed analog circuits. Establishing that any high-speed circuit works as intended involves lots of testing. Historically, the primary tool for such testing has been the analog oscilloscope. Unfortunately, most of these instruments are limited in bandwidth to 400 MHz or less, so their use on modern logic is becoming increasingly inappropriate. Modern wideband-digital oscilloscopes provide an alternative and can help you troubleshoot and characterize fast digital circuits.

Digital oscilloscopes operate by successively sampling waveforms with an analog-to-digital converter (ADC). Two classes of instruments are available: single-shot and repetitive. Although the single-shot oscilloscope generally offers great versatility, it employs high-speed ADCs and fast memories. These state-of-the-art components are expensive, and, as a consequence, single-shot oscilloscopes usually carry higher price tags than repetitive units of equal bandwidth.

Figure 41.1 shows a typical single-shot instrument. It employs an extremely high-speed ADC that samples once at each clock cycle. This scope is designed to record single-shot waveforms – those that occur only once or infrequently. For example, this type of instrument best suits ana-

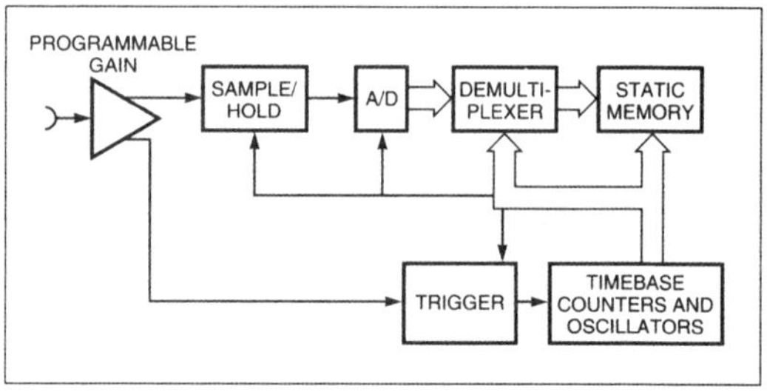

Figure 41.1 With a single-shot sampler, the maximum sample rate is typically > 4 to 10 × the scope bandwidth

lyzing circuits used for PROM burning or studying the effects of power-supply turn-on.

Luckily, in most electronic circuits, the majority of signals are repetitive. If a signal is repetitive, the scope reconstructs the signal by acquiring it multiple times using a slower ADC and then interleaving the samples. The method requires very good timing resolution (normally provided by a time-to-digital converter) but allows the use of relatively inexpensive ADCs and conventional memory. Instruments that use this technique provide most of the advantages of digital oscilloscopes without exacting the price premiums of single-shot instruments.

Measuring high-frequency repetitive signals

The two techniques for repetitive waveform sampling are random-interleaved sampling (RIS), which some vendors call random-repetitive sampling, and sequential (SEQ) sampling. Block diagrams for both types appear in Figure 41.2.

Random-interleaved sampling takes advantage of the fact that the signal and the sampling clock are asynchronous. By accurately measuring the time from the sample clock to the trigger, a scope that has only a moderate-speed ADC can achieve a sample rate that is, in effect, very high. For example, by using a high-resolution (< 10 ps) time-to-digital converter, equivalent sampling rates of 20 G samples/s (effective sample intervals of 50 ps) are possible. Besides allowing you to use a relatively inexpensive ADC for capturing the details of high-speed waveforms, this type of repetitive sampling has an important advantage: It allows full viewing of the signal's history prior to the trigger point.

In contrast, SEQ sampling adds incrementally longer delays to the sample clock for each repetition of the signal. This method of sampling is typically used in very-high-bandwidth (> 10 GHz) oscilloscopes. Even though SEQ sampling provides very fine timing resolution, it does not normally provide pretrigger information. In fact, to account for internal timing, SEQ sampling scopes require a signal delay as long as 10 to 50 ns (introduced by an internal cable).

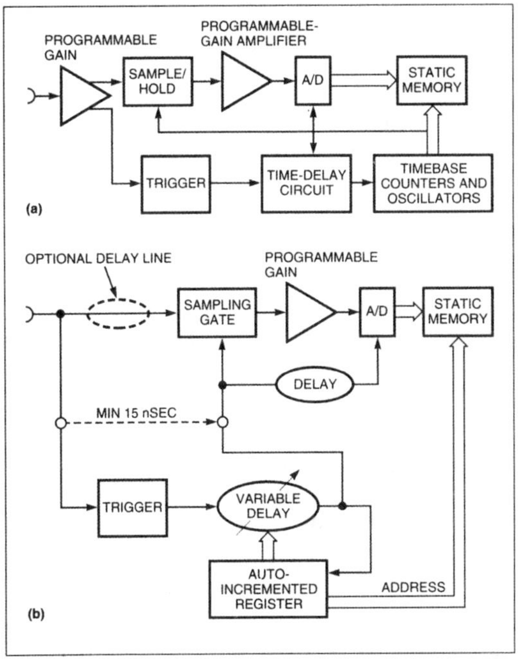

Figure 41.2 With a random-interleaved sampler (a), the scope bandwidth (in Hz) is typically many times the actual sample rate (whose units are samples/s). At least at high sweep speeds, however, the effective sample rate is many times the bandwidth. With a sequential sampler (b), although the architecture is quite different, the same statements apply

Bandwidth and sample rate

The bandwidth of a digital oscilloscope is the analog bandwidth of its front-end electronics; that is, the analog bandwidth of the amplifier/attenuator and the track-and-hold circuit. Bandwidth indicates the ability of the circuit to follow a fast edge (or a high-frequency signal) without substantially filtering it. An oscilloscope that has insufficient bandwidth will display a waveform that has reduced high-frequency content; high-frequency signal amplitudes will be reduced and pulse edges slowed down.

The sample rate of an oscilloscope also sets a limit on the fastest signal that an instrument can follow. The highest frequency that can be digitized without serious errors (the so-called Nyquist frequency) is just under half the sampling rate. In practice, accurate waveform reconstruction is difficult or impossible when the signals contain frequencies above ¼ or ⅕ the sampling rate. Sampling signals at an insufficient rate produces poor timing resolution. At worst, such 'undersampling' can result in aliasing, which completely distorts the displayed waveform.

Note that the constraints imposed by bandwidth and sampling rate are independent of each other and affect the measurement in different ways. Ideally, you would use a scope with a wide bandwidth and a high sampling rate to characterize and troubleshoot modern circuits. To adequately analyze signals that have the subnanosecond edges, such a scope would have at least 1 GHz of analog bandwidth and be capable of taking 5G samples/s. Instruments of this type exist, but they carry a price tag of close to $50,000.

As a more cost-effective alternative, a good laboratory might have a pair of scopes, one a wideband RIS instrument and the other a somewhat lower-bandwidth single-shot unit. This combination carries a much smaller price tag than the previously described fast sampling, wide bandwidth, digital oscilloscope. Moreover, with two scopes on hand, the laboratory personnel can perform two measurement tasks at once – something they couldn't do with a single, higher-priced scope.

Bandwidth – how much is enough?

A high-bandwidth instrument preserves pulse shape even on very fast signals. In addition, high bandwidth helps to maintain the integrity of fast edges (allowing better timing measurements) and can reveal effects buried in the edges. But how much bandwidth is enough? A reasonable way to address the question is to determine the magnitude of pulse slowing caused by the oscilloscope and to set a limit.

For comparison, three different oscilloscopes, a LeCroy 9414, a 9310, and a 9324, have bandwidths of 150 MHz, 300 MHz, and 1 GHz, respectively. Each scope captured and measured pulses having various rise times. The results were then compared to those obtained on a scope with 20-GHz bandwidth. Figure 41.3 shows the results. As the rise time approaches 2 or 3 ns, the 150- and 300-MHz scopes start to introduce significant errors (> 10%). In contrast, the 1-GHz oscilloscope can comfortably handle measurements in the 1 ns range.

One way to judge a scope's response is to determine the usable rise-time response (URR). The URR is the shortest input rise time for which the scope overestimates rise times by no more than 5%. For signals whose rise times are slower, the scope makes accurate measurements. For signals that have faster rise times, the instrument will cause significant distortion.

The URR for each of the oscilloscopes is shown in Table 41.1. Another way to judge the response capability of an oscilloscope is from its bandwidth. In the case of the LeCroy oscilloscopes, the URR is related to the bandwidth as follows:

$$\text{Bandwidth (MHz)} = 350/\text{rise time (ns)}.$$

The bandwidth of the fastest acceptable input edge in Table 41.1 has been computed in this way. Before using a scope to make measurements on modern circuits, you should examine the relationship between the input-signal bandwidth and the scope bandwidth:

$$\text{Bandwidth}_{\text{OSCILLOSCOPE}} \geq 3 \times \text{bandwidth}_{\text{SIGNAL}}.$$

Note that the relationship above applies only to instruments whose frequency-response curves

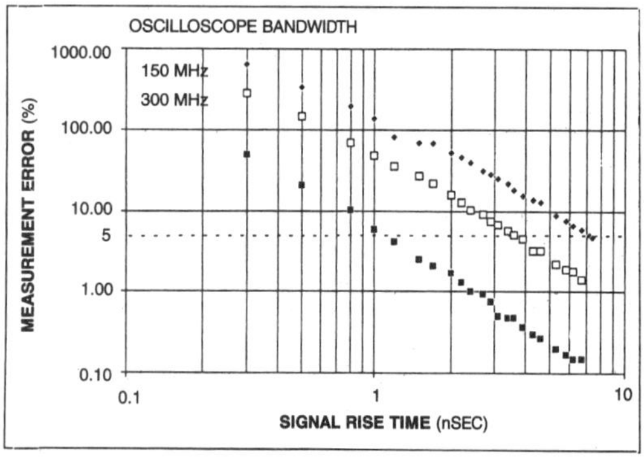

Figure 41.3 Scopes with 150-MHz, 300-MHz, and 1-GHz bandwidths and well-behaved frequency responses measure rise time with a 5% error when the actual rise time is 7.3, 3.6, and 1.1 ns, respectively. From these times, you can calculate the usable frequencies shown in Table 1. Each frequency is a bit less than ⅓ of the scope's -3-dB bandwidth

Table 41.1 URR and signal bandwidth vs scope bandwidth[1]

Oscilloscope bandwidth (-3 dB)	**URR** (nsec)	**Signal bandwidth** (MHz)
1 GHz	1.1	318
300 MHz	3.6	97
150 MHz	7.3	48

Note: 1. For scopes that have well-behaved frequency response.

(Bode plots) exhibit single-pole lowpass response. This is the case for the instruments' measurements in Figure 41.3. In practice, when compared with oscilloscopes that have equivalent bandwidth, but whose frequency-response curves are complex, instruments that have single-pole frequency response can measure faster rise times accurately. Oscilloscopes that have complex response curves must have bandwidths that exceed the signal bandwidth not by 3 times, but by 4 or 5 times.

Rise time: As shown above, signal rise time is accurately measured as long as its value is greater than the URR of the scope. Rise times are typically measured from the 10 to the 90% point on a waveform. According to IEEE Standard 181–1977, this is defined with respect to the full swing of the waveform even if there is an inflection point on the leading edge. In the fastest circuits, it is standard to measure the 20%–to–80% rise time. This makes the rise-time specification insensitive to inflections near the top of the pulse.

Overshoot: Measurement of the signal overshoot requires proper attention to the bandwidth of the oscilloscope used to make the measurement. In addition, the overshoot of the oscilloscope itself is important. For input pulses whose rise times approach the limit of the oscilloscope, overshoot larger than 5% is not uncommon.

Measurements in the presence of noise: Sometimes you must characterize a circuit under adverse circumstances, for example, when it is not reasonable to attach a probe ground close to the signal test point. In such cases, the noise in the circuit can mask the measurement. See Figure 41.4(a).

Because noise is usually distributed across a signal's entire frequency range, viewing it with a wider-bandwidth oscilloscope will generally reveal more noise. A simple way to improve the signal-to-noise ratio is to use continuous averaging with a small weighting function of, perhaps, 1:3. This type of signal processing provides a small amount of noise rejection without any compromise in bandwidth.

An example of an averaged waveform appears in Figure 41.4(b). An additional benefit of averaging is that the resulting averaged waveform has a greater dynamic range than does a single wave-

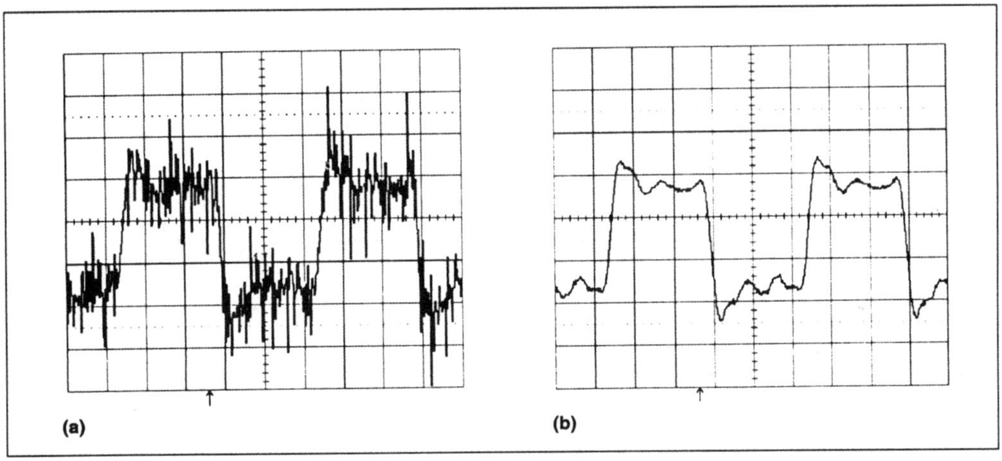

Figure 41.4 A simple DSP filtering algorithm cleans up a noisy waveform (a), so that the signal details are clearly visible (b), yet the bandwidth is not degraded. (Scales: time – 5 ns/div; voltage – 200 mV/div.) The waveform in (b) is the average of 231 acquisitions

form. The added dynamic range can be very useful when measuring small effects, such as overshoot on large signals.

Today's digital systems make extensive use of ICs that become faster with each new release. Modern IC families, such as FAST TTL, CMOS FACT, BiCMOS, and ECL, provide rise times, transit times, and toggle frequencies to several hundred MHz. Table 41.2 gives the maximum toggle frequency and the rise and fall times associated with the most important logic families. The phenomena that affect circuit performance are often at frequencies 3 to 10 × as high as the toggle rate. For example, the transition time (the time to go from one of the two possible states to the other) is much faster than the speed implied by the toggle rate.

Transition times and edge effects can have serious consequences on circuit performance. A user who is unconcerned with these effects can overlook a large class of problems, which include glitches, race conditions, ground bounce, oscillations, reflections caused by bad bus terminations, clock skew, metastable states, crosstalk, and more. Following are a few examples that show the analog nature of today's digital signals – and some of the problems associated with circuit-board debugging.

Table 41.2 Toggle and edge speeds for various logic families

Logic family	Logic type	toggle rate (MHz)	times (nsec)
TTL	Standard (74) Low-power Schottky (74LS) High-speed (74F)	30 30 70	10 10 4
CMOS	Standard (74C) High-speed (74HC) Advanced (FACT, etc)	10 40 150	20 8 <1
BiCMOS	Standard (BCT) Advanced (ABT)	70 200	2 1
ECL		300	0.8
ECLinPS		2000	0.3

The first example shows the effect of a 5-pF parasitic capacitance on the integrity of an ECLinPS pulse. Figure 41.5(a) shows the schematic diagram of the circuit under test. Figure 41.5(b), (c), and (d) show the signal at point A, as seen by instruments with 150-MHz, 300-MHz, and 1-GHz bandwidths.

The pulse viewed with the 150-MHz oscilloscope appears to be very clean and relatively noise free. The signal has a rise time of 2 ns and a fall time of 3 ns. Both edges appear to be smooth, although there is a small amount of overshoot on the leading edge.

When viewed with a 300-MHz oscilloscope, the pulse shows a somewhat different character. Even though the edge rates have not changed appreciably, there is a significant inflection in both edges. It is also evident that the noise of the signal is somewhat greater.

When viewed with the 1-GHz oscilloscope, the anomalies in the signal are dramatic. The two small peaks at the beginning and at the end of the pulse are reflections produced by the 5-pF capacitance. These peaks actually alter the rise and fall time of the main pulse and affect the switching time of the following stage. Even worse, they can generate a double firing or a metastable state in the following stage. Although the reflections are apparent with a scope having 1-GHz bandwidth, they are only subtle effects when viewed with a 300-MHz oscilloscope and are completely hidden with the 150-MHz instrument.

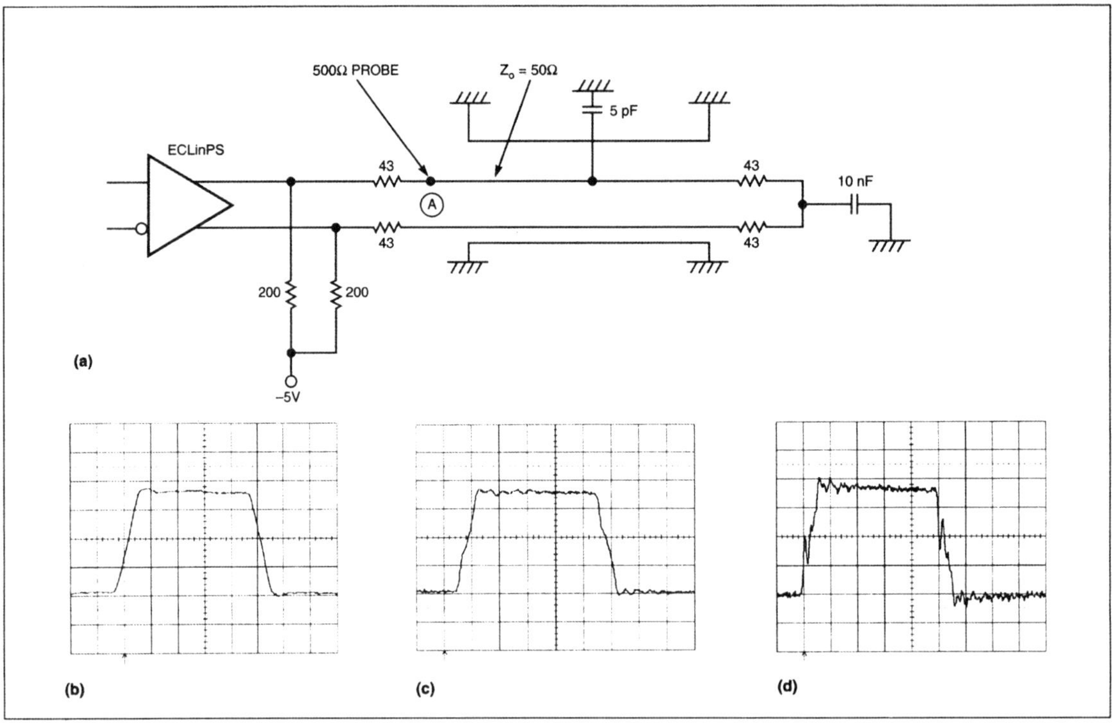

Figure 41.5 A 5-pF parasitic capacitance to ground (a) introduces reflections into a transmission line driven by an ECL gate. Scopes with bandwidths of 150 MHz and 300 MHz, (b) and (c), don't reveal any problem, however. With a 1-GHz scope (d), you see that the transitions are nonmonotonic and could cause double triggering if the signal were received by, say, a flip-flop. (Scales: time – 2 ns/div; voltage – 10 mV/div.)

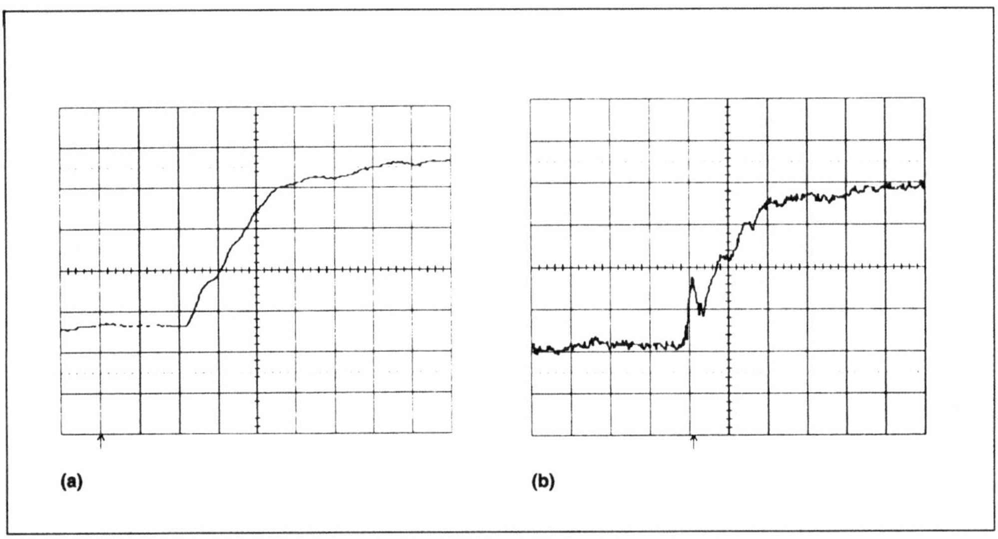

Figure 41.6 The signal rise at the end of a transmission line on a pc board looks rather clean when you view it with a 300-MHz scope (a) but potentially troublesome when you look at it with a 1-GHz scope (b). (Scales: time – 2 ns/div; voltage – 40 mV/div)

The second example shows the possible effect of a pc-board bus termination that is poorly matched to the impedance of the trace. To illustrate the importance of using an oscilloscope of adequate bandwidth, a particularly poor termination of a long (10-cm) bus was measured. The results of measurements taken with a 300-MHz oscilloscope appear in Figure 41.6(a). Results taken with a 1-GHz scope appear in Figure 41.6(b).

Figure 41.6(a) shows an almost smooth rise and does not indicate that the circuit under test has any significant problems. However, Figure 41.6(b) clearly shows the reflections produced by the incorrectly terminated bus. The first reflection is big enough to cross the threshold of the following circuit and can therefore produce double firing. Proper bus terminations not only avoid this problem but also reduce unwanted electromagnetic emission and crosstalk. Note that if older generations of digital logic were involved, they would have been immune to such narrow glitches, but with modern logic, the reflections would certainly cause problems. Proper measurement technique becomes a vital tool for debugging these types of design faults.

The fast edges of modern logic devices can often propagate through the parasitic capacitance between a trace pair on a pc board. This phenomenon, known as crosstalk, can have catastrophic effects. In severe cases, crosstalk can produce glitches that are large enough to cross some circuits' logic thresholds. These glitches can cause unpredictable failures such as unwanted logic pulses appearing in a data path or, more subtly, timing errors that result in device misfiring. Detection of glitches and accurate measurement of their amplitude and width are of major importance.

Figure 41.7 shows an example of how the apparent glitch amplitude can be altered by instruments with different bandwidth. In this example, the glitch shown is produced by the output of a 2-input AND gate. The cause of the glitch is incorrect timing of the two input signals which, in turn, results from marginal timing and jitter produced by crosstalk. Figure 41.7(a) and (b) may make you think that the glitch has insufficient amplitude to produce failures. Only the 1-GHz-bandwidth oscilloscope shows the problem to be worse than expected (Figure 41.7(c)).

Assume that the glitch shown in the previous

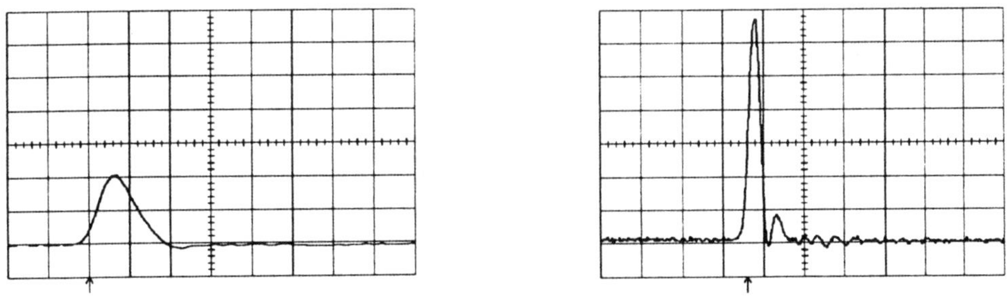

(a)

(c)

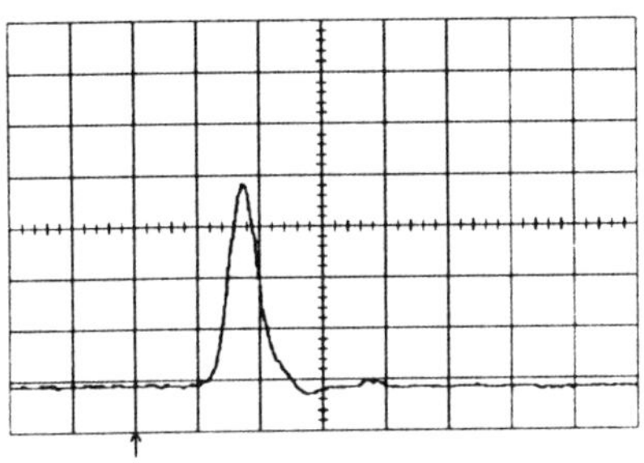

(b)

Measured pulse parameters vs scope bandwidth

Scope bandwidth (MHz)	150	300	1000
Pulse amplitude (mV)	132.4	247.5	397.6
Pulse width (nsec)	2.08	1.86	0.622
Rise time (nsec)	1.20	0.73	0.456
Fall time (nsec)	1.85	1.84	0.336

Figure 41.7 A narrow glitch looks too small to worry about when observed with a 150-MHz scope (a). With a 300-MHz scope (b), there is still a question about whether the glitch has sufficient amplitude to cause concern. The 1-GHz scope (c) removes all doubt. (Scales: time – 2 ns/div; voltage – 60 mV/div.) The table shows the parameters of the same pulse as measured by the three scopes

figures is inadvertently applied to one input of a 2-input AND gate and, at nearly the same time, another pulse arrives at the other input. There is a real chance that the following circuits will be sensitive to negative transitions. If they are, catastrophic failure will occur. Figures 41.8(a) and (b) show what you would see if you monitored the output of the gate using scopes with 150- and 300-MHz bandwidths. Again, no major problem is apparent. Although Figure 41.8(b) shows an inflection in the edge, it appears to be too small to cause any concern. Only when the signal is viewed with the 1-GHz oscilloscope does the real magnitude of the problem become apparent. Figure 41.8(c) shows the true size of the inflection and how it was largely hidden by the other instruments.

Probes and probing

In high-frequency circuit characterization, the proper use of probes is crucial. The simple act of touching a probe tip to a high-frequency circuit can significantly alter the behavior of the circuit. For each type of measurement, it is important to choose the right probe and to ground it well. Keep the ground lead as short as possible and, whenever possible, use a spring-clip ground pin.

Three factors are important when selecting a probe: its bandwidth, its resistance, and its capacitance. The effect of inadequate probe bandwidth is obvious; it degrades the composite probe/oscilloscope bandwidth.

One solution to the bandwidth issue is to use a low-impedance probe (typically 50 Ω), which is commonly a transmission line with a series resistor at the probe tip. Such a probe provides excellent bandwidth (several GHz); however, it can cause significant resistive loading, which results in a loss of signal amplitude. This attenuation is often a problem when using logic families, such as conventional TTL, which have 1-kΩ pull-up resistors; it is not usually a problem when using ECL.

You should be aware that passive probes with a high input impedance (HiZ) usually impose significant capacitive loading on a circuit. This loading can be a major bandwidth-degrading factor and, as a result, can cause signal distortion. For example, the capacitance of a 1-MΩ probe is typically around 15 pF. This means that with a 1-kΩ source impedance, bandwidth degradation would limit rise-time measurements to the order of 33 ns. Passive HiZ probes are generally restricted to applications where the signal frequency is less than 50 MHz.

Using an active probe is a high-performance approach to circuit probing. Such probes locate a bipolar or field-effect transistor at the probe tip to act as the input stage of a buffer amplifier. Active probes provide high bandwidth, high impedance, and low capacitance, but they also cost more than passive probes and are more sensitive to damage from excessive voltage.

Modern digital circuits operate at such high speeds that you must treat them with proper analog technique. In particular, attention to layout and termination are crucial. To test and debug modern circuits correctly, it is important to use test instruments that preserve signal fidelity. In particular, you should avoid using oscilloscopes that have inadequate bandwidth. Low-bandwidth instruments can easily mask important problems that can lead to unpredictable and unreliable designs.

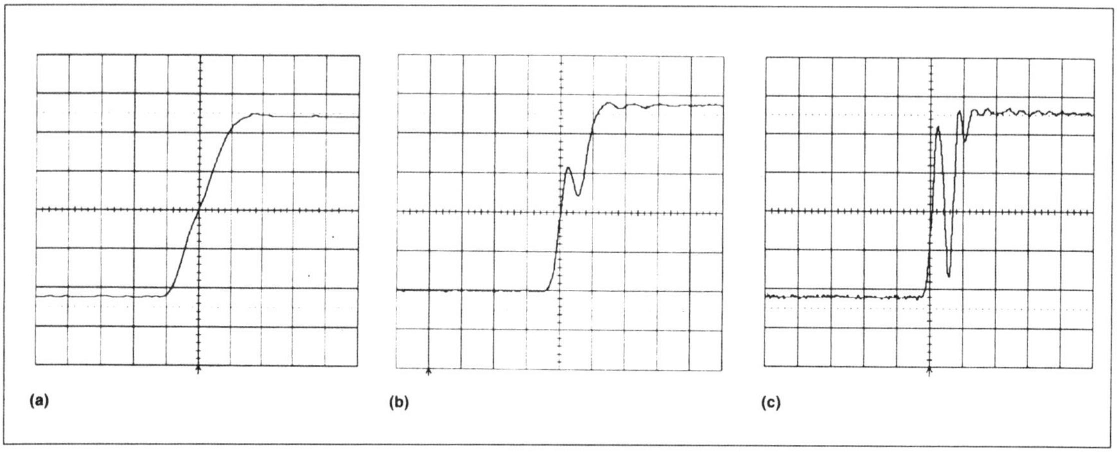

Figure 41.8 This signal's rising edge looks perfectly clean and monotonic when viewed with a 150-MHz scope (a). A 300-MHz scope (b) reveals what looks like a fairly harmless oscillation, but a 1-GHz scope (c) shows a 1-ns-wide pulse of nearly full amplitude, clearly big enough to cause the following circuits to misbehave. (Scales: time – 2 ns/div; voltage – 100 mV/div)

42 In >50-MHz digital design, measurements are a must

Use traditional RF engineers' tools in high-speed logic circuit design

If a board full of high speed digital circuitry is to work properly, the design must take into account factors such as the impedance of tracks on the PC board and similar factors, which could be ignored in the days of older slower logic: you just can't buck the laws of physics! Consequently, in addition to high-speed oscilloscopes, instruments from the RF engineer's armoury, such as network analysers and time domain reflectometers, must be employed.

DAN STRASSBERG
Senior Technical Editor

You've heard it dozens of times: state-of-the-art digital design is becoming more and more of an analog problem. But maybe you think that only purists hold to such a view ... that surely some rules of thumb will let you extend familiar design approaches into the territory beyond 50 MHz. Don't count on it!

In all likelihood, putting your head in the sand will cost you dearly. First, your design is unlikely to operate reliably. Second, it probably won't meet EMC (electromagnetic compatibility) requirements – for example, those of the FCC (Federal Communications Commission). Third, adopting 'conservative' approaches in an effort to ensure reliability will raise the cost of parts, making the product uncompetitive.

But don't think that designing high-speed systems is a black art practiced only by microwave engineers. A huge body of field and wave theory supports this area of technology. Although you must have some grasp of that theory to interpret the data, time-domain techniques allow you to visualize the measured performance of critical system elements in a way particularly well suited to how digital designers think (see 'Rho, rho, rho your board gently through the time domain'). The specific techniques are time-domain reflectometry (TDR) and time-domain transmission (TDT). Some people classify these two techniques as subsets of TDNA (time-domain network analysis).

The function of interconnections isn't hard to describe: an interconnection should get a signal from point A to point B without changing it, without allowing it to interfere with other signals, and without picking up interference from other signals. For signals at moderate voltage and current levels at frequencies from dc to tens of MHz, interconnections of most kinds usually do pretty much what they're supposed to. But in digital systems clocked at 50 MHz and more, interconnections can alter signals in devastating ways. The main culprits are reflections, delays, crosstalk, and amplitude losses. Time-domain measurement techniques help to localize and characterize several of these kinds of signal degradations.

Although some suppliers of analog simulation software might disagree, only by building and evaluating a technology board can you find out early in the design cycle whether you've made appropriate decisions on issues such as connectors and pinouts, insulation thickness, pc-board line widths, and ground-plane configuration. Technology boards usually contain no active circuits, but they do contain transmission lines, connectors, IC packages, and other interconnection elements whose performance is crucial to system performance.

According to Dr Ed Sayre, PE (Professional Engineer) of North East Systems Associates (NESA), consultants on interconnect and EMC issues and suppliers of TDR/TDT software (see 'Looking ahead'), most project teams can learn what they need from just one iteration of a technology board. If the board is designed with the requisite care, only rarely will the team need to design and evaluate a second board.

One reason that a technology board is so useful is that, if properly designed, it solves one of the stickiest problems in applying time-domain-measurement techniques. Specifically, a well designed technology board will include coaxial connectors that permit easy connection to TDR/TDT instruments. Most such instruments are designed for use in 50Ω environments. A technology board should provide coaxial connectors (often SMA type) that connect to 50Ω cables. A 'real' board from your system isn't likely to have these. A consequence of jury-rigged connections to the board under test is that in your TDR displays, the discontinuities associated with the connections to the board can hide the ones you want to investigate.

Proper fixturing is crucial to ensuring that TDR measurements contain the information you seek and to guarantee that significant effects are not masked by phenomena peculiar to your test setup. EEs who use TDR repeatedly cite two firms as sources of fixturing: Cascade Microtech and Inter-Continental Microwave. Inter-Continental specializes in test fixtures and probes for accessing pc-board features such as IC packages and strip line.

In addition to items such as these, Cascade supplies a broad line of probe stations and probes that have well controlled high-frequency characteristics. If you have to make connections within or near tiny structures – multichip-module (MCM) packages, for example – probes and probe stations, such as Cascade's, are a must.

Fixturing will not be an inconsequential part of your instrumentation cost. Prices range from around $4000 for a test fixture without probes from Inter-Continental to about $10,000 for a Cascade probe station complete with positioner and microscope. Probe prices from both firms range from $250 to roughly $1200. The Cascade station allows measurements on a variety of boards without added fixturing. Cascade also supplies calibrated impedances to verify the accuracy of its probes. For $1000, you get a 50Ω pass-through, a short circuit, an open circuit, and a 50Ω load accurate to within 0.3% traceable to NIST (the National Institute of Standards and Technology).

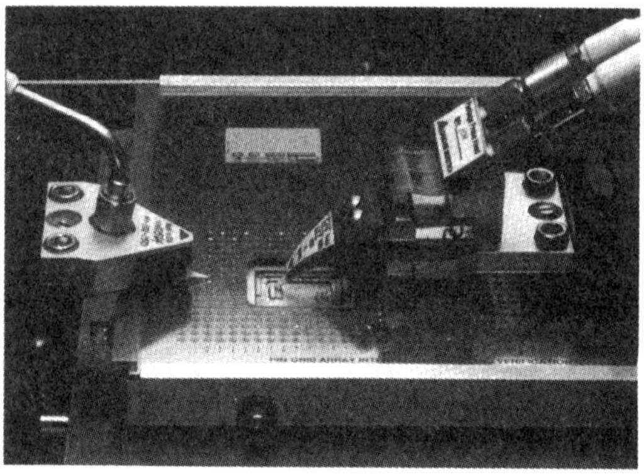

Figure 42.1 These probes from Cascade Microtech allow probing within IC packages to characterize the behavior of the package leads with sub-nanosecond edges. By adding probe points within the package, you can extend the characterization to include the bonding wires that connect a chip to the package leads

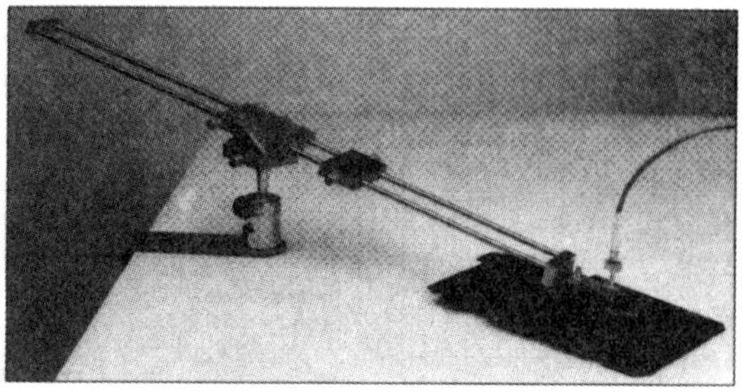

Figure 42.2 Controlled-impedance fixturing is essential to obtaining meaningful results from TDR measurements. Here is an example of probing a board using an Inter-Continental Microwave probe holder and one of the firm's TDR probes

Fixturing is just one area that you should approach with caution, as is the use of vendor-supplied test data. Component vendors often furnish elaborate sets of test results, including TDR waveforms. Don't depend on them unless you're sure that they relate to your application. In particular, beware of information that is unimportant or irrelevant at your system's speed. Vendors sometimes supply simulation models. Even here, more detail isn't always better, because it can cause excessive simulation time or require computing resources beyond what you can afford. Nevertheless, models containing excessive detail are not likely to mislead you; curves that show data beyond your region of interest can.

For example, suppose you have to choose a back-plane connector for a system whose fastest edge speed will be 400 ps. You've narrowed your selection to two candidates – one moderately priced, and the other much more expensive. You have manufacturer data showing that the lower-priced connector exhibits unacceptable crosstalk with 100-ps edges, whereas the more expensive unit performs quite well under the same conditions. If you choose the higher-priced connector without testing the less expensive one at 400 ps, your 'cautious' approach could cost your company a lot of money. The less expensive connector might well perform quite acceptably in your system.

It is important that system-architecture decisions and technology evaluations proceed concurrently and interactively. The people responsible for these two activities must talk often. Partitioning logic among boards or ASICs without regard to technology issues can be a recipe for disaster. For example, a partitioning scheme that doesn't allow an adequate number of ground pins is likely to exhibit crosstalk or ground-bounce problems whose solutions can cost far more than the pins that were saved.

Many large companies – and even some medium-sized ones – have in-house groups that can help designers with technology decisions. But, in beating the drum for consulting firms such as his own, Sayre points out that although most of the technology groups he's familiar with have very competent staffs, they don't always agree with design-team agendas and, because of other commitments, they can't always meet project timetables.

On the other hand, because consultants usually are paid from project budgets, their loyalties tend to align with the team's. Moreover, consultants work very hard to nurture a spirit of trust, because their future work depends on it. As a result, Sayre says, consultants usually enjoy smooth relationships with design groups.

Sometimes, however, these good relationships can backfire. For example, a consultant and a team might agree on an approach that, though appropriate for a specific project, differs from the corporation's long-term technology strategy. An in-house technology group would almost certainly steer the design team to a different approach.

Design at the cutting edge of technology isn't the only place you'll find TDR. Although the first commercial TDR instruments appeared in the early 1960s, until rather recently, TDR's main application has been pretty far from high-speed digital-system design. Low-cost TDR instruments, many of them handheld, have long been used for inspecting cables and for locating faults in cable installations – in local-area networks (LANs), for example.

TDR as used by the suppliers of the multilayer boards used in high-speed digital systems also differs in emphasis from TDR as used by the designers of those systems. Board suppliers use TDR measurements in quality control. The panels from which multilayer boards are made contain areas called 'coupons', which are miniature technology boards. A board vendor measures the characteristic impedance of a transmission line on the coupon to show that a group of parameters, including line width, dielectric constant, and insulation thickness, are in tolerance.

The emphasis in these measurements is on absolute values. Designers, on the other hand, usually are more interested in relative measurements and on locating discontinuities.

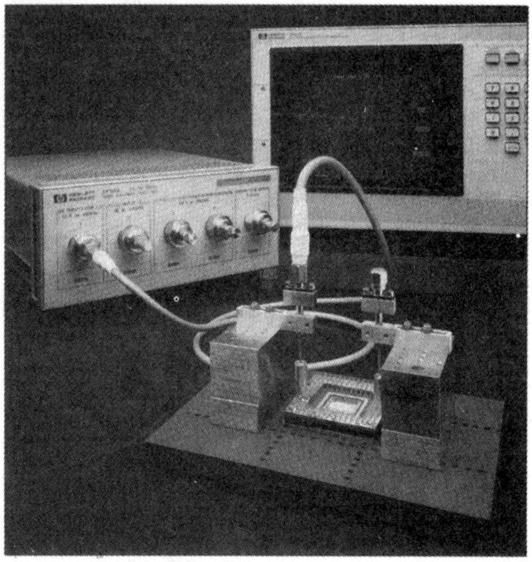

Figure 42.3 Hewlett-Packard's 54120T-series TDR systems consist of a sequential equivalent-time-sampling DSO and an external unit

You might think that a field whose underlying principles are based on a discipline as old as linear-network theory would be cut and dried. Surprisingly, TDR still provokes lively discussions. One debate centers on the best way to make TDR measurements on differential lines, such as on pairs of wires used to transmit high-speed signals through ribbon cable. When Tektronix introduced the 11800 series of scopes, it introduced a TDR plug-in that implements differential TDR by applying pulses of opposite phase to the

Figure 42.4 Tektronix's IPA 310 interconnect parameter analyzer is a PC-based turnkey system that develops Spice models. The heart of the system is the firm's 11800B DSO which accepts plug-ins that adapt it to making TDR/TDT measurements

two wires of a pair. Intuitively, the technique is appealing because it mimics the way you would drive the pair in actual use. However, you don't *have* to make the measurement in this way.

You can think of a single-ended signal as being composed of a common-mode component and a differential component. If you subtract the common-mode TDR response from the single-ended response, what's left is the differential-mode response. Hewlett-Packard makes the measurement in this manner. You can use NESA's Scope54 software (see 'Looking ahead') to extract the differential TDR response. Not too long ago, HP and Tek were engaged in heated arguments over which technique was superior[4,7,8,10,11]. It's hard to tell whether any users' opinions were changed by this controversy, but clearly, the companies' opinions were not.

Network analyzers, too

You should also be aware that TDR isn't the only way to examine discontinuities or to characterize the high-frequency behavior of interconnections. Another way to gather the data is with a frequency-domain instrument. You can use a network analyzer to measure the S-parameters of interconnects. DSP techniques can convert the S-parameter data into the time domain. The resulting displays look just like TDR traces. Indeed, the capability to convert frequency-domain data into the time domain is built into some network analyzers.

Whether TDR- or network-analyzer-based techniques are superior is the subject of more heated discussion. The partisans seem to agree, however, that although network analyzers can provide greater detail, they are somewhat slower. Network-analyzer advocates maintain that viewing a component's impedance or transmission characteristics in the frequency domain makes the reasons behind the device's behavior more obvious. Champions of TDR disagree and add that, although network analyzers provide extra detail, high-performance TDR instruments provide more than enough resolution for nearly any practical purpose.

In this debate, you should not overlook an economic issue: a TDR instrument costs somewhat less than a network analyzer whose frequency range extends to the – 3-dB frequency of the TDR's scope. TDR systems with 20-GHz bandwidth from both HP and Tektronix are priced from roughly $30,000. An HP vector network analyzer with time-domain capability and a frequency range extending to 20 GHz costs more than twice as much. In addition, high-performance TDRs include wideband scopes. Often, the scope proves invaluable. Even with built-in DSP, a network analyzer can't do the job of a scope. The battle cry of the TDR partisans, then, is 'If you buy a TDR instead of a network analyzer, your savings will pay for a very fine logic analyzer'.

Rho, rho, rho your board gently through the time domain

Time-domain reflectometry (TDR) is often likened to radar. Both systems use signal reflections to pinpoint the locations of things. In radar, the things are objects – along the ground, in the air, or in space. In TDR, the things are discontinuities in the characteristic impedance of signal-interconnection media. The media include cables, printed-circuit traces that behave as transmission lines, IC packages, and even connectors.

A TDR instrument makes a single connection to the medium under test. At the connection point, the instrument injects a voltage step and detects the resulting reflections. For time-domain transmission (TDT) measurements, you connect the instrument to the medium at two points. The signal (still a voltage step) enters at one point and is detected at the second point, along with reflections from discontinuities beyond the detection point and between the excitation and detection points.

In fact, the TDR systems used on high-performance pc boards and components are sampled-data systems built around sequential equivalent-time-sampling digital oscilloscopes, the widest-bandwidth DSOs available. Such scopes reconstruct waveforms from multiple signal repetitions. Because the signals repeat, the TDR 'step' must return to its baseline between iterations. In other words, the step is, in fact, a train of pulses.

Suppose a lossless transmission line having a characteristic impedance of Z_0 and a length of L is terminated in a resistance of Z_0. If at the unterminated end, you apply a zero-rise-time voltage step of amplitude A from a source whose internal resistance is Z_0, you will immediately see a step of amplitude $A/2$ at the sending end. The sending-end voltage will then remain constant. After a delay of L/V, where V is the propagation velocity, the step will appear at the receiving end, also with an amplitude of $A/2$. Once the step appears at the receiving end, the voltage there will hold constant.

If you remove the receiving-end termination and repeat the test, you will still see the $A/2$ step initially at the sending end. However, after a delay of $2L/V$, which is the time required for the step to travel to the receiving end and back, a second step will add to the first, so that the signal amplitude becomes equal to A. At the receiving end, you will see a step of A, delayed with respect to the initial sending-end step by L/V.

If you now replace the open circuit at the

receiving end with a short circuit, you will still see an initial sending-end step of $A/2$. After a delay of $2l/v$, however, the sending-end voltage will drop to zero. At the receiving end, the voltage will be zero at all times.

These measurements lead to the definition of the reflection coefficient, ρ (rho); $\rho = E_R/E_1$ (reflected voltage/incident voltage). In the preceding examples, $E_1 = A/2$. In the case of the Z_0 termination, $E_R = 0$, so $\rho = 0$. With the open-circuit termination, $E_R = A/2$, so $\rho = +1$. With the short-circuit termination, $E_R = -A/2$, so $\rho = -1$.

You should now begin to see that you can use ρ in determining the value of impedances. $Z = Z_0 (1 + \rho)/(1 - \rho)$. With the Z_0 termination, $\rho = 0$, making $Z = Z_0$. With the open-circuit termination, the denominator becomes zero, showing that Z is infinite. With the short-circuit termination, the numerator equals zero, so $Z = 0$. Figure 42.5 shows TDR waveforms produced by complex impedances and the formulas used to obtain the element values from the waveforms.

When hit with a voltage step, a capacitance initially behaves like a short circuit. (The voltage across a short is always zero; you can't change it.) Similarly, because you can't supply infinite current, you can't change the voltage across a capacitor in zero time. Therefore, the TDR response from a shunt capacitance exhibits a reflection that initially looks like that from a short circuit and then gradually returns to a steady-state value.

When hit with a voltage step, an inductor initially behaves like an open circuit. (Just as you can't force current through an open circuit, you can't change the current through an inductor in zero time because you can't apply infinite voltage across it.) A TDR response from an inductance exhibits a reflection that initially looks like that from an open circuit and then gradually returns to a steady-state value.

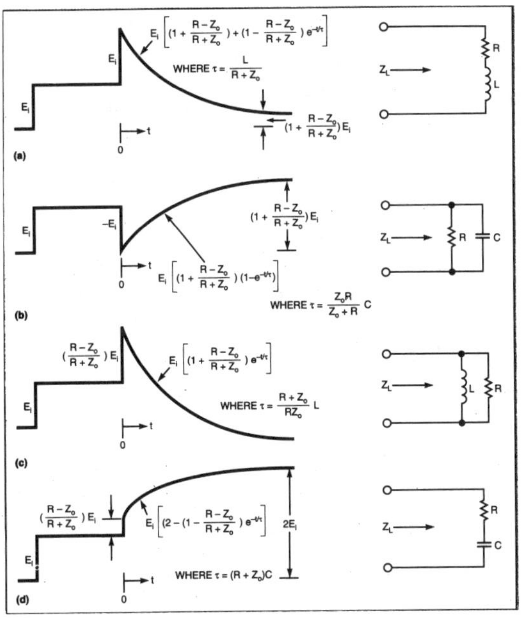

Figure 42.5 If you connect complex loads across a transmission line having a characteristic impedance of Z_0, the displays will look similar to these. The series R-L case is shown in (a); the parallel R-C case appears in (b); the parallel R-L case is in (c); and the series R-C case is in (d). The formulas let you calculate the element values from the step amplitudes and time constants

This entire theory is built around the concept of linear, time-invariant (LTI) networks. The math related to TDR assumes that superposition applies (the definition of linearity) and that component values do not change as a function of time. Generally speaking, passive components conform to these rules (ferromagnetic elements such as iron- and ferrite-core inductors and transformers are exceptions), but active components do not.

Nevertheless, you can still make meaningful TDR measurements on circuits that contain non-linear elements. And although TDR measurements on passive networks, especially interconnections, are the most common, TDR is often used on active circuits under power. A practical consideration when using TDR on powered circuits is the possibility of damaging the sampler. You must use caution because the samplers in high-performance TDR systems can't withstand large voltages.

You may wonder why TDR requires specialized equipment – why you can't use a general-purpose pulse generator and a wideband oscilloscope to make TDR measurements. The answer lies in resolution. To resolve detail critical to understanding the behavior of high-performance digital circuits, you need a pulse generator that produces very fast edges (< 100 ps) and a scope that offers very wide bandwidth (> 20 GHz). If the resulting TDR waveforms show detail so great that it causes confusion, you can use DSP techniques to filter the data after you've acquired it. If the detail is missing to begin with, no amount of signal processing will recover it (see Figure 42.6).

It is impossible to generate zero-rise-time voltage steps or pulses. Similarly, neither the sampler nor the test setup have infinite bandwidths. You can't distinguish between a pair of discontinuities if the round-trip travel time between them is less than half the system rise time (the square root of the sum of the squares of the voltage step, sampler, and test-setup rise times). In other words, a TDR system can't resolve discontinuities separated by a distance smaller than that corre-

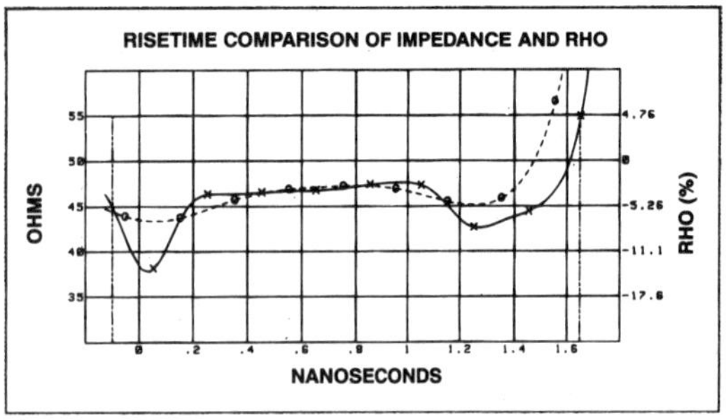

Figure 42.6 These two curves, produced by NESA's Scope 54 software package, show that impedance variations and reflections decrease as you degrade the system rise time. The software filtered the data after capture. For the "x" data points, the rise time was 200 ps; for the circled points, it was 500 ps. Note that the results appear in terms of both Z and ρ. When you make the Z scale linear (as it is here), the ρ scale becomes nonlinear

sponding to ¼ of the system rise time. For illustration, Hewlett-Packard's 54120T-series TDR systems have a system rise time of 45 ps, corresponding to 14 mm in air. Therefore, the systems' theoretical resolution is 3.5 mm; HP's spec is 7 mm. With other dielectrics, the slower signal velocity improves the spatial resolution.

One way of improving the resolution is by speeding up the pulse transition time. For this purpose, Picosond Pulse Labs' 4015B generator (suggested US distributor price: $7500) is designed to work with the 54120T series and is also usable with other time-domain transmission TDR instruments. The generator uses the 54120T's 35-ps rise-time pulse as a trigger and produces a larger pulse that has a 15-ps transition time. The pulse emanates from a remote pulse head that you can place closer to the unit under test than the generator itself. This feature eliminates the pulse-slowing effects of the cables you would otherwise need.

The discussion of resolution points out another aspect of TDR: by themselves, TDR measurements can't find the actual positions of discontinuities. To pinpoint locations, you need to know the propagation velocity. You can, of course, determine the propagation velocity by dividing the path length by the propagation time. TDT measurements are perfect for measuring propagation time, but to measure the path length you need a high-tech instrument such as a ruler or a tape measure. Once you know the propagation velocity, some TDRs let you display measurements in units of distance rather than of time.

Looking ahead

There can be little doubt that as more and more companies begin to build systems based on μPs clocked at > 50 MHz, the importance of time-domain measuring techniques for board and system designers will grow. Vendors of components, especially IC packages and connectors, will continue to refine the high-frequency models they supply. And CAE software for modeling the high-frequency behavior of circuits will continue to improve. Nevertheless, the day is still far off when an engineer can sit at a workstation to design a complex pc board that will operate at > 50 MHz and expect the board to work flawlessly and meet all applicable EMC standards the first time power is applied.

Two current products illustrate what to look for in future time domain measurement products: Tektronix's IPA 310 interconnect parameter analyzer and North-east Systems Associates' Scope54 software. The IPA 310 is a $75,000 PC-based turnkey system that extracts Spice models. The system is built around Tek's 11800B DSO with TDR plug-ins. Scope54, which costs $7500 and works with HP's 54120T-series TDR systems, runs on MS-DOS PCs. Although it does not extract models, Scope54 performs a variety of functions that make TDR measurements simpler and more repeatable. (Examples are post filtering of data and converting displays between calibration in ρ and Z. The package also facilitates storing and recalling waveforms and measurement setups.)

Because of the great variety of instruments and CAE tools already in use, there is a good chance that within a year or two you will be able to buy unbundled, modular software that will work with both Tek and HP TDR instruments, and perhaps network analyzers as well. The software will operate on TDR data only in the ways you want. If you use a network analyzer, the software will first convert frequency-domain data into the time domain, unless the network analyzer has already performed the conversion. With such software, if you want a feature like model extraction, you'll be able to get it; if you don't want it, you won't have to pay for it.

Time-domain reflectometry products and services

Amherst Systems Associates
Amherst, MA
(413) 596–5354

Cascade Microtech
Beaverton, OR
(503) 626–8245

GigaTest Labs Inc
Cupertino, CA
(408) 996–7500

Gilbert Engineering
Glendale, AZ
(602) 245–1050

Hewlett-Packard Co
Santa Clara, CA
(800) 452–4844

High Design Technology
Sunnyvale, CA
(408) 730–0383

Integrity Engineering Inc
St Paul, MN
(612) 636–6913

Inter-Continental Microwave
Santa Clara, CA
(408) 727–1596

LeCroy Corp
Chestnut Ridge, NY
(914) 425–2000

Mikon Consulting
Santa Clara, CA
(408) 727–5697

North East Systems Associates
Stow, MA
(508) 897–8787

Picosecond Pulse Labs Inc
Boulder, CO
(303) 443–1249

TechKnowledge
Portland, OR
(503) 289–2637

Tektronix Inc
Beaverton, OR
(800) 426–2200

References

1. Tilden, Mark, 'Measuring controlled-impedance boards with TDR', *Printed-circuit fabrication*, February, 1992.
2. Hewlett-Packard Co, *'TDR fundamentals'*, Application note 62, April, 1988.
3. Hewlett-Packard Co, *'Improving time-domain network-analysis measurements'*, Application note 62–1, April, 1988.
4. Hewlett-Packard Co, 'TDR techniques for differential systems', Application note 62–2, October, 1990.
5. Hewlett-Packard Co, *'Advanced TDR techniques'*, Application note 62–3, May, 1990.
6. Tektronix Inc, *'11800/CSA803 using filtering to control rise time'*, Application note, 1993.
7. Tektronix Inc, *'Crosstalk characterization of a ribbon cable using the 11800 series'*, Application note, 1993.
8. Tektronix Inc, *'Characterizing a differential amplifier using the 11800 series'*, Application note, 1993.
9. Tektronix Inc, *'Z-profile algorithm'*, Technical brief, 1993.
10. Tektronix Inc, *'Characterization of a differential line using the 11800-series differential TDR capabilities'*, Technical brief 47W-7519.
11. Tektronix Inc, 'Differential ohms measurement with the 11800 series', Technical brief 47W-7520.
12. Hewlett-Packard Co, *1993 high-speed digital symposium and exposition*, proceedings.

43 Measuring transient voltages between separate ground points

When ground is not ground, circuit design gets difficult

'Circuit ground' as a solid reliable zero-voltage reference point is a great idea, but often it is little more than that – just a fictitious idea. Audio buffs have always known this, battling with hum and other problems caused by ground loops. At the UHF frequencies which must be taken into account in high-speed logic design, even points on the same ground-plane on a printed circuit board can be momentarily at quite different potentials. This article tells you how to measure such potential differences.

Art Porter
Hewlett-Packard Co

Textbooks apply the phrase 'circuit ground' to a reference potential from which you can measure relative voltages. In practice, however, no two points on a circuit's ground plane are exactly at the same potential, and they generally exhibit a nonzero impedance between them. Overlooking this fact in the presence of the transient ground currents common in today's high-speed digital circuits can lead to circuit problems that are difficult to analyze and troubleshoot. Three scope methods offer pros and cons to characterize these potential differences by measuring the voltage between separate ground points.

Digital circuits respond to the potential difference between an input signal and the ground potential measured on the die. If there is sufficient inductance between the die and the package pins, the transient voltage between the die's ground potential and the ground potential at the pin can be significant. The notorious ground-bounce effect in high-speed CMOS circuits is a good example.

The problem is further exacerbated by inductance in the trace that connects the ground pin to a ground plane on the pc board. (Figure 43.1). An additional potential difference is due to the inductance of the ground return between the pc board's ground plane and the power supply's ground. Essentially every ground conductor has inductance, and rapidly changing ground currents will induce a transient voltage in the conductor according to $E_{ind} = L \times di/dt$.

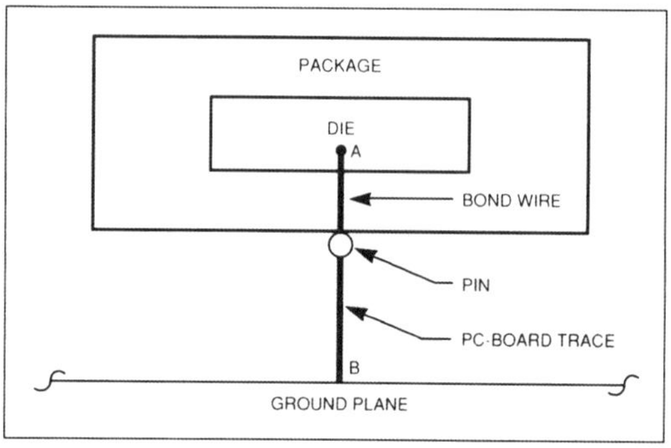

Figure 43.1 The ground potential at point A on a die can be different from the potential at point B on a pc board's ground plane. The inductance of the bond wire from the die to the package pin and the inductance of the pc-board trace from the pin to the ground plane induces a transient voltage when there are rapidly changing ground currents

Suppose you're scratching your head because a gate or a flip-flop changes state when there is no appropriate change of input signal. The enigma may be resolved if you consider the IC's ground as one of the input signals. Remember the IC responds to the potential difference between an input pin and its ground pin; voltage changes on either pin affect the circuit the same way. To ascertain if a glitch or metastable state is caused by an induced ground potential, you need to be able to measure the voltage difference between two separate points on the ground plane.

To measure the voltage across a low-resistance path when there is no absolute ground reference, you must make a differential measurement between the two points. One method is to use a dual-channel oscilloscope and two probes of the same type. You connect one probe to one point on the ground path in question and the other probe to another point on the ground plane. You then subtract the signals for each channel using the scope's A – B feature.

To make reliable measurements, you must minimize the time skew between each channel, including the probes. To de-skew the probes, you can follow the recommended procedures in the operator's manual, but generally only digitizing oscilloscopes let you adjust time skew between channels. If you use an analog oscilloscope, you must use probes that have well-matched time delays. The Tektronix Model P6135A matched pairs of probes are good examples.

The gains of the two channels and the two probes must also match. Most digitizing scopes have a probe attenuation factor that lets you match the probe and channel gains. You connect both probes to a common signal, set the scope for A – B operation, and adjust the probe attenuation factor on either channel for minimum deflection on the display. On an analog scope, you can use the vertical sensitivity vernier on either channel to balance the gains. Some digitizing scopes have a continuously variable vertical sensitivity, which lets you match the gains.

If you use 1- or 10-MΩ style probes that need to be frequency compensated, be sure to carefully adjust the compensation on each probe. Any mismatch in the probe's compensation can lead to severe degradation of the common-mode rejection at high frequencies. Connect the probe grounds using a short wire and then connect the center of this wire to a ground point in the system under test.

Because the system ground point can affect the test results, you should choose it with some care. Ideally, the system ground point should not experience a current transient that coincides with the induced voltage transient you are trying to measure. A sheet-metal ground or a board-mounting screw near the measurement point is often a good choice. It's also a good idea to intertwine the two probe cables to minimize electromagnetic coupling to the probe shields. (Figure 43.2).

A low-resistance divider probe, such as the HP54006A or the Tektronix P6150, is generally the best probe for making dual-channel differential measurements. The HP54006A probe has a flat frequency response and a bandwidth of 6 GHz. Although the divider probe has a minimum input impedance of 450 Ω, resistive loading is negligible because the measurement is being performed on a low-impedance circuit.

There are some inherent inaccuracies in dual-channel measurements, but a few procedures can

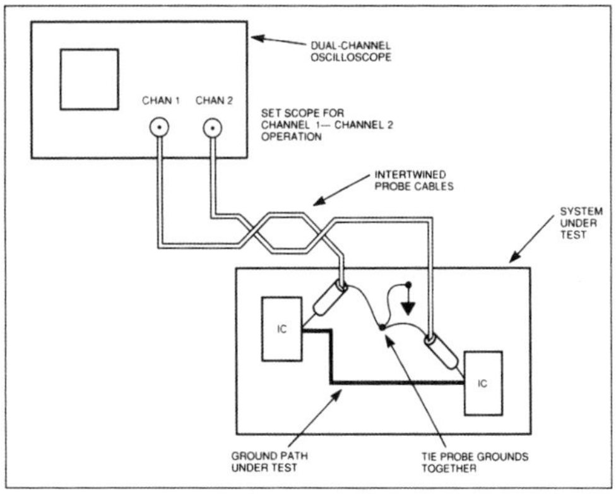

Figure 43.2 When using a dual-channel scope, connect the two scope ground probes and connect a short wire at the center point to a convenient ground in the system under test. Intertwining the probe cables minimizes electromagnetic pickup

help you minimize them. One useful technique to evaluate errors caused by ground loops or induced transient voltages at the system ground point is to short the tips of the probes together and disconnect them from the system under test. Observe the scope to see if there are any deflections on the display. If so, move the system ground to another point and try to find a point that minimizes the deflection on the display. In some instances, it may be best not to connect the probe shields to the system ground point. The probe tips must be shorted together to determine this, however.

Other potential sources for measurement error are ground loops formed by the probe shields and the scope ground(Figure 43.3). All ground loops are susceptible to pickup voltages caused by circulating electromagnetic fields. If you use 1- or 10-MΩ compensation-type probes, the input capacitance from the probe tips to the shield and the scope's input-channel capacitance creates ac ground loops. You can minimize the susceptibility of these loops to electromagnetic pickup by intertwining the probes together.

During measurement, each probe is subject to complex electromagnetic and electrostatic fields. Because the two tips are at different locations, the field coupling is different. The signal you observe on the display is not only the actual voltage difference between the measured points, but also the difference in the electromagnetic and electrostatic coupling at the probe tips.

Inaccuracies caused by coupling differences are hard to evaluate and difficult to control, but usually are small enough to ignore. You can test for this coupling effect by moving the probe cables around and observing changes in the scope display. The further apart the measurement points are, the more pronounced the problem is.

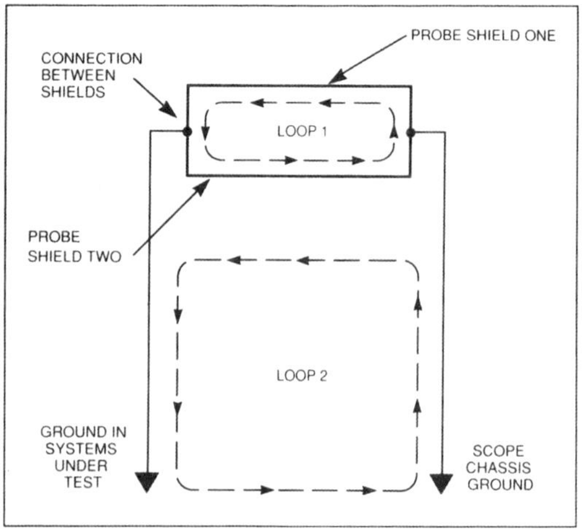

Figure 43.3 In a dual-channel measurement the two probe shields form a loop which can pick up induced currents caused by electromagnetic fields. The connections of the probe shield to the scope-chassis ground and to the ground in the system under test form an additional loop that is susceptible to pickup

Another difficulty can arise when the scope can't trigger on the difference signal. Usually triggering the scope on one of the signal channels overcomes this difficulty.

A differential probe, such as the HP 1141A, gives you a second method for making differential voltage measurements. Differential probes provide high sensitivity and excellent common-mode rejection for frequencies as high as 100 MHz. In addition, differential probes eliminate the concern of having two probe shields forming a large loop. You don't have to frequency compensate, time de-skew, or match the gains when using differential probes. Thereby you eliminate these sources for error. Differential probes also let you trigger the scope on the differential signal.

The differential-probe method presents some other difficulties, however. For one, the probe tips on a differential probe are physically close together. If you must make a differential measurement at two points that aren't close together, you must extend the lead lengths of the probes, which adds significant inductance or capacitance. Extending the lead lengths increases the size of the loop that is subject to electromagnetic pickup.

The bandwidth of a differential probe is considerably less than that of a resistive divider probe. For example, the HP 1141A bandwidth is 200 MHz, compared to the 6-GHz bandwidth for the HP 54006A divider probes. You still must determine where or whether to connect the differential probe's ground to the system under test. Similar to the dual-channel method, you must short the inputs of the probe together and determine the optimum grounding point by minimizing the residual signal on the display.

If you have the luxury of owning a digital-storage scope, a third method for making differential measurements is often the most accurate. This method uses only one probe to store digitized data from two points into waveform memory and then perform waveform mathematics to calculate a difference waveform. The procedure is:

1. Measure the waveform at one point and store this data into waveform memory.
2. Measure the waveform at the other point using the same probe and the same scope channel. Don't move the probe's ground connection to the system under test. Store the data into another portion of waveform memory.
3. Use the scope's waveform math capability to subtract the data for the two waveforms stored in memory. The scope will automatically calculate the difference and display the waveform.

If you use the third method, you should adhere to the same precautions concerning the probe shield's connection to a system ground point as in the first two methods.

There are several reasons why the third method is more accurate than the other two. For one, you use the same probe and scope channel for collecting data from two points. Therefore, there are no errors caused by mismatches in gain, frequency response, or time delays. Because there is only one hook-up from the probe shield to a system ground point, there is no loop formed by two probe shields as there is in the first method mentioned. Also, because the probe tip only measures one point at a time, the ground strip under test doesn't complete a loop, which could pick up induced currents from changing electromagnetic fields.

If you use the third method, you should be aware of a couple of its idiosyncrasies. Unlike the first two methods, the probe loading for the system under test is not balanced. The probe only loads one point at a time. However, because the ground trace under test has a low impedance, the unbalanced load is not a significant problem. In addition, if the physical distance between the two points is long or the system ground point is far removed from the trace under test, the wire connecting the probe shield to the system ground will be long. The inductance of the long lead can cause ringing and electromagnetic pickup. Theoretically, the ringing and pickup should cancel when the scope subtracts the two waveforms. However, the induced voltages caused by the electromagnetic fields surrounding the probe and the ground lead will change when you move the probe. Therefore, the waveforms may not exactly cancel.

Regardless of which method you use, you should be aware of some details that are common to any circuit measurements. Anytime you make a measurement on a circuit, the electrical characteristics of the measuring apparatus become part of the system under test. Circuit loading due to the probe's input impedance is an example, but generally is not a factor when the circuit impedance is low. Two immediate concerns, however, are the currents injected by the probes into the circuit under test and changes in the ground-circuit configuration caused by grounding the probes.

The scope probes inject current into the circuit from the electromagnetic fields surrounding the probes. Because the ground path under test is a current path between the probe tips, the fields induce currents into the path under test. You can minimize the induced currents by using a differential probe as described in the second method or eliminate the current path by using one probe as described in the third method. In addition, if there is a connection between the scope chassis and the ground for the system under test, ambient electromagnetic fields can induce currents in this ground path.

Another reason for current injection is a potential difference that may exist between the scope chassis and the ground for the system under test. Potential difference causes a current to flow between the scope and the system under test through the probe shields and the probe's ground wire. In addition, some current may also flow through the probe tip to the circuit under test. If you use a low-resistance divider probe, such as the HP 54006A, the 450 Ω tip resistor limits the current to a negligible value. However, if you use a 1- or 10-MΩ compensation-type probe, current can flow into the circuit through the probe's compensation capacitor, which has a low impedance at high frequencies.

When you connect the scope's ground to the system under test via the probe-shield ground, make sure you don't modify the current path for the ground trace under test. Make sure you don't select a ground point that is directly in the path you are measuring. If the system has separate analog and digital grounds, you should choose an analog ground point when you are measuring a ground potential in the digital section and a digital ground point when measuring a ground potential in the analog section.

Although following these guidelines will increase your confidence in the accuracy of a measurement, it's impossible to set up exact recipes to follow. The best method depends on the particular situation. You should understand the basic principles that cause ground potentials to exist in your specific ground system. As a reality check, you should make measurements using at least two of the methods. If you obtain inconsistent results, one of the methods is probably invalid. Remember, however, arbitrary high accuracy cannot always be attained.

References

1. 'Differential measurements on wideband signals', HP Application Note 1221, HP publication number 5091–3863E.

2. 'Selecting oscilloscope probes for high-speed digital circuit measurements', HP product note 54720A-3, HP publication number 5091–3758E.

44 Noncontact devices hone measurements

Useful introduction to pyrometry

Infrared-radiation thermometry measures the temperature of hot bodies by the intensity of radiation they emit. Explains the concepts of black-body radiation, emissivity and 'grey bodies'.

John Gallant
Associate Editor

Infrared-radiation thermometry accrues the many benefits of noncontact temperature measurement. One of the most important benefits of this type of measurement is that, by detecting only the IR energy emitted by a hot body, the temperature reading reflects the localized temperature of the radiant body and not the ambient temperature of the surrounding medium. Therefore, IR-radiation thermometers can monitor the temperature of hot spots, small targets, and moving bodies more accurately than contact devices, which respond to ambient temperature. In addition, because the IR-radiation temperature probe doesn't contact the radiant target, it doesn't damage fragile bodies.

Unlike IR-radiation thermometers, thermocouples and resistance temperature devices must come in contact with either the hot body or the surrounding medium. These transducers measure heat via conduction or convection, so they respond to the temperature of their surroundings, which may be different from the temperature of the target. Another disadvantage of contact sensors is that they can damage fragile bodies.

IR-radiation detectors have shorter thermal time constants than contact sensors. IR-radiation detectors have thermal time constants of a few hundred milliseconds, whereas thermocouples and resistance temperature detectors have thermal time constants of several seconds. Because they have faster time constants, IR-radiation thermometers respond more quickly to temperature variations in the target and give accurate readings for moving objects with wide temperature ranges.

Noncontact temperature probes cost almost ten times more than contact probes, but their ability to perform precise temperature measurements can reduce the cost of process control. IR-radiation thermometers contain an optical section that focuses the radiant energy from a target onto the detector. Detectors located 20 cm away from the heated body can detect radiant energy from a target a few square millimeters in size. This capability lets IR-radiation thermometers measure the temperature of small areas such as semiconductor components, textile fibers, precision-machined parts, and solder joints, independent of the ambient temperature. Therefore, you can control the temperature of these parts with more accuracy.

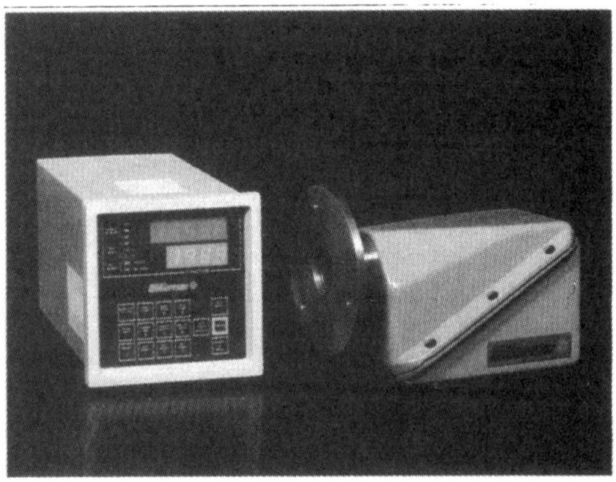

Figure 44.1 Dual-wavelength thermometers are insensitive to emissivity variations. Williamson's Tempmatic 8000 Series measures temperature by computing the ratio of IR energy in two adjacent wavelength bands

Because an IR-radiation probe doesn't contact the target, the probe maintains a lower temperature than the hot body. Therefore, an IR-radiation thermometer has no upper temperature limit. In contrast, the hot junction of a thermocouple must reach the temperature of the hot body, which places an upper temperature limit on the thermocouple based on the hot junction's melting point.

Thermopiles are popular

At the heart of the IR-radiation thermometers is the detector. Most IR-radiation thermometers use

Figure 44.2 The Vanguard II is a 2-wire IR-radiation transmitter with a 4- to 20-mA output signal. The Vanzetti Systems unit has a 50-ms response time, an accuracy of ± 1% of the reading, a repeatability of ± 1°C, and operates over 0 to 550°C

a detector with a thermopile, a photovoltaic IR diode, or a photoconductive IR semiconductor. The thermopile is the most popular detector used in IR-radiation thermometers. Although the thermopile is less sensitive to IR energy than the other detectors, it exhibits a flat spectral response extending from the near IR region (0.7 μm) to the extreme IR region (greater than 50 μm). This feature allows vendors to offer optional IR filters that select wavelength bands to match specific temperature ranges. The other two IR-radiation detectors are more sensitive, but they have spectral responses that peak at wavelengths of less than 10 μm.

A thermopile consists of a large number of thin-film thermocouples connected in series. The junctions of successive thermocouples connect to form a string of sensors. The ends of the string connect to the leads of the detector, which is sealed in a TO-style can. A thermopile doesn't require a bias voltage or current for operation. Because it behaves as a pure resistor, it generates Johnson noise only – no 1/f noise.

Figure 44.3 Certain handheld IR-radiation thermometers are as compact as photoelectric sensors. The Thermo-Hunter from Capintec Instruments weighs 350 g and measures 33 × 77 × 77 mm

Dexter Research Center offers a variety of thermopile detectors used in commercial IR-radiation thermometers. The Model 2MC thermopile detector features 36 thermocouple junctions, a sensitivity of 10 volts per watt, a 40-ms thermal time constant, and a resistance of 6 kΩ. The detector costs $38 (100).

EG&G Judson offers a variety of photovoltaic diodes and photoconductive semiconductors for IR-radiation detection. The IR devices, which are made from germanium, indium arsenide, indium antimonide, and mercury cadmium telluride, operate over a wavelength range of 1 to 22 μm. The company also offers the J16Si dual-wavelength detector, which consists of a silicon photodiode mounted with a germanium photodiode in a single TO-style package. The silicon and germanium diodes respond to wavelengths from 0.4 to 1 μm and 1.1 to 1.8 μm, respectively. The device provides two output signals from a single 500-to-1200°C source for use in dual-wavelength temperature sensing. Each device costs $230 to $240, depending on the size of the active area.

Vendors slash their prices

IR-radiation thermometry has progressed on two fronts in the last five years. The first front is cost reduction. In 1985, a typical IR-radiation thermometer with an LCD cost $2500 to $3000. Today, a similar unit costs half that figure. In fact, you can find some units for the OEM market that cost less than $1000.

The devices in Williamson's Transtemp 1000 Series have a base price of $1300 each. Models in the series operate over – 40 to + 2500°C using thin-film thermocouples that respond to wavelengths from 0.7 to 14.0 μm. The units come in a NEMA-4 enclosure and have an LCD. An operator can adjust the thermal time constant from 0.2 to 5.0 s and can digitally adjust the emissivity setting from 0.1 to 0.99. (For a detailed explanation about emissivity, see 'The Importance of emissivity'.)

The units operate from a 10 to 40 V dc power supply. You supply power to the unit using the same two lines that provide a 4- to 20-mA output signal; hence the unit is known as a 2-wire transmitter. The linear, 4- to 20-mA output signal connects to a recorder or a computer through a current loop. The units also specify an accuracy of 0.75% of full scale and a repeatability of 0.5% of the reading value. Options include a water-cooling, air-purging jacket for protecting the lens.

Figure 44.4 Portability and a built-in μP that performs automatic emissivity compensation are two features of Exergen's D-501F IR scanner. The unit operates from a single 9 V battery and updates the output signal 10 times/s

Instruments in the Transtemp 700 Series boast the same specifications as those in the 1000 Series and cost $700 each. Instead of NEMA-4 housing, however, the units have an IR-radiation probe that connects to a circuit board via a cable. Both series have an optional peak detector with an adjustable hold function. The circuit lets you adjust the decay time constant on the peak detector when you're monitoring moving targets with variable emissivities.

Exergen's Microscanner D-501F IR-radiation thermometer measures temperatures from – 50 to + 550°C and has a spectral response from 2 to 20 μm. The unit comes in a waterproof housing and specifies an accuracy of ±0.5% of the target temperature for a target with 0.98 emissivity. Repeatability is ±0.25% of the reading value. The thermometer, which costs $649, operates from a 9 V alkaline battery. It has a μP that automatically adjusts the thermometer for the emissivity of the target. The unit provides a linear voltage output that's updated every 100 ms with 0.1° resolution. It also contains a peak detector with a hold function and an audible alarm indicator for monitoring maximum and minimum temperature readings.

Capintec Instruments' Thermo-Hunter Built-in Series also costs less than $1000, with prices starting at $900. Models in this series operate over – 50 to + 1000°C. The units use either thin-film thermocouples or lead-selenium photodiodes as detectors, which have thermal time constants of 500 and 80 ms, respectively. Because the lead-selenium detectors operate in the 1- to 4-μm range, they can measure temperature through glass. A delay function allows the units to measure the mean temperature in an adjustable time frame of 0 to 7 s. The units provide a linear-voltage-output signal.

Fiber optics boosts sensing

In addition to their reduced prices, IR-radiation thermometers are also more sophisticated than their predecessors of five years ago. These devices now offer fiber-optic wave guides, single-lens reflex focusing, portability, advanced signal processing, and dual-wavelength sensing – features that previously weren't economically feasible.

By assembling quartz optical fibers into fiber bundles, optical wave guides can transmit energy in the near infrared region – a region that was once opaque to IR energy. The optical cables are flexible, so IR-radiation thermometers equipped with fiber-optic cables can look around obstacles to measure temperatures from 300 to 2000°C. The IR fiber-optic cables connect to an optical system located at the viewing end of the cable. The wave guide transmits the energy to a detector and amplifier remotely located from the heated surface.

Applications for fiber-optic temperature sensing abound. For example, fiber-optic sensing lets you obtain a much more precise temperature measurement of a molten metal heated by induction coils. You can place the viewing end of the cable between the coils, permitting the thermometer to measure the molten-metal temperature independent of the coil temperature (Figure 44.5).

Vanzetti's LTD Series linearized IR-radiation detector/amplifiers are designed for use with flexible fiber-optic cable. Specifications include an accuracy of ± 1% of the reading, a repeatability of ± 0.1% of the setpoint, and a thermal response time from 100 to 5 ms. The device chops the IR energy at a 400-Hz rate, permitting synchronous demodulation to discriminate against electromagnetic noise at other frequencies. The LTD 4035-S operates over a wavelength range of 0.8 to 1.05 μm and costs $1755.

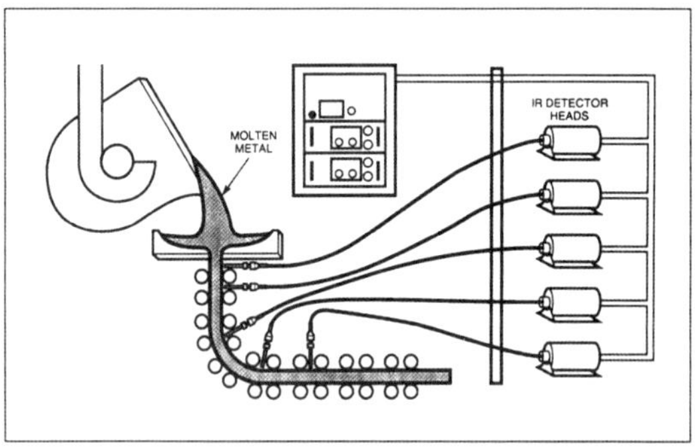

Figure 44.5 An optical system in the viewing end of a fibre-optic cable can fit between the coils of an induction heater to measure the surface temperature of a molten metal, independent of the ambient or coil temperature

The Land Fiberoptic is another series of IR-radiation detector/amplifiers designed for use with flexible fiber-optic cables. Units use either a silicon or a germanium detector cell. The FP11 model has a silicon cell that responds to wavelengths from 0.7 to 1.0 μm. The model specifies an accuracy of ± 0.25 % of the absolute temperature in °K + 2°C, a repeatability of ± 0.15% of the absolute temperature in degrees Kelvin, and a thermal time constant of 5 ms. Prices range from $1600 to $2000.

Advances in CMOS technology have made handheld noncontact thermometers a reality. Raytek and Ircon are only a few of the companies that offer standard handheld products that operate from batteries. Ircon's Ultimax is a handheld unit that contains a number of user-friendly features. It operates in a similar manner to a single-lens reflex camera. The thermometer contains a dichroic mirror that transmits visible energy and reflects IR energy, so you can view an object through the mirror while the IR energy from the object reflects off the mirror to an IR-radiation detector.

The Ultimax contains a μP for signal processing; this chip responds to pushbutton controls for setting the emissivity, thermal response time, and peak-detector decay time. Models operate over – 50 to + 3000°C, covering six different IR regions. Response time ranges from 0.5 to 2.0 s depending on the model, and the repeatability is within 1.0°C of the reading. In addition to providing an LCD, the thermometer has a linearized 0-to-1 V analog output and an RS-232C port for sending logged data to a computer or printer. The unit, which costs $2250, weighs 1.9 lbs and uses four AA alkaline batteries that operate for 20 continuous hours.

Two colors cancel emissivity

All of the IR-radiation thermometers discussed so far are single-wavelength devices that determine temperature by measuring the brightness of a target. If the target has a low emissivity, the instrument will provide a low temperature reading because the brightness is low. Therefore, to ensure accurate readings, single-wavelength thermometers must have an emissivity adjustment.

In contrast, dual-wavelength (two-color) thermometers, which appeared less than five years ago, are insensitive to emissivity. Therefore, they can provide accurate temperature readings in the presense of dust, steam, smoke, and other particles in the air. These thermometers take the ratio of the responses to two different IR wavelengths. If a target's emissivity is the same for both wavelengths (a good assumption when the wavelength bands are adjacent), the emissivity effect cancels. The output of the dual-wavelength thermometer at high temperatures is approximately

$$E_O = \frac{W_{\lambda 1}}{W_{\lambda 2}} = \left(\frac{\lambda_2}{\lambda_1}\right)^5 e^{\left[-C_2\left(\frac{\lambda_2 - \lambda_1}{\lambda_1 \times \lambda_2}\right)\frac{1}{T}\right]}.$$

Furthermore, dual-wavelength thermometers are insensitive to any changes in the target's field of view. If an obstruction or the target's orientation result in a low brightness reading, the effect occurs at both wavelengths; therefore, they both cancel when the radiometer takes the ratio of the two readings.

Both Land Instruments' System 3 Series and Williamson's Tempmatic 8000 Series contain thermometers based on the dual-wavelength principle. The thermometers have a front-end, motor-driven cylinder that has mounts for two selective IR filters. The control circuitry synchronizes the motor's position so that a single detector can alternately receive IR energy in the two bands. The succeeding circuitry takes the ratio of the two readings. Certain models in the Tempmatic Series operate over a temperature range of 150 to 2400°C. Units in the Tempmatic Series start at $2600, which includes the dual-wavelength processor and a visual-aiming feature. Land Instruments' Ratio Thermometers operate over 250 to 4200°C and cost $5000 each, including the dual-wavelength processor.

To help you gain a more thorough understanding of the complexities of IR-radiation thermometry, manufacturers provide technical support and numerous application notes for their products. In addition, they offer a variety of accessories, including black-body radiation calibrators; NEMA-qualified housings; air-purging and -cooling jackets; optional, thermocouple-compatible output ports; PID (proportional-integral-derivative) controllers; power supplies; meters; fiber-optic cable; and an assortment of optical systems.

The importance of emissivity

All bodies with temperatures above absolute zero (– 273°C) possess thermal energy. A certain amount of this energy, which is caused by vibrating molecules, can radiate from the surface of the body in the infrared region of the electromagnetic spectrum. IR-radiation thermometers detect this radiant energy to determine the body's temperature.

For a specified temperature, the maximum amount of radiation a body can emit is called black-body radiation. Black-body radiation is at a known level for every wavelength, and it depends only on the temperature of the emitter.

The level of black-body radiation at each wavelength for a specific temperature is given by Planck's Law:

$$W_\lambda = C_1\lambda^{-5}\left(e^{\left(\frac{C_2}{\lambda T}\right)} - 1\right)^{-1},$$

where W_λ is the emitter's spectral radiance in W/m^2 for each wavelength; λ is the wavelength in meters; T is the emitter's temperature in degrees Kelvin; C_1 is Planck's first radiation constant, which equals 3.74×10^{-16} W · m²; and C_2 is Planck's second radiation constant, which equals 1.4388×10^{-2} m·°K.

Black-body radiation is only theoretical, however, and it can only be approximated for physical measurement. It exists in the space between the molecules of materials. When the radiation reaches the surface of the material, it's partially reflected, so only a fraction of the radiation emerges from the surface. The fraction emerging from the surface is known as the material's emissivity, a specification that plays an important role in IR-radiation thermometry.

A body's emissivity is the ratio of its emitted radiation to the emitted radiation of a black body at the same temperature. The emissivity is a value from 0 to 1. A gray body has an emissivity less than unity, which is constant for all wavelengths. Three black-body-radiation profiles are shown by the solid curves in Figure 44.6. Gray-body radiation is shown dotted for a temperature of 1200°F. The peak of the radiation curves shifts to shorter wavelengths as the temperature increases. In fact, very hot bodies emit energy in the visible spectrum.

A conventional IR-radiation thermometer measures temperature remotely by focusing the radiant energy onto an IR-radiation detector using an optical system. Vendors calibrate the thermometer's temperature indication as if the radiant energy comes from a black body. Because real bodies emit less energy than black bodies, the thermometer must have an emissivity adjustment to calibrate the instruments to the observed radiant energy. The total black-body radiation is given by the Stefan-Boltzmann Law, which is the integral of Planck's Law for wavelengths from 0 to ∞. The Stefan-Boltzmann Law for a gray body that adjusts the thermometer to the target's emissivity is

$$W_{TOTAL} = \epsilon T^4 \sigma,$$

where ϵ is the emissivity of the gray body; T is the absolute temperature in degrees Kelvin; and σ is the Stefan-Boltzmann constant, which equals 5.63×10^{-8} W/m²/°K⁴. The Stefan-Boltzmann Law shows a fourth-power relationship between the total IR radiation and the body's absolute temperature.

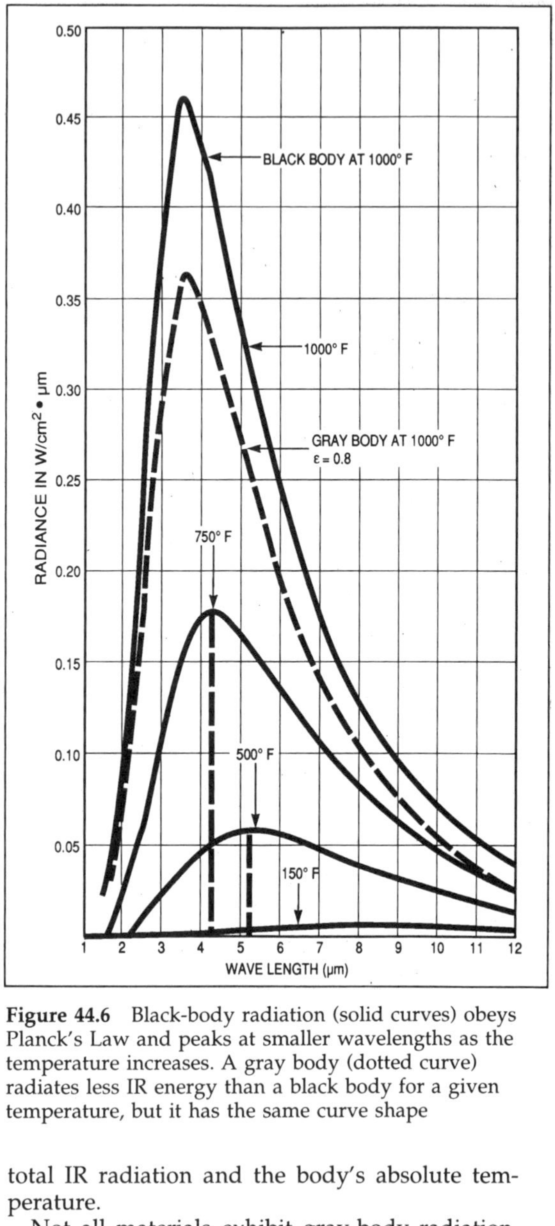

Figure 44.6 Black-body radiation (solid curves) obeys Planck's Law and peaks at smaller wavelengths as the temperature increases. A gray body (dotted curve) radiates less IR energy than a black body for a given temperature, but it has the same curve shape

Not all materials exhibit gray-body radiation, however. Many materials have strong and weak radiation regions throughout the IR spectrum. Therefore, to match the radiance of a specific material to the IR-radiation detector's sensitivity, IR-radiation thermometers include wavelength-selection filters in the optical system.

The body's emissivity isn't the only factor that affects the temperature reading. The optical system must be positioned so that background IR radiation from sources other than the target doesn't enter the system's field of view. In addition, because the optical system doesn't contact the target, IR radiation travels through a medium before reaching the detector. Contaminants in the atmosphere can affect the medium's transmission response and subsequently can decrease the detected IR radiation. To avoid significant errors in the accuracy of the temperature reading, you must calculate an effective emissivity, which compensates for medium-path losses.

Another source of temperature error is the detector's characteristics. Because a body's radiance increases exponentially with temperature, the detector's voltage-vs-temperature curve is nonlinear. Signal-processing circuitry often includes a look-up table that provides a linearized output signal. Thus, deviations from a straight-line curve affect the measurement accuracy.

The IR-radiation detector's ambient temperature can also affect the calibration, particularly when measuring low temperatures (below 200 °C). Because the detector is also a hot body, it emits IR radiation, which subtracts from the IR radiation emitted by the target. Therefore, to provide a calibration reference, low-temperature thermometers require an additional temperature transducer to measure the detector's ambient temperature.

The thermometer's ability to compensate for uncertainties determines the absolute accuracy of the temperature reading. However, many applications require repeatability more than absolute accuracy. It's often sufficient if an IR-radiation thermometer produces consistent readings, even though the absolute readings are in error by a few degrees.

Manufacturers of IR-radiation products

Capintec Instruments Inc
6 Arrow Rd
Ramsey, NJ 07446
(201) 825–9500
FAX 201–825–1336

Dexter Research Center Inc
7300 Huron River Dr
Dexter, MI 48130
(313) 426–3921
TLX 757075

EG&G Judson
221 Commerce Dr
Montgomery, PA 18936
(215) 368–6900
FAX 215–362–6107
TLX 846120

Exergen Corp
251 W Central St
Natick, MA 01760
(800) 422–3006
in MA, (508) 651–8146
FAX 508–651–8140

Ircon Inc
7301 N Caldwell Ave
Niles, IL 60648
(800) 323–7660 in IL, (312) 967–5151
FAX 312–647–0948

Land Instruments Inc
2525 Pearl Buck Rd
Bristol, PA 19007
(215) 781–0700
FAX 215–781–0723

Raytek
1201 Shaffer Rd
Santa Cruz, CA 95060
(800) 227–8074
in CA, (408) 458–1110
FAX 408–458–1239
TLX 171890

Vanzetti Systems
111 Island St
Stoughton, MA 02072
(617) 828–4650
FAX 617–341–2084

Williamson Corp
mBox 1270
Concord, MA 01742
(508) 369–9607
FAX 508–369–5485
TLX 923343

45 ISO 9000 standard cultivates customer satisfaction

Company qualification to ISO 9000 brings benefits – at a price

Applied properly (and your company won't achieve recognition under it any other way) ISO 9000 touches every aspect of company performance. This useful staff-written article outlines the benefits – and pitfalls – awaiting companies going for ISO 9000 approval.

Brian Kerridge
Technical Editor

In the beginning there were inspectors, and the inspectors stamped the products to tell customers that the products were good. Then came production controllers to monitor manufacture; and the inspectors still stamped the products to tell customers that the products were good. Then arrived quality-management systems; they influenced designers, and production controllers, and the inspectors that stamped, and so on.

Now comes ISO 9000: it checks your quality-management system and awards you international recognition to tell customers that your products are good.

In truth, being qualified to ISO 9000 is no more guarantee that your products are good than was the case in the days of the rubber-stampers. But, qualifying does mean that your company concentrates effort to examine and analyze every detail of its objectives, structure, and operation. Qualifying does mean your company will have thoroughly documented its findings in an all-embracing company quality manual. It also means opening your doors to regular external audits to prove adherence to that master tome. It's on this basis then, that customer satisfaction with your products or service, while still not being a certainty, becomes a highly likely result.

The complete ISO 9000 standard exists in several parts, each pertaining to different operations within your company. The section that concerns your design function appears in ISO 9001. (For background information on other ISO 9000 sections and general status of the standard, see 'ISO 9000 status and signposts'.)

When you first examine the ISO document, don't expect to see a neat little list showing what you have to do to qualify. Similar to many standards, the text is deliberately general to cater to a range of situations. You can be designing anything from bricks to jumbo jets, and the ISO standard applies equally. In fact, the entire ISO 9001 document covers barely seven pages, and of that, only part of one page focuses on design control. Essentially, the subsection describes a very broad framework within which you must establish, document, and maintain procedures to control and verify your designs.

But despite the standard's brevity, don't think you'll sort out what to do over a quick chat with the boss. Typically in the electronics business, companies are taking one to two years from first deciding to go for qualification to becoming qualified. During that time, companies hold several weekly meetings at different levels to study the standard's requirements. Virtually everyone in your company contributes toward qualification, since almost every employee's work influences product quality. Therefore, expect to involve from presidents to cleaning staff in putting together the vital procedures that ultimately appear in your company's quality documents. Remember, the combined manual becomes the centerpiece of your quality-management system; it embodies your company's mission statements and explains the how, why, and where you fit in with your company's objectives and philosophy. This manual eventually will serve as a permanent reference work for all staff.

Going for standard compliance is not a task for the uncommitted or the poor. On top of the overhead costs of endless internal meetings and the work involved in compiling manuals, your company will need a budget of around $25 k for direct external costs in reaching qualification. After qualifying, you can add another $8 k/year to maintain conformance. If you're still feeling keen, consider that companies find it necessary to appoint a project leader to coordinate the company-wide process up to qualification. Also, companies generally hire the services of a management consultant, especially to lead them through the early stages of the process. A summary of the complete process appears in 'Essential steps on the path to ISO 9000 approval'.

With all this activity running through the company, you can be sure that your design department's work will not escape the spotlight. Don't look upon this inspection as a threat, but more as an opportunity – through your own involvement – to improve the way you work as well as the conditions under which you work. Remember, during the run-up to qualification, probably more people in your company than at any time, and at all levels, will focus attention on what you do and how your work fits in with theirs. While others are receptive to change, it's also a unique opportunity to get things done in ways that most benefit your customers.

In essence, your department will have to design itself a system for designing. And, just like modelling any design, you'll need to study inputs, know what output you're aiming for, and establish a transfer function for what falls between.

It's in establishing that transfer function where you have the most to contribute. You'll have to write down some design rules for yourself; know how your design activity will be organized; see where the various job functions fit into that organization; and derive job specifications, including responsibilities for those functions.

You'll need to outline a process for designing your company's products from start to finish; break down the process into manageable phases; and specify goals for those phases. You'll also need to develop a review procedure, including checklists to demonstrate that you're ready to move on to the next phase, and outline what level of resources to use along the design path.

ISO 9001 is hot on documentation control. You'll have to decide what level of documentation you produce and maintain as you pass through your design phases. Unofficial documentation no longer exists: no more back-of-the-envelope design work, no carrying important design information around in your head; instead, you'll need to keep a legible log of what you're working on.

Your design tools don't escape the scrutiny of an ISO quality-management system either. You'll need to place all design test-and-measurement instruments on a recalibration schedule and specify what level of performance to check each of your instruments. If you use a handheld DMM in design, but only on dc voltage ranges, then that unit must bear a label restricting its use in design on other ranges.

With the transfer function mostly behind you, now look at your design input and output. These areas are where your work directly interfaces with other departments – and where establishing ground rules needs careful negotiating. One of your critical inputs comes from marketing colleagues who supply your design requirements specification. You'd like them to tie down every last detail of what you have to design, but they want to stay flexible to respond to new requirements as they develop. You'll both have to thrash out what level of detail is right for your company. But at least you'll need a performance specification, with time constraints and cost estimates.

Other important inputs arrive in the form of new components or processes, and you'll need to outline how your company accommodates such new things.

At the other end of your process, you want to stay loose, but your manufacturing friends want the earth. Again, you'll need to compromise on the number of prototypes, the level of documentation, how you will implement design changes, and so on.

Overall, you'll end up with a set of procedures that define employees' roles and responsibilities. The outcome is likely to simplify your working practices, to encourage self-discipline, and to force decisions where previously things were woolly. Other benefits include clearer objectives, less chance of designing into blind alleys, and greater visibility of your contribution throughout the company.

Force Computers GmbH of Munich, Germany, (ISO 9001 approved since March of 1992) sees major benefits of conformance for customers. Jürgen Baumann, director of engineering at Force, says that with ISO 9001, the company is better able to deliver what it promises and serves its customers in a more controlled and quality-conscious way. Since qualifying, Force finds that the company operates more smoothly. Several company procedures in operation before qualifying were pruned out, simplifying interfaces between departments, streamlining product design, and enabling the company to plan better.

Baumann admits there are significant increases in the amount of paper people now keep on their shelves, but staff refer to those documents on a daily basis. Baumann says designers at Force had already initiated guideline procedures for design, so incorporating them within the ISO standard was a relatively simple task. He says an engineer's principal aim is to design successful products, and working to ISO 9001 simply promotes that objective.

Lynn Hitchcock, ISO 9001 project leader at Heurikon, reports equal enthusiasm among designers at the prospect of embracing the ISO standard. (Heurikon, computer systems division of Computer Products Inc, is working toward the standard and should qualify toward the end of 1993). Hitchcock says Heurikon designers identified significant areas for improvement by analyzing their existing design process. Loopholes were evident, particularly at the front and back ends of the process. Again, studying interfaces between departments yields the most return for your effort.

ISO standard awards recognition

ISO 9000 apart, chances are that your company already has documented procedures in place. It may be that using your procedures, products are just as good as if you were ISO-qualified. The difference, however, is that qualifying for ISO 9000 adds value to your procedures – and products – because of international recognition. As more and more companies achieve qualification (currently 1300 in the US, 450 in Canada, and 23,000 in the UK), a shadow of doubt remains over products from companies that hang back. Eventually, nonconformance will mean losing business. In Europe, there's already strong evidence that buyers prefer vendors with ISO approval. Even if you don't lose business directly, artful buyers never miss the chance to penalize you for a perceived shortcoming. Really small companies have little option but to contemplate such penalties. The sheer cost of qualifying probably commits them to life outside the standard.

But for companies that can afford to qualify, it's tempting to overlook the primary objectives of the standard and simply join the club and frame the certificate. Common advice from assessors and consultants alike says that your quality-management system must serve the needs of the business first and the standard second. Companies that lose sight of this guiding principle will see a poor return on their investment. Even with ISO objectives firmly in place, return on investment is not a forgone conclusion. To make the system win for you, every member of your company must maintain a high level of commitment, and you must review and update your procedures continually.

Essential steps on the path to ISO 9000 approval

- Executive managers commit to ISO 9000 quality management
- Appoint a steering committee
- Appoint and train a project leader
- Appoint a consultant to conduct a 'situation review' and help steering committee produce an action plan
- Detail your action plan: i.e. who produces what by when
- Communicate (and sell) the plan to all staff
- Produce the quality manual
- Document your procedures
- Select an assessor (known as registrars in the US, and accredited certification bodies in the UK)
- Phase in your procedures
- Conduct internal self-assessment
- Submit quality documents for external pre-assessment (called system audit)
- Undergo on-site assessment for qualification (called compliance audit)
- Qualification awarded
- Proceed with regular on-site audits
- Continue to improve procedures

ISO 9000 status and signposts

The ISO (International Standards Organization) 9000 document exists in several sections[1]. ISO 9001 embodies ISO 9002 and ISO 9003 and is the standard to use if you intend to qualify your company's design/development, production, and inspection/testing activities. There are other sections; ISO 9000 and ISO 9004 are guideline documents, and ISO 9000–3 specifically addresses the matter of qualifying software development.

As with most international standards, each country uses its own numbering system, although the document's contents are identical. So in Europe, the US, and the UK, ISO 9001 is known as EN 29001, ANSI/ASQC Q91, and BS 5750 part 1, respectively.

At the local level, there are additional sections and support documents that don't appear internationally. In the UK, British Standards Institution publishes BS 5750 part 4. This document is an additional guide to using ISO 9001 and usefully elaborates on the standard itself. For example, it lists items you should have in your design project plans; it lists questions to ask when making changes during your design process; and it outlines requirements for a documentation-control system.

Another highly recommended document comes from the TickIT office of the UK government's department of trade and industry. Entitled 'Guide to Software Quality Management System Construction and Certification using EN 29001' £10), the 175-pg booklet covers in great detail how to apply ISO 9001 to software. Even if software is not your product, this work is well worth studying because most of the contents apply equally well to hardware products. The booklet includes guides for purchasers, suppliers, and assessors, and lists valuable pointers to standards, information sources, and recommended reading material.

In the US, CEEM Information Services is a valuable source of help with ISO 9000. This company offers training programs, videos, and monthly newsletters, and maintains databases of qualified companies and assessors worldwide.

When it's time to select an assessor, you'll need to contact the authority that controls assessors in your country (ASQC in the US or NACCB in the UK). These authorities publish directories of assessors with their corresponding range of disciplines. Currently, the directories list approximately 30 assessors each in the US and the UK, although not all cover electronic products. For example, in the UK, only four of the 30 assess software companies.

A consultant advises ...

Paul R Greenwood, Quality Director – International, with quality management consultants Beard Dove Ltd, offers advice to companies embarking on an ISO 9000 program:

Thinning-out your quality manual

First and foremost, try not to produce a single weighty tome. Although some companies do combine all documents into one manual, staff are less likely to use it in that form. Purchasing people simply don't want to know how to make pc boards, so don't obscure their buying procedures with irrelevant information from other departments. You can successfully trim down your quality manual by writing-in only policy and general practices – a framework document. In this way, you transform your manual into a steering document, with pointers to other specific procedures and work-instruction documents distributed around your operation. One advantage of this modular approach is that updating becomes less awesome. Also, when customers request a copy of your manual (as aggressive buyers often now do), you can satisfy the request without revealing what you may regard as confidential or proprietary data.

Stamping on bureaucracy

Don't go overboard in aiming for ISO 9000. It's a mistake to tie down every last detail across your company's operation. If you do, you'll simply throw your staff into a documental straightjacket. Chances are you'll sail through the assessment first time with this approach, but the overhead on your operation is simply not justifiable or workable. You should set the level of detail commensurate with the training and experience of your staff. For example, unskilled staff require detailed instructions, and skilled or professional staff need guiding procedures.

Choosing a long-term ally

Exercise great care when selecting an assessor, because you're taking the first step in forming a long-term business relationship. Prepare a short list of viable candidates, get them to make you a presentation, inquire into their business reputations, and get them to tell you why you should appoint them. Also, check out how the assessors go about their work. Some assessors will want to conduct a complete reassessment or review every three years, others will be content to continue with a high level of routine checks. Some may wish to pay you surprise visits; others work to fixed schedules. After qualification, some assessors insist you don't make changes without their scrutiny; others show more flexibility. Lastly, remember you're the paymaster, and bear in mind that assessor's charges can vary almost two to one. Overall, don't allow assessors to intimidate you.

For more information

Registrar Accreditation Board in US:

American Society for Quality Control (ASQC)
Milwaukee, WI
(414) 272–8575

Beard Dove Ltd
London, UK
71 387–1313

Center for Energy and Environmental Management (CEEM)
Fairfax, VA
(703) 250–5900

Disc TickIT Office
London, UK
71 602–8536

Force Computers GmbH
Munich, Germany
89–60814–0

Heurikon Corp
Madison, WI
(608) 831–0900

National Accreditation Council for Certification Bodies (NACCB)
London, UK
71 233–7111

Reference

1. Eugene R Hnatek 'ISO 9000 for the Test Engineer', *Test & Measurement World* (international testing supplement), October, 1992, p. 7.

46 Europe lays down EMC Law

Guide to Euro-EMC legislation

Here is one of those useful editorial articles for which *EDN* is so well known. While the impact of the various directives has not to date been as dramatic as many forecast, nevertheless the squeeze is on as the end of various transition periods for the introduction of the different measures approaches (Additional notes at the end update this article to the state-of-play as at February 1994).

Brian Kerridge
European Editor

Ohm's Law, Kirchoff's Laws, and now Europe's EMC Law. All electronic circuits produce electromagnetic emissions (EMI) at some level. Equally, all circuits become susceptible to EMI at some level. Europe's Electromagnetic Compatibility (EMC) Law sets out exactly what levels are acceptable in both cases for your product. In the future, when designing products for Europe, you will have to consider the EMC Law alongside other circuit constraints.

As part of Europe's move toward a single market structure, politicians and technocrats decided to formalize and harmonize EMC regulations. The result is European Directive 89/336/EEC (European Economic Community), entitled *Approximation of the Laws of the Member States relating to Electromagnetic Compatibility* issued by the Council of European Communities. Products will require a 'CE' mark, as well as other documentary evidence, as proof of conformance.

In principle, as from January 1, 1992, it is a criminal offense to contravene the regulations. In practice, you still have a breathing space of two or three years in which to prepare.[1]

If you declare compliance and your product fails to conform, ultimately it must be removed from sale and your stock impounded. If you fabricate documentation in order to imply compliance, then your company will face legal action against its European signatory.

The range of products affected by the new ruling is vast. According to an interpretation of the Directive by Chris Grimes, Chief Consultant on EMC with Siemens Plessey Assessment Services, 'If an electron flows in your product, then the Directive is applicable to it'. The EMC

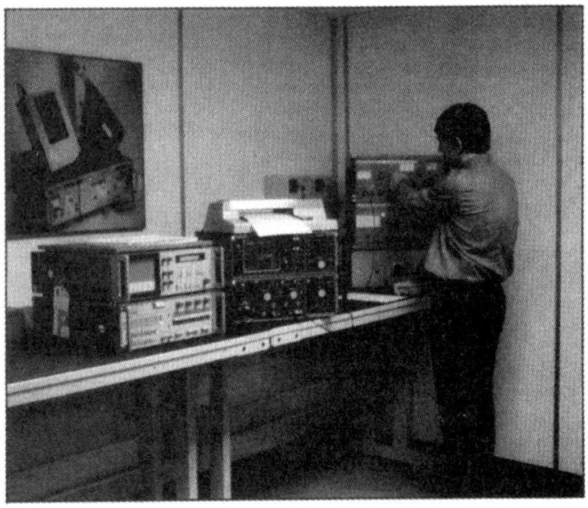

Figure 46.1 Setting up a lab for EMC-emission and -immunity testing is expensive. Test equipment alone costs about £200,000, and you'll need a shielded and damped room to put it in

Directive encompasses regulations on emissions and immunity for conducted and radiated electromagnetic sources, line distortion, and electrostatic discharge (ESD). CENELEC (European Committee for Electrotechnical Standardization) is the European authority producing the new series of technical documents (distinguished by European Standard prefix 'EN' (from its title in German)). Generally, the new documents specify harmonizations of existing specifications from CISPR (International Special Committee on Radio Interference), or the national standards of the European countries. Table 46.1 identifies the EN documents produced so far and the provisional EN documents (prENs) under development, cross-referenced to established specifications. When deciding which specifications apply to your product, look for product-specific versions

Table 46.1 European standards relating to EMC

	European standard[1]	Subject	Equivalent standard
Emission	prEN 50081-1	Generic domestic, commercial, and light industrial equipment	None
	prEN 50081-2	Generic industrial equipment	None
	EN 55011	Industrial, scientific, and medical equipment	CISPR 11
	EN 55013	Radio and TV receivers	CISPR 13
	EN 55014	Household appliances	CISPR 14
	EN 55015	Lighting equipment	CISPR 15
	EN 55022	Information technology equipment	CISPR 22
	EN 60555	Line disturbances	IEC 555
Immunity	prEN 50082-1	Generic domestic, commercial, and light industrial equipment	None
	prEN 50082-2	Generic industrial equipment	None
	EN 55020	Radio and TV receivers	CISPR 20
	prEN 55101-2	Information technology equipment—ESD	None
	prEN 55101-3	Information technology equipment—RF radiation	None
	prEN 55101-4	Information technology equipment—RF disturbances	None
	HD 481[2]	Industrial process, measurement and control equipment	IEC 801

Notes:
1. European standard designation is EN; provisional status is prEN.
2. No EN number yet, still at harmonization status.

first, such as in the case of information technology equipment. If more than one specification applies, adopt the toughest. In the case where no specification exists for your product, the Directive still requires your product to conform to the EMC law. (How to deal with that situation appears later.)

Transition period eases pain

But don't think all issues are cut and dried. Although legislation will be in place from day one, some of the technical specifications will come later.[2] Partly because of this delay, a transition period will enable a smoother switch from National to harmonized European specifications. The end date of the transition period is under consideration, but may extend to 1995.[3] In the transition period, you can either wait, or attain compliance from day one using a mixture of whatever new specifications exist together with prevailing national specifications. Naturally, there is some confusion, and even experts in the same European countries cannot entirely agree on what you need to do to conform. Differences also exist between the EEC's member states over what technical specifications should apply. Even wording of the Directive itself proves troublesome. The Council intends to publish a further document by the end of this year to clarify some issues, such as what 'taken into service' and 'placed on the market' mean.[4]

Make no mistake, though – despite these discordant notes, the EMC Directive exists and will soon apply throughout the EEC's 12 member states and the four EFTA (European Free Trade Association) countries. By the end of the transition period at the latest, all electronic products must conform, and only harmonized EN specifications will count. Compliance will be mandatory for products in current production, regardless of country of origin.

When politicians, specmakers, and marketeers have had their say, it will be up to you, the design engineer, to decide how to make the product conform.

That your product must ultimately conform is certain. How you reach conformance is less certain. The issues are complex, and subject to various interpretations. At the very least, your company will need to assign one person the responsibility of studying, interpreting, and tracking the developing situation.

Figure 46.2 shows EDN's interpretation of the various routes to compliance. Several organizations are already offering specialist help and advice. (See 'Who can help?')

Three ways lead to compliance

In overall terms, the route to placing a compliant product on the European market is straightforward. First, you adopt one of three possible methods to convince yourself that the product conforms technically. Then, you apply a 'CE' mark to the product and ship it with a declaration-of-compliance certificate.

The certificate must name a company signatory. Ideally, that person should be a resident European native. In the case of manufacturers outside

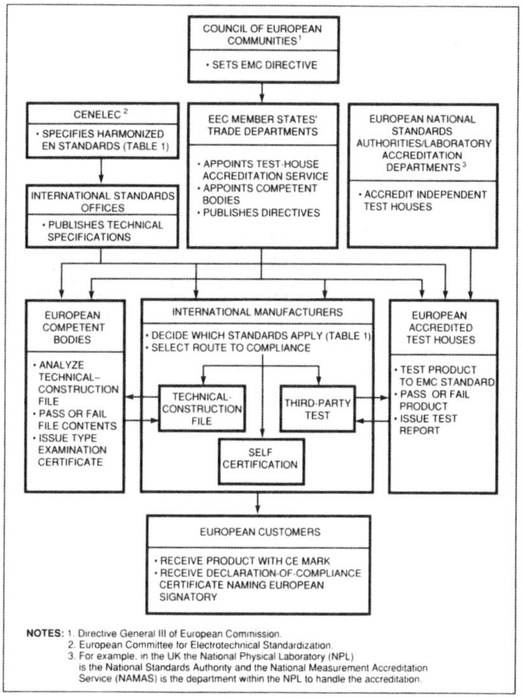

Figure 46.2 EDN's guide to compliance with Europe's EMC Directive

the EEC, that person would normally be in the employ of your representative or distributor.

The three methods of reaching a level of confidence that your product complies are:

- self-certification
- third-party certification
- technical-construction file.

Self-certification appears to be the most direct route, as it is informal and involves minimal interactions outside your company. As the Directive stands, this route allows you freedom to do whatever convinces you that your product complies. When you feel certain of compliance, then your path is clear to apply the CE mark and get on with the selling.

The danger with this route is if someone officially challenges your product's conformance. How able will you be to support your belief that the product does indeed conform? If you are unable to convince the EEC trade authorities, then your product must be withdrawn and your company stands the risk of being blacklisted. In extreme cases, your European signatory could end up explaining things in court.

Nonetheless, companies outside Europe will favor this route, especially where the convenience of local EMC test facilities exist. In the case of a challenge, when test results from a reputable source are available, the danger of a product ban diminishes.

Third-party certification is the route that offers optimal assurance that your product conforms. Essentially, you subcontract the work to an accredited laboratory. The laboratory tests, reports, and certifies product compliance. A third-party test house must be accredited, otherwise the laboratory is not authorized to issue certificates. In the UK, NAMAS, a department of NPL, accredits such laboratories. DAE is the accrediting authority in Germany.

At present, accredited labs exist only in Europe. By the end of 1991, NAMAS expects there will be around 30 accredited labs for EMC testing in the UK. Germany will have around 15.[5] Ideas to allow third-party testing in laboratories outside Europe are in the early stages. In the case of self-certification, it is not essential for your laboratory to be accredited.

Paperwork proves powerless

The technical-construction-file route is the most uncertain way to have your product declared compliant. In the first place, there is no clear idea yet of what should be in a technical-construction file. The attraction of this route is that the need for testing is not mandatory. The general intention is that the file contains sufficient information to convince a 'competent body' that your product complies. A competent body is another independent agency, authorized by governments to vet technical-construction files and issue approval certificates. By the end of 1991, each EEC member state's trade department will issue a list of accredited test houses and competent bodies.[6]

Figure 46.3 The CE mark signifies that your product complies with all applicable Directives of the EEC. The letters must be semicircular, greater than 3 mm in height, and the E-bar must be > 80%

Geoff Orford, at NAMAS, believes the technical-construction-file route, as the directive has outlined it, has little merit. He says that most EMC Standards are simply recipes for testing, and therefore you cannot hope to demonstrate compliance without some tests.

Grimes says the value of the technical-construction-file route shows up when manufacturers have a range of similar products. He recommends manufacturers to third-party test the worst-case product in the range, and then seek compliance for the rest of the range with technical-construction files. The technical-construction-file route also theoretically allows early compliance with the directive in the case of temporary absence of standards. Although, on what basis a competent body will make a judgment remains uncertain.

Grimes says he expects a technical file to include a technical report, a detailed block diagram, photographs, a wealth of EMC test data, and the test equipment's history.

A CE mark signifies that your product complies with all applicable EEC Directives. Some toys, for example, have to comply with three earlier Directives, and are among the first products already displaying CE marks. The first CE marks showing adherence to the EMC Directive should appear by the end of 1991, according to Tony Bond, head of the EMC office of the UK's Department of Trade and Industry.[7] He believes manufacturers will want to demonstrate compliance sooner rather than later in order to beat competitors.

As the transition period passes, Bond expects purchasing managers to show increasing preference to CE-labeled goods. By the end of the transition period, only goods with the CE mark will be acceptable. Bond warns against procrastinating in complying with the Directive. He points out that it's more expensive to massage a finished product through to compliance than it is to design in EMC features from the outset.

He also thinks manufacturers in the UK will provide a natural watchdog service over other suppliers. He says the system will be 'complaints driven' – competitors will step forward to expose one another.

In Germany, the Ministry of Telecoms has 50 offices spread throughout the country to police contraventions of EMC and other Directives. These offices routinely purchase and test products. If your product is found lacking, you will have to pay heavily for the test work, and be given a fixed period in which to conform. You may also be punished for importing the goods in the first place. Essentially, if you get to this stage in Germany, you're in big trouble.

All routes demand heavy toll

Attaining compliance will be expensive. Even discounting your additional design effort and manufacturing costs, the extra work in assembling documentation, and the likely need for third-party testing, will cost at least $2500 per design. Naturally, companies flinch at this level of penalty, especially coming at a time when business is not exactly booming. The authorities, recognizing this situation, claim the transition period introduces a degree of flexibility into the approval procedure. But still there is not really a low-cost route.

At a recent forum of EMC-Directive experts in the UK, a delegate from a small company producing custom designs asked how he could afford to meet the Directive and survive. The experts had no satisfactory answer. The only advice from the panel (a test house representative), was to discuss the problem with a test house. It seems that companies in this predicament have a difficult decision to make: They can either dodge the directive and risk getting caught, or go out of business.

If you consider self-certification using your own set-up, the cost is exorbitant. Tom Leahy, technical manager with Schaffner EMC Ltd, estimates a minimum of £280,000 is necessary in order to do emission and immunity testing. Of that figure, £100,000 buys you an electromagnetically shielded and damped room of around 25 m^2, which he says is essential to obtain repeatable test results. Leahy sees that the main problem with setting up your own facility is in locating experienced engineers to make the measurements. He reports mainly multinationals following this route, and only a few of the smaller companies.

For third-party EMC testing, TRL Ltd charges a daily rate of £750, which is typical for the industry. Mark Heaven, EMC consultant with TRL, estimates that compliance tests on the average product will take about two or three days. This period assumes the work goes smoothly, and the product passes. Before you submit a product to TRL, the company likes to consult with you on critical aspects of the design. These consultations minimize the chance of failure in the test house, and therefore limits your expense. According to Heaven, 80% of all EMC failures result from cables. Offending cables include line-cords, data-links (such as RS-232C and IEEE-488), and signal leads. Apertures in the product's enclosure have the second largest effect on a product's failure.

Heaven says fixing EMC problems in a finished product is often a losing battle, and costly. Unless you carry out substantial redesign, involving changes to the pc-board layout, all you can do is install more shielding or filtering. The best way is to be conscious of EMC requirements throughout the design process. In that way, you can minimize, or even eliminate, additional product and manufacturing cost. Also, you optimize your chance of passing the test routine for the first time. Some products cannot avoid cost penalties, however – notably, enclosures for information-technology products. Additional shielding using electroless plating on plastic surrounds increases component cost as much as three to ten times.

Wolfgang Sammet, EMC specialist with Siemens, also emphasizes the importance of good design practice from the beginning. He says particular attention to grounding and short signal paths on your pc-board layout pays dividends. He also meets many problems at a system level. It is quite common for a collection of modules that pass an EMC test as individual units to fail when connected together. This situation highlights the problem of 'configuration control', which Sammet sees as the major obstacle in the future. Several EMC specialists suggest Henry W. Ott's *Noise Reduction Techniques in Electronic Systems* (Wiley-Interscience, 1988) as a comprehensive and readable text to help you design around EMC problems.

Germany leads Europe on the EMC specification front. German National Standards Institution (VDE) specifications for commercial equipment have been around for decades. Siemens's commitment to the EMC business is exceptional. The company has extensive test facilities in Munich and markets a variety of EMC components for products and installations. Siemens has run EMC training seminars for design engineers for the last 20 years. Currently, the program takes place six times a year at international venues. Despite these advances, it is Sammet's view that a lot more needs to be done. In particular, he thinks electricians and technicians, as well as design engineers, need to be aware of, and understand how their work affects EMC.

The impact and complexity of the new EMC legislation is likely to influence all involved to move cautiously through the early months of 1992. Nobody knows exactly what to expect.

So far, every sign from the authorities suggest a commonsense approach, particularly in the UK. The objective of the EMC Law is to be protective more than pernicious. If you can demonstrate that you've taken a responsible attitude, then your problems should be few. The authorities anticipate that most companies will respond in this way, just as they have with safety regulations.

Some anomalous and difficult areas persist, however. It's not likely that they will disappear until the new regulations have passed a burn-in period.

Lack of commonality throughout the EEC is one such problem. In the transition period, for example, it will be possible for different specifications to apply in different countries, but all these specifications will lead to qualifying for the same CE mark. For example, as German VDE specifications are the toughest, it will be attractive to qualify elsewhere. Even after the transition period, it's likely that there won't be common rules for authorizing competent bodies or accrediting test houses. At present, NAMAS in the UK has mutual agreements with France and the Netherlands. The Western European Laboratory Accreditation Cooperation is engaged in providing common rules, but progress is at snail's pace.

Looking further afield, there is a strong case for establishing equivalent competent bodies and accredited test houses outside Europe. This move would make it much more straightforward for international manufacturers to use the third-party test route to comply. However, the likelihood of this happening in the next five years seems remote. Yet another organization, the International Laboratory Accreditation Cooperation, supports the idea, but implementing any program is way off.[8]

The most persistent headaches will result from difficulties of configuration control. For example, if a pc plug-in card complies in pc type A, but fails in pc type B, you will have to decide how many other pc types you should try it in. If you are a systems integrator, your headaches multiply. All your systems may have a different configuration, and the system may pass at one site, but fail at another. If you're a small company making custom designs, it's unlikely you can pass the cost of testing to your customer for every single product.

Europe's EMC Law is here to stay and is certain to disturb the tranquility of your R&D lab. In design-cycle terms, the 1995 drop-dead date is not so far away. Tune in to Europe's EMC emissions, before Europe designs in immunity to your products.

Help on Europe's EMC Law

For a copy of Directive 89/336/EEC:[2]
Alan Armstrong Ltd
2 Arkwright Rd
Reading RG2 0SQ, UK
(734) 751771
FAX (734) 755164

For European Norm documents:
BSI Sales
Linford Wood
Milton Keynes MK14 6LE, UK
(908) 221166
FAX (908) 322484

For general information on the EMC Directive:
Bundesamt für Post und Telekom
Herr Lehning, Referat 124
Postfach 8001
Templestrasse 2–4
W-6500 Mainz 1, Germany
(6131) 18 1200
FAX (6131) 18 5600

Department of Trade and Industry
Tony Bond
Manufacturing Technology Div 4E
151 Buckingham Palace Rd
London SW1W 9SS, UK
(71 215) 1408
FAX (71 215) 1529

For accreditation information:
Deutsche Akkreditierungsstelle Elektrotechnik
Herr Dr Facklam, Geschaftsstelle
Stresemannallee 19
W-6000 Frankfurt, Germany
(69) 6302 380
FAX (69) 6302 3167

NAMAS Executive
Geoff Orford
National Physical Laboratory
Teddington TW11 0LW, UK
(81) 943 7140
FAX (81) 943 7134

International EMC Standards and Test house facilities:
ERA Technology Ltd
Gordon Jackson
EMC Dept
Cleeve Rd

Leatherhead KT22 7SA, UK
(372) 374151
FAX (372) 374496

Siemens Plessey Assessment Services Ltd
Chris Grimes
Segensworth Road
Titchfield, Fareham PO15 5RH, UK
(329) 844440
FAX (329) 853234

TRL Technology Ltd
Mark Heaven
Alexandra Way
Ashdown, Tewksbury GL20 8NB, UK
(684) 850438
FAX (684) 850406

EMC Products:

Schaffner EMC Ltd
Headley Park Area 10
Headley Rd E
Woodley, Reading RG5 4SW, UK
(734) 697179
FAX (734) 699846

Siemens Matsushita Components GmbH
Product Marketing EMC Components
Balanstrasse 73
W-8000 Munchen 80, Germany
(89) 4144 4269
FAX (89) 4144 2575

What does it all mean?

The following list of acronyms are in common usage in the Standards and EMC world:

ANSI: American National Standards Institute
BSI: British Standards Institution
CE: European Community, used for CE mark of compliance (from its title in French)
CENELEC: European Committee for Electrotechnical Standardization (from its title in French)
CISPR: International Special Committee on Radio Interference (from its title in French)
DAE: German Laboratory Accreditation Service (from its title in German)
EEC: European Economic Community (Belgium, Denmark, France, Germany, Greece, Ireland, Italy, Luxembourg, The Netherlands, Portugal, Spain, and the United Kingdom)
EFTA: European Free Trade Association (Austria, Finland, Sweden, and Switzerland)
EMC: Electromagnetic Compatibility
EMI: Electromagnetic Interference
EN: European Standard (from its title in German)
ESD: Electrostatic Discharge
ETSI: European Telecoms Standards Institute
FCC: Federal Communications Commission
IEC: International Electrotechnical Commission
ILAC: International Laboratory Accreditation Cooperation
NAMAS: National Measurement Accreditation Service (Department of NPL)
NPL: National Physical Laboratory (in UK)
prEN: Provisional EN, not fully ratified
RFI: Radio Frequency Interference
VDE: German National Standards Institution (from its title in German)
WELAC: Western European Laboratory Accreditation Cooperation

Editorial update

The following notes provide an update as to the state-of-play at the end of 1993.

1. The European Directive 89/336/EEC (as amended in 1992) is now in effect. However, as some specifications and/or definitions are still to be finalized, in the interim individual member states can offer companies the choice of compliance with the Directive, or with existing national standards. This relaxation will cease at the end of 1995, the Directive being mandatory from 1 January 1996.

2. This state of affairs persists and is the reason for the relaxation described in note 1.

3. The end date of the transition period is currently fixed for 31 December 1995.

4. See the 1992 amended version of the Directive.

5. As of November 1993, NAMAS lists 36 accredited test houses in the UK. In Germany, there are 11 establishments approved as 'competent bodies'. (Note that not all accredited houses are approved as competent bodies for the purpose of granting final approval as fully meeting the Directive.)

6. Most now issued.

7. Equipments bearing the CE mark, denoting full compliance with the Directive, are now appearing. For example, current IBM computers bear this mark, indicating that, as delivered, they comply. But what happens if you fit someone else's plug-in card is another matter.

8. In the USA, there is a lab attached to a VDE approved (German) organization, whilst another is similarly attached to a UK body. These can offer advice and preliminary testing, but final assessment (or approval of a technical construction file) must be carried out in Europe, as there are currently no competent bodies outside Europe.

(I. H.)

Index